ENVIRONMENTAL BIOTECHNOLOGY

*Sustainable Remediation of Contamination
in Different Environs*

ENVIRONMENTAL BIOTECHNOLOGY

*Sustainable Remediation of Contamination
in Different Environs*

Edited by

Rouf Ahmad Bhat, PhD
Moonisa Aslam Dervash, PhD
Khalid Rehman Hakeem, PhD
Khalid Zaffar Masoodi, PhD

AAP | APPLE
ACADEMIC
PRESS

First edition published 2022

Apple Academic Press Inc.
1265 Goldenrod Circle, NE,
Palm Bay, FL 32905 USA
4164 Lakeshore Road, Burlington,
ON, L7L 1A4 Canada

CRC Press
6000 Broken Sound Parkway NW,
Suite 300, Boca Raton, FL 33487-2742 USA
4 Park Square, Milton Park,
Abingdon, Oxon, OX14 4RN UK

© 2022 by Apple Academic Press, Inc.

Apple Academic Press exclusively co-publishes with CRC Press, an imprint of Taylor & Francis Group, LLC

Library and Archives Canada Cataloguing in Publication

Title: Environmental biotechnology : sustainable remediation of contamination in different environs / edited by Rouf Ahmad Bhat, PhD, Moonisa Aslam Dervash, PhD, Khalid Rehman Hakeem, PhD, Khalid Zaffar Masoodi, PhD.
Other titles: Environmental biotechnology (2023)
Names: Bhat, Rouf Ahmad, 1981- editor. | Dervash, Moonisa Aslam, editor. | Hakeem, Khalid Rehman, editor. | Masoodi, Khalid Z., editor.
Description: First edition. | Includes bibliographical references and index.
Identifiers: Canadiana (print) 20210393912 | Canadiana (ebook) 20210393920 | ISBN 9781774638309 (hardcover) | ISBN 9781774638316 (softcover) | ISBN 9781003277279 (ebook)
Subjects: LCSH: Bioremediation.
Classification: LCC TD192.5 .E56 2023 | DDC 628.5—dc23

Library of Congress Cataloging-in-Publication Data

Names: Bhat, Rouf Ahmad, 1981- editor. | Dervash, Moonisa Aslam, editor. | Hakeem, Khalid Rehman, editor. | Masoodi, Khalid Zaffar, editor.
Title: Environmental biotechnology : sustainable remediation of contamination in different environs / edited by Rouf Ahmad Bhat, PhD, Moonisa Aslam Dervash, PhD, Khalid Rehman Hakeem, PhD, Khalid Zaffar Masoodi, PhD.
Description: First edition. | Palm Bay, FL : Apple Academic Press, 2023. | Includes bibliographical references and index. | Summary: "This book provides a review of innovative and novel biotechnological techniques that can be implemented to assess, analyze, and mitigate harmful pollutants and wastes that result from agricultural and industrial operations. It helps to meet the much-needed demand for improvement of low-cost technologies that tackle pollution problems scientifically for the safeguard of the environment, focusing on bioremediation solutions that also create useful and renewable forms of energy. The biotechnological interventions discussed in the volume include approaches involving genomics, proteomics, transcriptomics, metabolomics, and fluxomics. In addition, biological agents such as microalgae, bacteria, fungi, and bacteriophage, which can also prove to be helpful in the elimination of wastes, are explored. Topics in Environmental Biotechnology: Sustainable Remediation of Contamination in Different Environs include the associated consequences and hazards from agricultural and industrial waste and a variety of bioremediation measures, including the use of bioaugmentation, biosensors, challenges of biofuel production, and more. The book is directed to researchers, scientists, industrialists, farmers, agricultural waste management authorities, as well as to faculty and students, and aims to help implement these novel technologies for environmental stability"-- Provided by publisher.
Identifiers: LCCN 2021060311 (print) | LCCN 2021060312 (ebook) | ISBN 9781774638309 (hardback) | ISBN 9781774638316 (paperback) | ISBN 9781003277279 (ebook)
Subjects: LCSH: Bioremediation.
Classification: LCC TD192.5 .E93 2023 (print) | LCC TD192.5 (ebook) | DDC 628.5--dc23/eng/20211220
LC record available at https://lccn.loc.gov/2021060311
LC ebook record available at https://lccn.loc.gov/2021060312

ISBN: 978-1-77463-830-9 (hbk)
ISBN: 978-1-77463-831-6 (pbk)
ISBN: 978-1-00327-727-9 (ebk)

ENVIRONMENTAL BIOTECHNOLOGY

*Sustainable Remediation of Contamination
in Different Environs*

Edited by

Rouf Ahmad Bhat, PhD
Moonisa Aslam Dervash, PhD
Khalid Rehman Hakeem, PhD
Khalid Zaffar Masoodi, PhD

AAP APPLE
ACADEMIC
PRESS

First edition published 2022

Apple Academic Press Inc.
1265 Goldenrod Circle, NE,
Palm Bay, FL 32905 USA

4164 Lakeshore Road, Burlington,
ON, L7L 1A4 Canada

CRC Press
6000 Broken Sound Parkway NW,
Suite 300, Boca Raton, FL 33487-2742 USA

4 Park Square, Milton Park,
Abingdon, Oxon, OX14 4RN UK

Library and Archives Canada Cataloguing in Publication

Title: Environmental biotechnology : sustainable remediation of contamination in different environs / edited by Rouf Ahmad Bhat, PhD, Moonisa Aslam Dervash, PhD, Khalid Rehman Hakeem, PhD, Khalid Zaffar Masoodi, PhD.
Other titles: Environmental biotechnology (2023)
Names: Bhat, Rouf Ahmad, 1981- editor. | Dervash, Moonisa Aslam, editor. | Hakeem, Khalid Rehman, editor. | Masoodi, Khalid Z., editor.
Description: First edition. | Includes bibliographical references and index.
Identifiers: Canadiana (print) 20210393912 | Canadiana (ebook) 20210393920 | ISBN 9781774638309 (hardcover) | ISBN 9781774638316 (softcover) | ISBN 9781003277279 (ebook)
Subjects: LCSH: Bioremediation.
Classification: LCC TD192.5 .E56 2023 | DDC 628.5—dc23

Library of Congress Cataloging-in-Publication Data

Names: Bhat, Rouf Ahmad, 1981- editor. | Dervash, Moonisa Aslam, editor. | Hakeem, Khalid Rehman, editor. | Masoodi, Khalid Zaffar, editor.
Title: Environmental biotechnology : sustainable remediation of contamination in different environs / edited by Rouf Ahmad Bhat, PhD, Moonisa Aslam Dervash, PhD, Khalid Rehman Hakeem, PhD, Khalid Zaffar Masoodi, PhD.
Description: First edition. | Palm Bay, FL : Apple Academic Press, 2023. | Includes bibliographical references and index. | Summary: "This book provides a review of innovative and novel biotechnological techniques that can be implemented to assess, analyze, and mitigate harmful pollutants and wastes that result from agricultural and industrial operations. It helps to meet the much-needed demand for improvement of low-cost technologies that tackle pollution problems scientifically for the safeguard of the environment, focusing on bioremediation solutions that also create useful and renewable forms of energy. The biotechnological interventions discussed in the volume include approaches involving genomics, proteomics, transcriptomics, metabolomics, and fluxomics. In addition, biological agents such as microalgae, bacteria, fungi, and bacteriophage, which can also prove to be helpful in the elimination of wastes, are explored. Topics in Environmental Biotechnology: Sustainable Remediation of Contamination in Different Environs include the associated consequences and hazards from agricultural and industrial waste and a variety of bioremediation measures, including the use of bioaugmentation, biosensors, challenges of biofuel production, and more. The book is directed to researchers, scientists, industrialists, farmers, agricultural waste management authorities, as well as to faculty and students, and aims to help implement these novel technologies for environmental stability"-- Provided by publisher.
Identifiers: LCCN 2021060311 (print) | LCCN 2021060312 (ebook) | ISBN 9781774638309 (hardback) | ISBN 9781774638316 (paperback) | ISBN 9781003277279 (ebook)
Subjects: LCSH: Bioremediation.
Classification: LCC TD192.5 .E93 2023 (print) | LCC TD192.5 (ebook) | DDC 628.5--dc23/eng/20211220
LC record available at https://lccn.loc.gov/2021060311
LC ebook record available at https://lccn.loc.gov/2021060312

ISBN: 978-1-77463-830-9 (hbk)
ISBN: 978-1-77463-831-6 (pbk)
ISBN: 978-1-00327-727-9 (ebk)

This book is dedicated to Jammu and Kashmir (J&K),
universally acknowledged as the "heaven of Earth."

About the Editors

Rouf Ahmad Bhat, PhD, is an Assistant Professor at Cluster University, Srinagar, Jammu and Kashmir, India, specializing in limnology, toxicology, phytochemistry, and phytoremediation. Dr. Bhat has been teaching graduate and postgraduate students of environmental sciences for the past three years. He is an author of more than 53 research papers and 35 book chapters and has published more than 20 books with international publishers. He has presented and participated in numerous state, national, and international conferences, seminars, workshops, and symposiums. Dr. Bhat has worked as an Associate Environmental Expert in a World Bank-funded flood recovery project and also environmental support staff in several Asian Development Bank (ADB)-funded development projects. He has received many awards, appreciations, and recognitions for his services to the science of water testing and air and noise analysis. He has served as an editorial board member and reviewer for several international journals published by Elsevier, Springer Nature, Taylor and Francis, SAGE and Wiley. Dr. Bhat continues to write and experiment diverse capacities of plants for use in aquatic pollution.

Moonisa Aslam Dervash, PhD, is actively involved in teaching the graduate and postgraduate students of environmental science for the past 1 year in Sri Pratap College Campus, Cluster University Srinagar, J&K, India. She has earned a number of awards and certificates of merit. Her specialization is in mesofauna and carbon sequestration. She has published scores of papers in international journals and has more than three books with national and international publishers. She is also the reviewer for various international journals.

Khalid Rehman Hakeem, PhD, is working as Professor at King Abdulaziz University, Jeddah, Saudi Arabia. After completing his PhD (botany; specialization in plant ecophysiology and molecular biology) at Jamia Hamdard, New Delhi, India, in 2011, he has worked as Assistant Professor at the University of Kashmir, Srinagar, for a short period. Later, he joined Universiti Putra Malaysia, Selangor, Malaysia and worked there as Postdoctorate Fellow in 2012 and Fellow Researcher (Associate Professor) from 2013–2016, respectively. He joined King Abdulaziz University in August, 2016, and was promoted to professorship recently.

Dr. Hakeem has more than 10 years of teaching and research experience in plant ecophysiology, biotechnology and molecular biology, medicinal plant research, plant-microbe-soil interactions as well as in environmental studies. He is the recipient of several fellowships at both national and international levels; he also has served as the visiting scientist at Fatih Universiti, Istanbul, Turkey, as well as at Jinan University, Guangzhou, China. Currently, he is involved with a number of international research projects with different government organizations.

To date, Dr. Hakeem has authored and edited more than 70 books with international publishers, including Springer Nature, Academic Press (Elsevier), CRC Press, etc. He has also to his credit 155 research publications in peer-reviewed international journals and 65 book chapters in edited volumes with international publishers.

At present, Dr. Hakeem is serving as an editorial board member and reviewer for several high-impact international scientific journals from Elsevier, Springer Nature, Taylor, Cambridge, Francis, and John Wiley Publishers.

Khalid Zaffar Masoodi, PhD, is Assistant Professor in the Division of Plant Biotechnology at Sher-e-Kashmir University of Agricultural Sciences and Technology of Kashmir, India. Dr. Masoodi is a recipient of a SERB Early Career Research Award (2016); Travel Award, SBUR, USA (2012); *Society of Endocrinology Journal Award* (2018), Glasgow, UK; and many more. Dr. Masoodi holds a distinction of qualifying CSIR-UGC net in Life Sciences, DBT-BET JRF, and GATE. His laboratory (Transcriptomics Laboratory also called as K-Lab, inspired by Klenow fragment at SKUAST-K) is well equipped with fluorescence imaging and RT-PCR technology and state-of-art instruments. His research focus is on multidimensional work in transcriptomics of stress biology (biotic/abiotic) in plants and drug discovery from medicinal plants against prostate cancer. Apart from that he is extensively working on DNA barcoding. He has written well-received papers published in Oncogene, Molecular Cancer therapeutics, Endocrinology, Plos One, Journal of Functional Foods, Physiologia Planetarium, and many more, and has over 350 sequences in Genbank. Currently he has three externally funded projects from SERB and NMPB that focus on high throughput screening of medicinal plants for prostate cancer and cisgenic apple. During 2019–21 Dr. Masoodi filed seven patent applications with the Indian Patent Office and one with US Patent Office. He has also deciphered the whole transcriptome of apple and discovered new genes involved in imparting scab resistance.

He holds a master's and PhD degree in Biotechnology. After completing his PhD, he worked in Prof. Zhou Wang's Laboratory at the University of Pittsburgh, PA, USA, as Postdoctoral Associate.

Contents

Contributors

Adeyemi Adesina
Department of Civil and Environmental Engineering, University of Windsor, Windsor, Canada,
E-mail: adesina1@uwindsor.ca

Hadiza Abdullahi Ari
Faculty of Sciences, National Open University of Nigeria, Nigeria

Elena Bonciu
University of Craiova, A.I. Cuza 13, Craiova, Romania

Monica Butnariu
Banat's University of Agricultural Sciences and Veterinary Medicine "King Michael I of Romania"
from Timisoara, Calea Aradului 119, Timis, Romania; E-mail: monicabutnariu@yahoo.com

Rubiya Dar
Centre of Research for Development, University of Kashmir, Srinagar, India;
E-mail: rubi07@rediffmail.com

Moonisa Aslam Dervash
Division of Environmental Sciences, SKUAST-K, Shalimar, Srinagar, J&K, India;
E-mail: moonisadervash757@gmail.com

Zulaykha Khurshid Dijoo
Department of Environmental Sciences/Centre of Research for Development, University of Kashmir,
Srinagar; E-mail: zulaykhazehra@gmail.com

Saleem Farooq
Department of Environmental Science, University of Kashmir, Srinagar, India

S. A. Gangoo
Faculty of Forestry, SKUAST-K, Benihama, Ganderbal, J&K, India

Mehvish Hameed
College of Agricultural Engineering, Sher-e-Kashmir University of Agricultural Sciences and
Technology of Kashmir, Shalimar Campus, Srinagar, J&K, India

Saima Hamid
Centre of Research for Development/Department of Environmental Sciences, University of Kashmir,
Srinagar, India; E-mail: cord.babasaima4632@gmail.com

Tijjani Sabiu Imam
Department of Biological Sciences, Bayero University Kano, Kano State, Nigeria

Shah Ishfaq
Department of Environmental Science, University of Kashmir, Srinagar, India

Muatasim Jan
Centre of Research for Ethnobotany, Govt. Model Science College, Jiwaji University, Gwalior, India

Rizwana Khurshid
Department of Management Studies, Iqbal Institute of Technology and Management, Srinagar, India

Khalid Z. Masoodi
Division of Plant Biotechnology, SKUAST-K, Shalimar, Srinagar, J&K, India

Tawseef Ahmad Mir
Centre of Research for Ethnobotany, Govt. Model Science College, Jiwaji University, Gwalior, India. E-mail: mirtawseef787@gmail.com

Saba Mir
Division of Plant Biotechnology, SKUAST-K, Shalimar, Srinagar, J&K, India

Mohammad Yaseen Mir
Centre of Research for Development, University of Kashmir, Hazratbal, Srinagar, Jammu and Kashmir, India. E-mail: yaseencord36@gmail.com

Javeed A. Mugloo
Faculty of Forestry, SKUAST-K, Benihama, Ganderbal, J&K, India

Hina Mushtaq
Department of Environmental Science, University of Kashmir, Srinagar, India

Irteza Qayoom
Department of Environmental Science, Sri Pratap College, Cluster University Srinagar, India; E-mail:qirteza@gmail.com

Mir Sajad Rabani
Microbiology Research Lab., School of Studies in Botany, Jiwaji University Gwalior, Madhya Pradesh, India

Gulab Khan Rohela
Biotechnology Section, Moriculture Division, Central Sericultural Research & Training Institute, Central Silk Board, Ministry of Textiles, Government of India, Pampore, Jammu and Kashmir, India

Asif B. Shikari
MCRS, SKUAST-K, Sagam, Srinagar, India

Wasia Showkat
Division of Plant Biotechnology, SKUAST-K, Shalimar, Srinagar, J&K, India

D. P. Singh
Department of Environmental Science, Babasahib Bhimrao Ambedkar University, Lucknow, India; E-mail: dpsingh_lko@yahoo.com

Ranjan Singh
Department of Environmental Science, Babasahib Bhimrao Ambedkar University, Lucknow, India

Dig Vijay Singh
Department of Environmental Science, Babasahib Bhimrao Ambedkar University, Lucknow, India

Adamu Yunusa Ugya
Department of Environmental Management, Kaduna State University, Kaduna, Nigeria; E-mail: ugya88@kasu.edu.ng

Atul Kumar Upadhyay
Department of Environmental Science, Babasahib Bhimrao Ambedkar University, Lucknow, India

Baba Uqab
Department of Environmental Science, University of Kashmir, Srinagar, India.
E-mail: uqabaalibaba@gmail.com

Riasa Zaffar
Department of Environmental Science, University of Kashmir, Srinagar, India

Abbreviations

ACC	acetyl-CoA
AChE	acetylcholinesterase
ACS	acyl-CoA synthetase
ADH	alcohol dehydrogenase
AMC	artificial microbial consortium
asRNA	antisense RNA
AWM	agricultural waste management
AWMS	agricultural waste management system
bm	brown midrib
BOD	biological oxygen demand
BRC	Berkley rapid composting
BTL	biomass to liquid
cBOD	biochemical oxygen
ccNiR	cytochrome C nitrate reductase
CMFC	cube microbial fuel cell
DDT	dichloro diphenyl trichloroethane
DMFC	double-chamber microbial fuel cell
DNAPLs	dense nonaqueous phase liquids
ECHA	European Chemicals Agency
EMC	environment maintenance consortium
EPA	Environmental Protection Agency
ERD	enhanced reductive dechlorination
ERFs	ethylene response factors
FAEE	fatty acid ethyl esters
FRET	fluorescence energy transfer
GFP	green fluorescent protein
HCB	hexachloro benzene
HCH	hexachlorocyclohexane
HCT	hydroxycinnamoyl transferase
HMs	heavy metals
IMG	heavy metal ions
ISEs	ion-selective electrode
ISFETs	ion-selective field-effect transistors

IUPAC	International Union of Pure and Applied Chemistry
NEC	nutrient exchange consortium
NMC	natural microbial consortium
OPH	organophosphorus hydrolases
PAHs	polycyclic aromatic hydrocarbons
PBDE	polybrominated diphenylether
PCBs	polychlorinated biphenyls
PCDDs	polychlorinated dibenzo-p- dioxins
PCDFs	polychlorinated dibenzofurans
PES	polyethylene succinate
PHB	polyhydroxybutyrate
PL	petroleum and lubricants
POPs	persistent organic pollutants
PVC	polyvinyl chloride
RME	rapeseed methyl ester
SDS	sodium dodecyl sulfate
SEC	signal exchange consortium
SeMC	semisynthetic microbial consortium
SFC	substrate facilitator consortium
SMFC	sediment microbial fuel cell
SPFM	sweet potato flour medium
SPR	surface plasmon resonance
SyMC	synthetic microbial consortium
TAGs	triacylglycerols
TCE	trichloroethylene
TCP	trichlorophenol
TSCA	toxic substances control act
TVOC	total volatile organic compounds
UNGHS	United Nations, called the Global Harmonized System
USEPA	US Environmental Protection Agency
VC	vinyl chlorides
VFAs	volatile fatty acids
WCBBS	whole cell-based biosensors

Foreword

While the world is remarkably progressing each day toward an advanced age of science and technology, the pace at which the pollutants are being produced and released into the environment is alarming. The dilution in the values of sustainability has produced pollutants that have impacted the air, soil, and water to such an extent that their restoration in the present scenario seems to be an impossible task.

Diverse pollutants having varied and serious effects have created a hostile environment that is not fit for a healthy life to thrive in. Continuous efforts have been made from time to time to combat these pollutants with the use of various techniques; however, due to various reasons pertaining to ecological sustainability and economic viability, these technologies have failed to touch the mark of success. With the advent of biotechnology, a breakthrough has been achieved and many success stories have been recorded in circumventing the pollution menace.

The book entitled, *Bioremediation and Biotechnology: Degradation of Pesticides and Heavy Metals in Contaminated Environs,* is a brilliant effort toward documenting the biotechniques and technologies that are and can be used to address the problems of pollution in a sustainable manner. These technologies are not only eco-friendly but also cost-effective. The latest research pertaining to biotechnology and bioremediation has been presented in a lucid manner. All the fifteen chapters of the book have dealt with different aspects of handling different types of pollutions, consequences of pollution through pesticides and heavy metals, development of mechanism or strategies to overcome metal exposure, and the role of modem innovative techniques for monitoring the pollution due to heavy metals and pesticides using sensors and biosensors.

With a mushrooming population in the world, there is a huge pressure on the agricultural sector for food production. However, chemical fertilizers are reducing the soil fertility as well as crop productivity; therefore, there is an urgent need to use alternatives to meet the need for nutrient sources for crops. This book provides an overview of global environmental regulations framed for management of pesticides for global environmental safety, human welfare, and sustainable development. This book is a valuable resource material for researchers of biotechnology and allied sciences. The authors and editors have put in a lot of effort to make a noteworthy contribution to the area of biotechnology and bioremediation. I wish them all the best for their future endeavors.

—Prof. Nazeer Ahmed
Former Vice-Chancellor,
Sher-e-Kashmir University of Agricultural Sciences & Technology,
Srinagar, Jammu and Kashmir, India

Preface

The high velocity of the movement toward a modern lifestyle, industrialization, and usage of novel technologies to ease a man's life has brought with it a negative impact on the environment. Environmental degradation is created by the amalgamation of a successfully extensive plus escalating population, persistently intensifying financial advances in addition to the function of asset grueling and poisoning technology. The environmental degradation has resulted in ozone depletion and global warming, which poses a threat to the planet and life of every organism present in it. Thus, there is an utmost need to tackle environmental pollution, and in this, environmental biotechnology has proved to be one of the major fields that resulted in positive outcomes. Majorly environmental waste can be attributed to the chemical industries, agriculture units, and food processing units. Estimates of agricultural wastes generated are limited but they are believed to contribute a considerable amount to the total waste materials in the developed world. These have been so far successfully tackled and dealt with environmental biotechnology.

The book is intended to provide various innovative and novel biotechnological techniques that can be implemented to assess, analyze, and mitigate pollutants and wastes resulting from agricultural and industrial operations owing to the fact that there is an increased and much needed demand for the improvement of low-cost technologies to tackle these problems scientifically to safeguard the environment. The authors, belonging to their respective areas of expertise, have put forward efficient technological methods and ideas to cater to the need of bioremediation by not only eliminating the pollutants but also to convert them into the useful and renewable forms of energy. The biotechnological interventions include genomics, proteomics, transcriptomics, metabolomics, and fluxomics. The biological agents such as microalgae, bacteria, fungi, and bacteriophage can also prove to be helpful in the elimination of wastes.

The book is aimed at researchers, students, scientists, industrialists, farmers, and agricultural waste management authorities to know and implement these novel technologies for the environmental stability. The technologies can be helpful for keeping a clear check on the degradation caused by agricultural and industrial chemical wastes which are the threat to water, soil, and air.

—Dr. Rouf Ahmad Bhat
Dr. Moonisa Aslam Dervash
Prof (Dr.) Khalid Rehman Hakeem
Dr. Khalid Zaffar Masoodi

CHAPTER 1

Assessment of Some Hazards Associated with Dangerous Chemicals

MONICA BUTNARIU[1*] and ELENA BONCIU[2]

[1]*Banat's University of Agricultural Sciences and Veterinary Medicine "King Michael I of Romania" from Timisoara, Calea Aradului 119, Timis, Romania*

[2]*University of Craiova, A.I. Cuza 13, Craiova, Romania*

Corresponding author. E-mail: monicabutnariu@yahoo.com

ABSTRACT

Hazard analysis is part of a broader concept, namely, risk management. It considers all aspects related to the operation of an industrial unit to prevent and properly treat an unwanted event. The hazard management includes hazard assessment, hazard mitigation measures, and emergency response plans. The update includes reassessment of hazard, implementation of emergency response measures, determination of hazard reduction modalities, analysis of benefit–cost ratio of implementation of appropriate measures, and implementation of hazard reduction measures at a level acceptable from the point of view of industrial practice. Hazard management involves the continually identifying of potential hazards and assessing the associated hazards. A periodic analysis of hazard reduction measures should be taken into consideration to ensure the capacity to control post-accident damages through emergency intervention plans.

1.1   EVALUATION OF HAZARD FACTORS REGARDING THE SOIL CONTAMINATION

1.1.1   *SOIL—THE PRIMARY FACTOR OF THE ENVIRONMENT*

Soil is the generic term that defines the entire set of natural fixtures that cover a large part of the surface of the land, consisting of mineral and organic matter, sometimes modified or even created by human, which contain living matter and can support vegetation directly under the open sky. As a factor of the environment, the soil fulfils a multitude of functions that can be classified into several categories, namely, ecological functions, industrial functions, and technical–economic functions (Cherubin et al., 2016).

The ecological functions of the soil are given as follows:

- The production of biomass, the soil being a support for plants, also serve as a reservoir of nutrients, water and air, traits that attribute an essential characteristic to it, namely, fertility.
- Filtration, buffering, and transformation are important functions by which the soil reacts to the action of pollution phenomenon for the protection of its own characteristics, for the prevention of pollution of the aquifer, and also of the agro-food products.
- Genetic reserve is a function by which the soil ensures the biodiversity specific to the edaphic environment.

As industrial and technical–economic functions of the soil can be as listed below:

- Use of soil as a support for different constructions.
- Use of soil for the storage of waste.
- Use of soil as a raw material (Di Falco and Zoupanidou, 2017).

It should be noted that in general, the industrial and technical–economic functions are in contradiction with the ecological functions.

1.1.2   *GENERAL ASPECTS ABOUT THE SOIL POLLUTION*

In general, the soil pollution means any disturbance that affects their quality from a qualitative and/or quantitative point of view.

The soil pollution can be classified according to several criteria, such as the nature, form and mode of dispersion of the polluting substances in the soil, the sources of production of the polluting substances, the effects produced by the metabolization/migration of the polluting substances in soil, etc.

In order to ensure a unitary approach and to comply with the norms of the European Union, the classification of soil pollution was regulated by the Soil Taxonomy System 2003 (types of pollution).

According to this normative, the soil pollution is divided into different categories, in which the types of pollution are listed as follows:

- Various soil pollution caused by industrial and agricultural activities.
- Soil pollution caused by slope processes and other physical processes.
- Soil pollution influenced by other natural and/or anthropic processes.

The following types of soil pollution are included in the different soil pollution category determined by industrial and agricultural activities:

- Pollution by daily excavation works (daily mining, ballast, quarries, etc.)
- Pollution by deposits, dumps, tailings ponds, floating tailings dumps, garbage dumps, etc.
- Pollution from inorganic wastes and residues (minerals, inorganic materials, including metals, salts, acids, and bases) from industry (including extractive industry).
- Pollution by substances carried by air.
- Pollution with radioactive materials.
- Pollution with organic waste and residues from the food and light industry and other industries.
- Pollution from waste, agricultural, and forestry residues.
- Pollution with animal waste.
- Pollution with human waste.
- Pesticide pollution.
- Pollution with contaminating pathogens.
- Pollution with salt water (from oil extraction).
- Pollution with petroleum products.

In the category of soil pollution by slope processes and other physical processes, the following types of soil pollution are included:

- Pollution by surface erosion, depth, landslides.

- Pollution by primary and/or secondary compaction.
- Sediment pollution caused by erosion (clogging).

The following types of soil pollution are included in the category of soil pollution by other natural and/or anthropogenic processes:

- Salted soils (saline and/or alkaline).
- Acid soils.
- Excess of water.
- Excess or deficiency of nutrients and organic matter.

The degree of soil pollution is assessed on five classes (low, moderate, strong, very, and excessive), either according to the percentage reduction of the crop in quantitative and/or qualitative terms, compared with the production obtained on the unpolluted soil, or by exceeding in different proportions, the defined thresholds.

It is noted that the units in the oil and gas extraction industry hold the most weight, followed by the large difference between the units in the extractive industry and the units in the waste recycling industry.

Considering the share of contamination of sites due to the units in the hydrocarbon extraction industry, the following are presented regarding the infestations with petroleum products, which are statistically due to:

- Diversity of human activities, approx. 65%.
- Processing industry—distribution, approx. 25%.
- Oil fields (exploitations), approx. 10% (Allen et al., 2013).

The industrial objectives in which these activities are carried out include oil and gas wells, separator parks, compressor stations, boiler batteries, depots, wastewater collection facilities, wastewater treatment plants, injection stations, transport pipes, water treatment plants, and slurry.

The sources of pollution for the environmental factors are the machines in which the basic activities of the scaffolding are carried out, such as extraction, collection, separation, treatment, storage, transport, as well as the machines in which the related activities are carried out, such as production, distribution of steam, treatment of wastewater, water injection, and slurry storage.

The machinery that can represent, depending on the operation, the specific sources of pollution for the soil/subsoil and aquifer environment factor include wells (extraction pipes, towers, and wells), separator parks (keyboards, separators, reservoirs, and pumps), compressor stations (cooling installations, cooling tower, and pumps), (boiler batteries—softening stations, and tanks), reservoirs (tanks, oil tanks), wastewater collection facilities (sewerage network, settling basins, and beams), treatment plants (oil separators), transport pipes (mixing, injection pipes), slurry battalions (slurry cells, retention dams, guard ditches, and siphon pipes).

The pollution of the soil, the subsoil, and the aquifer is resulting from the oil exploitation activities. The separation of the three phasic mixture associated with the activities of storage, distribution, and transport is complex and significant and is classified into:

- Oil pollution.
- Pollution with salt water.
- Mixed pollution, with crude oil and salt water (Riccardi et al., 2013).

The soil pollution with oil occurs during the activity of extraction, separation, and transport of crude oil and salt water during then leaks can occur that can reach the land and in the watercourses. The soils contaminated with oil exhibit a dense, compact crust of oil on the surface, which prevents the processes of infiltration–percolation of water in the soil and of exchange of gaseous substances between soil and atmosphere.

When the soil comes in contact with the oil, some changes occur in the physical, chemical, and biological properties of the soil. Several phenomena are taking place, namely, volatilization (evaporation) of light compounds (the most volatile hydrocarbons), which occurs in the first hours after the spill, stratification on the soil profile, the more polar components, especially the asphaltenes remaining at the surface of the soil and forming a compact film that prevents the exchange of gas with the atmosphere and which does not allow the normal circulation of water. The migration of oil on the soil profile, the depth at which it reaches depend on the intensity of the pollution and the texture and apparent density of the soil.

The ground soil and groundwater pollution with oil are due to oil leaks from certain oil installations (pipes, deposits, and extraction scaffolds).

The oil pollution of the aquifer comprises several phases that can be grouped into two distinct stages: (1) the pollutant migration stage (starting from the source) which has the effect of pollution by accumulation in the capillary area of the soil, located above the groundwater, in the form of a "floating" residual layer and the step of transferring soluble hydrocarbons from waste oil into groundwater, resulting in chemical pollution (Wang et al., 2020).

This type of pollution is significantly affected by the effects of the biodegradation of the pollutant stored in the vegetal soil or the ground cover of the groundwater aquifer.

The soil pollution with salted water occurs due to the fact that in the oil extraction industry, large quantities of salted water (reservoir water) are extracted along with the crude oil, and the salt content (NaCl) being sometimes high (over 1188 mg/dm^3).

Although the extraction inside the scaffolds of extracted matter is done in a closed system, there are still leaks of salt water, especially in the area of pipeline networks that transport this water from parks to treatment plants, and subsequently to injection stations by pipe breakage caused by corrosion. In this type of pollution, the content of soluble salts varies greatly depending on the amount of water discharged and its contact time with the soil.

The salted water seeps into the soil at different speeds depending on its texture.

At the impact of the salted water with the solid phase of the soil, cation exchange or adsorption processes take place whereby the chemical elements in the salt water are adsorbed in the colloidal soil complex. It is worth mentioning that soils that have been polluted with reservoir water exhibit a surface of gray–yellow soluble salts after the chemical processes and after drying the soil (Banks and Banks, 2019).

The mixed pollution with oil and salted water is the most serious type of pollution, the cumulative effects of the action of the pollutants being much more and are difficult to remove. It is very common in extraction scaffolds because, during the extraction process, the two phases can form emulsions because of intense mixing that occur.

The lands that present a mixed pollution generally have a total content of oil hydrocarbons (THP) and of changeable Na. In these cases, the very severe pollution causes the soils to become practically unproductive.

It is worth mentioning that of all types of pollution resulting from the oil extraction activities, the mixed pollution is very important.

The pollution caused by the transport of oil and oil products through pipes appears only in accidental damage caused to pipes, because the transport of oil and oil products through pipes is a technological process that takes place in a closed system, and which under normal conditions is made without emissions into the environment (Chen et al., 2016).

The main causes that cause damage and implicitly cause environmental pollution include accidental breakage caused by the corrosion of the pipes, the attack of the pipes (burglary) in order to steal petroleum products, and suffusion phenomena.

1.2 THE POTENTIAL HAZARD FACTORS AND PREVENTION OF SOIL CONTAMINATION WITH HAZARDOUS CHEMICALS AND MIXTURES

1.2.1 THE SOIL CONTAMINATION HAZARD WITH HAZARDOUS CHEMICAL SUBSTANCES AND MIXTURES

The European Chemicals Agency (ECHA) indicates the following categories of chemical substances and mixtures as being at high hazard for human health and the environment.

Biocides, which are products that contain or generate active substances and are used against harmful organisms in the home environment (disinfectants, rodenticides, repellents, and insecticides) or to protect natural or artificial products in industry and agricultural applications. Given that due to their intrinsic properties, the use of biocidal products can pose health risks, the EU has established a regulatory framework aimed at significantly increasing the safety of biocidal products used and sold in the EU, namely, EU Regulation 528/1012.

It provides for the release of biocides that are carcinogenic, mutagenic, toxic to reproduction, which disrupt the endocrine system or are hazardous to the environment (Gopinath et al., 2018).

Pesticides, which are products used to combat harmful organisms in agricultural crops, such as weeds and insects. However, their use can endanger the health and the environment. Many studies have revealed the link between pesticide exposure and its effects on the human body, such as cancer, fertility and reproductive problems, respiratory diseases, disruption of the hormonal (endocrine) system, and impaired immune system and nervous system. EU pesticide legislation imposes prohibitions on the use of certain hazardous chemicals in these products.

Endocrine disrupting substances are chemicals that interfere with hormones (the endocrine system) causing adverse health effects. A wide range of substances, both natural and human-made, are considered to cause endocrine disruption, including pharmaceuticals, pesticides, and industrial organic chemicals.

The European Community Strategy on endocrine disrupting substances, adopted in 1999, indicated the actions needed to counteract the impact of endocrine disrupting substances on human health and the environment. Thus, the endocrine disrupting substances were included in the lists of the REACH Regulation, and in January 2012, the European Commission launched an important study called "The Endocrine Disruptive Substance Evaluation Stage."

The study, which is based on thorough scientific documentation, was commissioned to inform the European Commission about identifying regulatory criteria for endocrine disrupting substances in accordance with the EU legislation (Ramón and Lull, 2019).

Heavy metals are elements that have a relatively high density and are toxic at low concentrations. Of these, cadmium, lead, and mercury are particularly toxic to the human body, and therefore, the EU has made considerable progress in addressing the problem of contamination with these metals. Thus, in 2005, the EU Mercury Strategy was launched which provides for restrictions on the sale of mercury-containing measuring devices, prohibitions on exports of metallic mercury from the EU and new safety rules for its storage.

This strategy was revised in 2010. The application of the EU Industrial Emissions Directive, adopted in 2010, will further reduce mercury emissions from industrial processes.

The EU strongly supports the international process of developing a global legal instrument on mercury under the aegis of UNEP. The levels of cadmium, lead, and mercury are strictly controlled in accordance with the restrictions of the Hazardous Substances Directive. Strict limit values are imposed for these metals on certain specific types of electronic equipment. Also, the End-of-Life Vehicle Directive further restricts the use of these metals. In addition, the use of cadmium in the manufacture of jewellery and plastics products has been banned under the REACH Regulation, while in the Water Framework Directive, cadmium and mercury are listed as priority hazardous substances, while lead is a priority substance (Gabbert and Hilber, 2016).

Persistent organic pollutants (POPs) are chemicals that remain intact in the environment for long periods of time, from where they are transported and enter living organisms, including humans, where they accumulate in their fatty tissue.

Due to their toxic characteristics, persistent organic pollutants have harmful effects on human health and the environment. The Stockholm

Convention is the world's leading instrument providing protection against persistent organic pollutants.

In this regard, the EU is committed to implementing the requirements of the Stockholm Convention and has encouraged the further listing of chemicals under this Convention.

Moreover, in the EU, the Persistent Organic Pollutants Regulation supports and strengthens the provisions of the Stockholm Convention.

The Persistent Organic Pollutants Regulation contains provisions on the production, placing on the market and use of such chemicals, managing the stocks and waste of persistent organic compounds and measures to reduce unintended emissions of persistent organic compounds.

Nanomaterials are chemicals or materials that are manufactured and used at extremely small dimensions (of the order of nm, i.e., 10^{-9} m).

Currently, many products containing nanomaterials are already in use and this market is expected to increase in the coming years. Nanomaterials have the potential to improve the quality of life and contribute to industrial competitiveness (Weber et al., 2019).

However, new materials can also raise health and safety issues.

These risks, and how they can be addressed through existing risk assessment measures in the EU, have been the subject of several scientific opinions issued by the European Commission Scientific Committee on Emerging and Newly Identified Health Hazards.

Combinations of chemicals and chemical mixtures, according to some studies, can have harmful effects on human health even when the individual chemicals are below the "safety level." In the EU, the regulatory risk assessment for chemicals is based on individual products.

However, there are concerns that this approach may not provide sufficient protection against the risks posed by simultaneous exposure to multiple chemicals and that another approach is needed.

Pharmaceutical residues have become an important source of environmental contamination. Many pharmaceutical chemicals are designed to be nondegradable and thus present a risk when they reach the environment where they persist.

The European Commission recognizes that environmental pollution from pharmaceutical residues is both a developing and a public health issue and is committed to addressing this issue.

In the following section, we have presented the systems of measures to prevent the contamination of soils and groundwater with the main groups of hazardous substances and chemical mixtures under the conditions of our country, namely, plant protection products (pesticides/biocides), fertilizers, heavy metals, compounds persistent organics, petroleum products/ residues (Mancini et al., 2016).

1.2.2   MANAGEMENT OF PLANT PROTECTION PRODUCTS (PESTICIDES/BIOCIDES)

The management of the products for the protection of the plants is regulated by law, the norms of approval and placing on the market, storage, and application. The use of technological means specific to them in the conditions of soil and water protection being presented in the Code of good practices in farm.

The document specifies the specific rules for approval and placing the products on the market, storage aspects, and application of plant protection products (Eudoxie et al., 2019).

1.2.3   FERTILIZERS MANAGEMENT

The management of fertilizers in farming which consists in storing, handling, and application of chemical fertilizers; the collection, storage, handling, and application of manure (manure and liquid manure); and the collection and storage of different effluents from livestock units must be carried under conditions of environmental protection (of soil and water) by applying the requirements of the Directive about waters protection against pollution with nitrates from agricultural sources (also called the "Nitrates Directive"), transposed into the Romanian legislation by GD no. 964/2000 which approved the Program of action for the protection of waters against pollution with nitrates from agricultural sources. For this purpose, the Code of good agricultural practices for the protection of waters against nitrate pollution from agricultural sources has been elaborated (Green et al., 2008).

It consists of 14 chapters and 13 annexes containing norms referring to General aspects regarding water and soil as renewable natural resources, prevention of environmental pollution as a means of protection and preservation of renewable natural resources, agriculture as a pollutant of the environment, particularly the soil and water and vulnerable areas or potentially vulnerable to pollution by nitrates from agricultural sources.

- Definitions, Agricultural systems.
- Fertilizers, potential sources of water and soil pollution.
- Fertilizers containing nitrogen.
- Fertilizers containing phosphorus.
- General rules for the storage and handling of chemical fertilizers.
- Storage and management of effluents and manure on agricultural holdings.
- Application of nitrogen fertilizers.
- Application of phosphorus fertilizers.
- Aspects of agricultural land management in terms of nitrogen dynamics.
- Prevention of pollution of surface and groundwater caused by fertilizers in the case of irrigation and watering.
- Fertilization plans and the register of fertilizer use in agricultural holdings.
- Datasheets and brochures included in the action program in vulnerable areas to nitrate pollution (De Waele et al., 2017).

1.2.4 THE HAZARD OF SOIL CONTAMINATION WITH HEAVY METALS AND ITS PREVENTION

Heavy metal soil pollution occurs mainly due to contamination of atmosphere by heavy metals.

The heavy metals emitted in the atmosphere in the form of sedimentable powders reach the surface of the soil by deposition or by the action of precipitation.

Rainwater can form streams, mobilize heavy metal deposits from the surface of the soil by transporting them in natural watercourses, or it can infiltrate into the soil by transporting heavy metals deposits into the soil, where they can be retained and fixed by the absorbent complex of the soil, either they can be absorbed by plants in solubilized forms due to the interaction with water or they can be smoothed in case of excess water, passing through the lower horizons of the soil and reaching the groundwater.

It should be mentioned that the pollution of the soil can also occur due to accidental causes (discharge from the tailings ponds of mining or extractive units, the action of precipitations on the dumps formed on the ground following the technological process of the mining or extractive units, etc.) (Chonokhuu et al., 2019).

The effects of heavy metals on the soil depend on the complex reactions between them and the components of the solid, liquid, and gaseous phases of the soil.

Reactions involving heavy metals in the soil include dissolution, adsorption, complexation, migration, precipitation, occlusion, diffusion (in minerals), binding to organic matter, absorption and adsorption by microorganisms, and volatilization.

These reactions are governed by the properties of the soil, namely, pH, redox potential, organic matter content, cation exchange capacity, clay and oxyhydroxide content of Fe and Mn. The presence of the heavy metals in soil has harmful effects on the biological productivity, more visible of the plants that absorb them, but also of the microorganisms in the soil, diminishing the amount of organic matter produced by them.

Heavy metals polluted agricultural soils can produce seemingly normal crops, but they can be hazardous for animal or human consumption. Because, heavy metals when accumulated in plants, enter the herbivorous/carnivorous–human food chain or through the direct consumption of affected plants by man.

Another danger is the possibility of heavy metals from the soil or from the soil reaching through the surface water or percolation into the surface waters, respectively in the groundwater, thus entering other trophic chains having the human to end.

The problem of heavy metal pollution is further complicated by the presence of other pollutants in the atmosphere (SO_2, NO_x, HF), resulting from technological processes, often the same ones that generate heavy metal emissions (e.g., combustion of fuels) and the presence of which in addition generates acid precipitation which contributes to the acidification of the soil, thus increasing the mobility of heavy metals (Živančev et al., 2019).

From the behavioral point of view and implicit risks to humans, the heavy metals can be classified as follows:

- Major bioaccumulation in water: Hg, Cd, Pb, Cu, Zn, Sr.
- Major bioaccumulation in soil: Cd, Zn, Ni, Sn, Cs, Rb.
- Sorption from the gastrointestinal tract: Cd, Hg, Zn.
- Penetration through the placenta: Cd, Hg, Pb, Zn.
- Penetration of the blood–brain barrier: Hg, Al, Pb.
- Reactions with thiol groups of proteins: Hg, Pb, Cd, Se.
- Alteration of the nucleic acid chain: Cd, Cu, Zn, Hg, Ni.

The human consumption of plants grown on heavy metal-polluted soils or of products obtained from the herbivores that consume these plants cause the heavy metals to reach the human body, with serious, sometimes dramatic consequences on health (Zeng et al., 2019).

It should be mentioned that in addition to heavy metals ingested through food, they can be inhaled (from powders in the atmosphere) as well as can accidentally be ingested (especially by children).

1.2.5   THE HAZARD OF SOIL CONTAMINATION WITH PERSISTENT ORGANIC COMPOUNDS AND ITS PREVENTION

As previously shown, POPs are chemicals that remain intact in the environment for long periods of time, from where they are transported and enter living organisms, including humans, where they accumulate in their fatty tissue.

Under the conditions of our country, from the agricultural point of view of the pollution of soils with persistent organic compounds, the pollution of agricultural soils with organochlorinated insecticides (Sruthi et al., 2017) was significant.

Organochlorinated insecticides, namely, those based on HCH (hexachlorocyclohexane) and DDT (pp'–dichlordiphenyl–trichloroethane) were introduced in use in the 1940s and have long been used to protect agricultural crops against vector insects of certain diseases. Organochlorinated insecticides HCH and DDT (isomers and metabolites) are very persistent, accumulating in the soil; the half-life in soils being 2 years for HCH and decades for DDT.

The widespread use of these pesticides, as well as their high persistence, have made their residues and metabolites evident in all elements of the environment.

There have been studies that have shown that in areas where soil is contaminated with DDT, there are residues of it and its metabolites in plants, in animals, and human adipose tissue (Gulan et al., 2017).

1.2.6   THE HAZARD OF SOIL CONTAMINATION WITH PETROLEUM PRODUCTS/RESIDUES AND ITS PREVENTION

Soil and aquifer contamination with petroleum products/residues, as well as the risks arising from soil contamination with such products have become serious problem. Given that the greatest danger of pollution with petroleum products/residues appears in the pipeline transport network, some preventive measures are presented below:

- Conducting of internal inspections in pipes with the help of the smart go-devil for checking the wall thickness.
- Strict compliance with the regulations regarding the periodicity of the pipeline controls, the periodicity of the pipeline reviews and repairs, the control of the pumping stations, the measures taken in case of faults or damages, the use of the depollution materials.
- Use of the SCADA system—for monitoring, control and data acquisition—for tracking the entire network of oil and pipeline transport pipelines (gasoline, liquid ethane, and condensate) equipped with the leak detection function to detect the area where the damage occurred and amount of pollutant escaped (Saravanan et al., 2018).
- The purchase of mobile slurry treatment devices recovered from the oil storage tanks, taking into account that the slurry recovered from the tanks is not neutralized (Akhigbe et al., 2019).

1.2.7 EUROPEAN POLICIES AND INSTRUMENTS FOR THE PROTECTION OF HUMAN HEALTH AND THE ENVIRONMENT AGAINST HAZARDOUS CHEMICAL SUBSTANCES AND MIXTURES

The European Union, an organization of which Romania is a part, has developed policies and instruments for the protection of human health and the environment against dangerous chemical substances and mixtures. The REACH Regulation for registration, evaluation, authorization, and restriction of chemicals is a key element of European Union regulations to make the use of chemicals as safe as possible.

The REACH Regulation addresses long-term exposures to chemicals and their implications for human health and environmental quality.

The REACH regulation creates a unique regulatory system that deals with industrial chemicals and seeks to eliminate the shortage of information on them by placing responsibility on the chemical manufacturers who are required to provide safety information on their substances and to manage the hazards from chemicals (Money et al., 2014).

The philosophy of the REACH regulation is that no chemical, in any form, should be placed on the market without proper documentation.

Each manufacturer and importer of chemicals who distributes quantities in excess of 1 ton per year is required to register them with the ECHA and to present information on their properties and uses and safe methods of handling them.

The REACH Regulation also calls for the progressive replacement of the most hazardous chemicals with appropriate alternatives. The first registration deadline ended on November 30, 2010 with approximately 5000 chemicals being registered with ECHA.

This marks an important step toward the safe management of chemicals. The CLP Regulation on the classification, labeling, and packaging of chemicals that aligned the existing classification system in the EU with the international system developed by the United Nations called the Global Harmonized System (UNGHS) of Classification and Labeling of Chemicals.

By classifying, labeling and packing chemicals according to the CLP Regulation, it is ensured that users and handlers of chemicals are aware of the dangers of their use and handling. All manufacturers and importers of hazardous chemicals have been obliged from 1 December 2010 to label the chemicals in accordance with the CLP Regulation and to notify the ECHA.

Thus, more than 100,000 chemicals were notified to ECHA, which allowed ECHA to make an inventory of the classification and labeling of chemicals and to provide users and chemists with the information needed to limit the risks and to select safer chemicals.

The HBM system for human biomonitoring involves collecting and analyzing samples of human tissues or body fluids to identify the presence and levels of chemicals in the human body. Chemicals are taken directly or indirectly from the environment. The HBM system is an important tool for assessing human exposure to environmental substances.

The EU Commission for Scientific Research has funded a research program that develops a human biomonitoring framework at EU level.

In this research project, we tested the feasibility of a coherent approach to human biomonitoring in Europe by identifying biomarkers of mercury, cadmium, phthalates, and tobacco smoke from the environment and in human hair and human urine.

The project will collect biological samples and information from 120 mother–child pairs from each participating EU country (Malkiewicz et al., 2009).

1.3   MANAGEMENT OF THE POTENTIALLY CONTAMINATED/ CONTAMINATED SITES. METHODS AND TECHNOLOGIES FOR SOILS DEPOLLUTION

1.3.1   METHODOLOGY FOR MANAGING THE POTENTIALLY CONTAMINATED/CONTAMINATED SITES

Because in the existing legislation these terms are not defined, and in order to eliminate the confusion regarding the use of these terms, it was considered necessary to define them, namely, the historically contaminated area is: "A contiguous site (land and/or aquifer layer) on which the anthropic activities have determined the presence of polluting substances in concentrations that present and/or may present, both for the existing site and for the surrounding areas, an immediate or long-term risk for the health of the population and the environment" (Vanni et al., 2016).

On a contaminated site, one or more sources of contamination can be found. The potentially historically contaminated area is: "A contiguous site (land and/or aquifer layer) that historical and/or present activities could have generated, both for the existing site and the surrounding areas, has a significant impact on the health of the population or the environment. This definition emphasizes contamination caused by past industrial practices that led to contamination as a result of improper storage/handling/ use of chemicals, soil and/or aquifer." The contaminated site management system comprises four phases:

- Identification and registration.
- Preliminary assessment.
- Detailed evaluation.
- The remedy.

A potentially contaminated site can be classified as a "suspicious" site following an analysis based on the following elements:

- Evaluation of the information presented by the owner of the site.
- Regular or irregular site inspections by local authorities and control bodies.
- Analysis of the effects of natural disasters on the site.
- The notification made by the owner to the regulatory bodies (Beccaloni et al., 2010).

A preliminary assessment should be carried out on all "potentially contaminated" sites. The purpose of the preliminary assessment is to determine whether the degree of contamination is real or not and whether a detailed site assessment is required. For this purpose, laboratory analyses will be carried out for the samples taken from the contaminated sites.

For the industrial sector, indicator parameters have been developed which are used to indicate any contamination, these being:

- Total oil hydrocarbons (HPT)
- Total organic halogens (HOT)
- Benzene, toluene, ethylbenzene, xylene (BTEX)
- Total volatile organic compounds (TVOC)
- Petroleum and lubricants (PL)
- Heavy metal ions (IMG)

Corresponding indicator parameters for each industry must be listed. There are three possible outcomes of the preliminary assessment of the contaminated site, namely,

- No further investigations are necessary—the site will be kept in inventory.
- Further investigations are required—a detailed evaluation of the site will be carried out.
- Remediation—the activities of cleaning and remediation of the contaminated site will be started immediately (Power et al., 2010).

Detailed assessment of the contaminated site requires the following steps:

- Characterization of the site and its contamination: on this occasion, information and data on the history of the site will be collected. The Conceptual Model for the Site (MCA) is being developed. All possible links are identified: source—transfer–receiver path; samples will be taken from the surface and from the soil profile and the concentration of contaminants will be determined following laboratory analyses. Samples will be collected from the groundwater to know if it is contaminated or not.
- Generic risk assessment: the measured contaminant concentration is compared with the soil quality standard to define the transfer paths, and the generic scenario will include inhaling fugitive particles,

inhaling volatiles, ingesting groundwater, dermal absorption. If the values in the standard are exceeded, the risk assessment for all potential open transfer paths will begin, all transfer routes will also be identified.

- Detailed hazard assessment: information about the specificity of the site is collected. Specific levels of risk are calculated. If the risk level is higher than the acceptable risk level, the site is rated as contaminated, needing to be remedied. For this purpose, the level of site cleaning is established, all the alternatives of cleaning technologies will be identified and the technology will be chosen appropriate to meet the remediation goals. In the remediation phase, the design and execution of the remediation work is carried out. The effects of the fix will be monitored both during the execution and postexecution periods (Dijkstra et al., 2018).

1.3.2 ECOLOGICAL REMEDIATION OF DEGRADED SOILS

With regard to the concept of ecological remediation of degraded soils, different terms have been proposed depending on the intensity of the ecosystems damage and the nature of the interventions that must be carried out for their restoration. In this sense, the following categories related to the respective concept can be defined, which are of interest to the process of ecological reconstruction:

- Ecological reconstruction (directed ecological recovery)—through which the reconstruction of a supra individual biosystem similar to the previous one is realized (e.g., restoration of the conditions of nutrition, of pH, of humidity, trophic level, and of the composition and structure of the biosystem);
- Ecological improvement—it is a much more intense action through which biosystems are realized that respect mainly the functionality and less structure and composition (e.g., salted soils improvement, sands improvement, modification of the hydrological and water regimes of the soil by drying or irrigation, plantations with species other than the zonally ones after the restoration of the soil, etc.);
- Ecological reconstruction—by which an artificial distribution of species is ensured in supraindividual biosystems according to arrangements considered optimal in which generally the function of

environmental protection (e.g., execution of works requiring large volumes of earth, such as terracing, selective levelling of earth materials and installation of biocenoses other than the initial ones) is considered important (Alvarenga et al., 2019).

From the ecological reconstruction point of view of the soils, it is sought that the fulfilment of the ecological functions of the soil in the ecosystem is to be ensured by the morphological, physical, chemical, and mineralogical functions, if not of the whole ecopedological profile, at least of an important part of it, in which the surface horizon and the immediately adjacent ones have a decisive role (Bohan et al., 2017).

Soil characteristics with an essential role in ecological reconstruction include edaphic volume, ion exchange capacity, humus quantity, soil reaction, and soil permeability.

1.3.3 THE CONTROL AND REDUCTION OF ADVERSE EFFECTS (ON HUMAN HEALTH AND/OR THE ENVIRONMENT) GENERATED BY SOIL POLLUTION WITH HAZARDOUS SUBSTANCES AND CHEMICAL MIXTURES, THROUGH APPROPRIATE REMEDIAL ACTIONS

In the case of sites contaminated with pollutants, either security procedures or remediation procedures can be applied.

While remediation ensures the quantitative and qualitative elimination or reduction of pollutants, the purpose of the securitization is to raise barriers to prevent the spread of pollutants over wider areas. Since the source of pollution generally remains, and the high barriers are subject to degradation and aging, securing is only a temporary measure, the remedy being the safest procedure to apply.

The remediation methods can be classified according to the following criteria: their place of application and the nature of the processes that are executed (Saupe et al., 2018).

Thus, according to the first criterion, ex situ and in situ processes are distinguished, and after the second criterion, thermal, physicochemical, and biological processes are distinguished. Ex situ treatment methods require the mobilization of contaminated soil followed by its treatment either on the same site (on-site remediation) or in a specialized off-site soil treatment facility (off-site remediation).

In situ treatment methods are performed directly in the contaminated site, without the need to mobilize soil from the site.

The thermal remediation processes are based on the transfer of pollutants from inside the soil in the gaseous phase through the contribution of thermal energy.

The pollutants are released from the soil by vaporization, migrate through the soil to the surface, from where they are captured and evacuated to the atmosphere after purification treatment.

Physicochemical processes are generally processes of extraction and/ or wet sorting.

The principle of ex situ extraction procedures of soils is to wash them in order to concentrate the pollutants in a residual fraction as small as possible, the water being the most commonly used extraction agent.

For the transfer of contaminants from the soil to the extractant, two mechanisms are important, namely, the creation of strong shear forces induced by pumping, mixing, vibrating or by using high pressure water jets, which break the particle agglomerations of polluted and unpolluted particles and disperse the contaminants in the extraction phase and cause dissolution of the contaminants by the components of the extracting phase (Richardson et al., 2010).

In situ extraction consists of the percolation of an aqueous extracting agent through the contaminated soil. The percolation can be done by surface grooves, horizontal drains, or deep vertical wells. The soluble contaminants in the soil dissolve in the percolating liquid that is pumped and subsequently treated on-site.

The biological processes are based on the action of microorganisms that have the ability to convert organic pollutants mainly into CO_2, water, and biomass or to immobilize pollutants by binding to the humic fraction of the soil.

Degradation is usually done under aerobic conditions, or more rarely under anaerobic conditions. In order to make the process more efficient, it is essential to optimize the development conditions of microorganisms (oxygen supply, pH, water content, etc.).

Stimulation of biological activity can be achieved by homogenizing the soil, active aeration, humidification or drying, heating, addition of nutrients or substrates, and inoculation with microorganisms (Yang et al., 2019).

Biological processes require much lower energy input than thermal or physical–chemical ones, but require longer treatment periods.

1.3.4 CURRENT TECHNOLOGIES FOR THE REMEDIATION OF POLLUTED SOILS

Representative technologies that are currently required to remedy soils contaminated with hazardous substances or chemical mixtures such as petroleum products/residues, heavy metals, persistent organic compounds, pesticides/biocides are presented below.

Soil remediation technologies for soils contaminated with petroleum products/residues commonly used today are called bioremediation technologies that include a number of systems or processes that utilize microorganisms to treat soil and groundwater for the purpose of degradation or decomposition of petroleum hydrocarbons (Ye et al., 2017).

The bioremediation process can be performed on-site or outside the site. The in situ process is used in cases where soil mobilization is an impractical solution and involves methods, such as biostimulation, bioventilation, biosparging or natural attenuation.

Biostimulation involves aeration and application of carefully selected micronutrients and biostimulants for indigenous microbial populations present in the substrate to degrade contaminants.

Biosparging is a rehabilitation technique in which a gas (most commonly air) and sometimes nutrients are injected into the contaminated area to stimulate the aerobic biodegradation process carried out by the biomass of native microorganisms. Air injection is carried out in the area saturated with soil pollutant or in groundwater.

Bioventing is a process commonly used for the biological treatment of hydrocarbon–contaminated soils. It consists of injecting relatively low air (oxygen) flow into the contaminated soil area to promote aerobic microbial biodegradation.

Natural attenuation or intrinsic remediation is the simplest form of bioremediation, which requires no intervention other than to demonstrate that indigenous populations exist and can have action to degrade the pollutants and monitor the degradation process.

For ex situ bioremediation, there are different already verified technologies that can be used for both depolarization of the unsaturated and saturated areas. These include the "biopile" method, the "land farming" method, and the composting or the soil treatment in the bioreactor.

The biopile method consists in the biological treatment of excavated soils placed in the pile on-site or off-site with the control of parameters, such as concentration in oxygen, soil moisture, content in mineral nutrients and microorganisms.

The method of land farming consists of depositing the soil contaminated with organic products on a surface prepared in advance, in a layer of reduced thickness on well isolated areas to protect the subsoil from any risk of infiltration.

The treatment in the bioreactor involves the introduction of the polluted soils in the form of sludge into a reactor equipped with agitation and aeration systems.

Composting is a biological process whereby dangerous organic contaminants are transformed by microorganisms under both aerobic and anaerobic conditions into stabilized nonhazardous products (Yang et al., 2019).

The remediation technologies for contaminated soils with heavy metals that are becoming more and more present are biological technologies which are "gentle" remediation technologies that maintain or restore the natural fertility of the soil, as opposed to "brutal" technologies which is characterized mainly by the manipulation of enormous quantities of soil or by the use of chemical reagents that cause the treated soil to lose its fertility.

For the rehabilitation of the soils contaminated with heavy metals, the most commonly used biological treatments in our country is the phytoremediation.

Phytoremediation uses plants to extract, fix and/or neutralize pollutants. This is an efficient, noninvasive, economically efficient, aesthetically pleasing, and socially accepted method for the remediation of polluted areas (Song et al., 2017).

Plants are the ideal means of remedying soil and water due to their genetics, biochemical, and physiological properties.

According to the main mechanisms involved in the process, phytoremediation can be classified as follows:

- Rhizofiltration—absorption, concentration, and precipitation of heavy metals by the roots of plants.
- Phytoextraction—technique that involves the whole organism of the plant in the process of taking contaminants from the soil.
- Phytotransformation—degradation of complex organic molecules into simpler molecules and incorporation of these molecules into plant tissues.

- Phytostimulation (plant-assisted bioremediation)—stimulates the degradation by bacteria and fungi by giving up exudates/enzymes in the root zone (rhizosphere).
- Phytostabilization—involves the absorption and precipitation of contaminants, especially metals, by plants, reducing their mobility and preventing washing to groundwater, or air or entering the food chain (Del Buono et al., 2020).

Of these methods, only phytoextraction and phytostabilization can be applied for remediation of polluted soils with heavy metals.

Phytoextraction is a phytotechnology that involves the cultivation of one or more species of hyperaccumulators of heavy metal plants, to which optimal development conditions are created to ensure the production of maximum plant mass in order to extract, accumulate, and remove a larger amount of metals.

The vegetal mass will be harvested, then it will be subjected to other metal extraction treatments or it will be dried and incinerated, and the ash is deposited in a controlled garbage dump. Remarkable plant species that accumulate concentrations of heavy metals up to 100 times higher are used than the normally grown plants.

Phytostabilization is used to restore the vegetation of the place, while decreasing the potential for migration of contaminants under the influence of wind and water erosion and of leaching to groundwater.

Another in situ "gentle" technology for the decontamination of metal-contaminated soils, which tends to impose more and more at present, is the electrochemical remediation technology (also called electrokinetic) (Radziemska et al., 2017).

The electrochemical remediation process can remove metals and organic compounds from soils with reduced permeability. This method uses electrochemical and electrokinetic processes to extract and then remove pollutants such as metals in ionized form.

The remediation technologies of soils contaminated with POPs that are applied more frequently are the bioremediation technology (in situ technologies presented above) as well as the thermal desorption technology that are discussed below. Thermal desorption is a thermal method that works on the principle of heating the polluted soil, so that the polluting substances evaporate, thus being able to move easily through the soil to the area of a system of wells that collects and evacuates them. Thermal desorption can be applied in situ or ex situ (Rajmohan et al., 2020).

In situ thermal desorption can be done by the following methods:

- Steam injection—steam is injected into the soil using wells dug in the polluted area. The steam heats the area and mobilizes, evaporates, and/or destroys the pollutants. The water and gases produced are collected in special wells, from where they are directed to treatment plants.
- Hot air injection—this process is similar to steam injection except that hot air is injected through wells instead of steam. Hot air heats the soil thus causing the evaporation of pollutants.
- Hot water injection—this process is similar to the two processes mentioned above. It involves introducing hot water instead of steam or hot air.
- Heating by electrical resistance—a process by which an electric current is applied to the ground through a network of steel electrodes. The heat generated by the electrical resistance of the soil when passing the electric current vaporizes the groundwater and the water from the soil creates the conditions for the evaporation of the pollutants.
- Radio wave heating—a process that involves placing an antenna that emits radio waves in a well. Radio waves heat the soil and thus causes the evaporation of pollutants.

In some cases, thermal desorption is followed ex situ, which is a more expensive process but the treatment process is much better controlled.

Typical ex situ thermal desorption systems work in several successive phases, namely, pre-treatment, desorption, and posttreatment of the solid material and the resulting gases.

- Pre–treatment involves the processes, such as sorting, dehydration, neutralization, and mixing.
- Desorption is carried out in thermal installations called desorption, whose classification is usually made according to the heating system used.
- Posttreatment of waste gases—depends on the specific factors of the equipment and may include combustion at high temperatures (over 1400°C) followed by purification and disposal, combustion at moderate temperatures (200°C–400°C) using catalysts or the purification and removal of the burnt gases (O'Brien et al., 2017).

1.4　THE MONITORING OF SOIL QUALITY

1.4.1　THE MONITORING OF SOIL QUALITY STATUS

Starting with 1977, according to the recommendations of the U.N.E.P., the Soil Quality Monitoring System was established as an integral part of the National Environmental Quality System. In 1992, in order to improve and modernize the existing surveillance process, a new Soil Quality Monitoring System was set up, which aims to achieve an integrated national system for monitoring the quality of agricultural and forest soils, which will be harmonized with similar European systems.

The Soil Quality Monitoring System is characterized by four basic elements:

- Spatial distribution.
- The density of the observation network.
- The set of indicators.
- The periodicity of the determinations (Teng et al., 2014).

According to Figure 1.1, the basic elements of the monitoring system are tracked on three levels of detail:

The first level is characterized by the following elements:

- At the macro scale, the spatial distribution of the soil sampling points (sites) lies at the intersections of a fixed observation network, covering the whole territory with a density of 16 × 16 km (in accordance with the recommendations of the document "Convention on Long Transboundary Air Pollution").
- At the micro scale, the selection and location of an observation point is made inside a square with a 400 m side, which has the center (the point of intersection of the diagonals) in the nodes of the observation network.
- From each observation point, a set of samples will be collected to determine the basic indicators which contains an average sample in disturbed structure, composed of 20–25 individual samples from the first 10 cm of soil depth, respectively three individual samples for each soil horizon and four samples in natural structure on the horizon.
- The periodicity of the determinations is of 4–10 years for the observation points without special problems and of 1–2 years for the

observation points with problems (in which processes of degradation of the quality of the soil take place, e.g., pollution) (Mukhopadhyay et al., 2016).

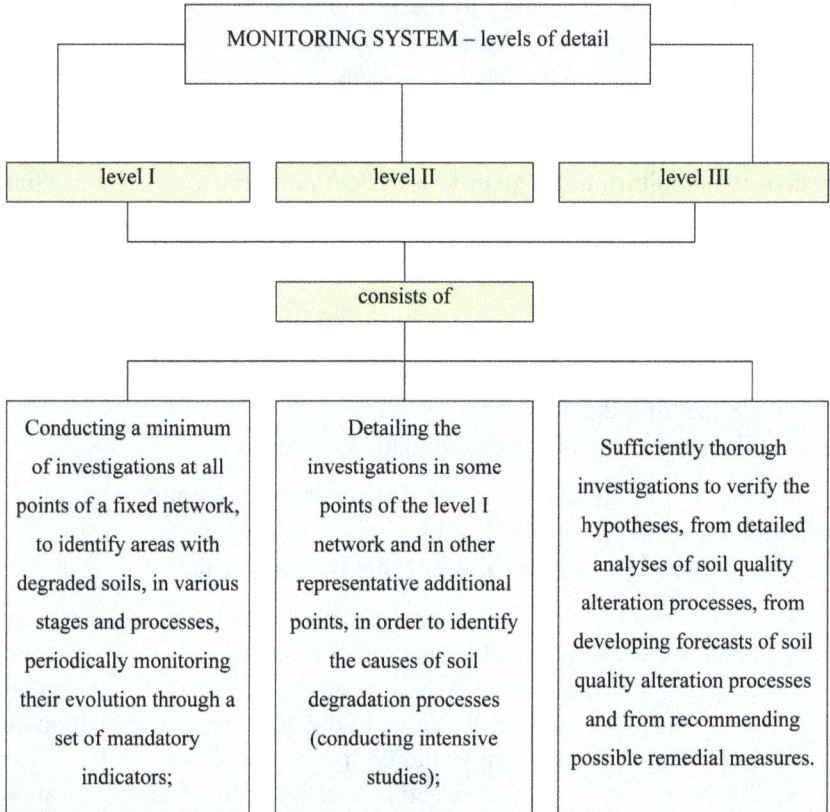

FIGURE 1.1   Detailed levels of the monitoring system.

1.4.2   *THE SOIL QUALITY MONITORING SYSTEM IN THE 8 × 8 KM PROFILES NETWORK*

The soil/land monitoring system for agriculture comprises the Program for the elaboration of pedological and agrochemical studies and the Program for the distribution of funds for the realization/updating the National soil quality monitoring system. The main objectives of the new soil quality monitoring system in the 8 × 8 km network include the following:

- Systematic monitoring of the qualitative and quantitative character-istics of the soils.
- Elaboration of forecasts regarding the evolution of soil quality.
- Warning of interested organisms on negative soil problems.
- Filling databases with complete soil profiles.

The new soil quality monitoring system, having the same levels of detail as the previous one is characterized by the following elements:

- At the macro scale, the network of 8 × 8 km profiles will have 2914 points for taking soil samples.
- At the micro scale, the methodology of site placement must ensure the representativeness of the site in relation to the weight of the soil units in the analyzed area, and its fixation in the territory will respect the rule of moving if necessary, namely, 100, 200, 300, 400 m from the normal point in the north, south, east or west order.
- The laboratory analyses that will be performed for the national soil quality monitoring system in the 8 ×8 km profiles network.
- The data obtained from the field and the laboratory will be processed taking into account the regulations regarding the assessment of environmental pollution, with the subsequent modifications and completions (Yu et al., 2014).

In the field works, the methodological specifications presented below will be respected:

- The location of the level I monitoring sites (8 ×8 km) on the ground will be based on the predetermined geographical coordinates using 1: 25,000 scale maps such as larger scale maps and sketches and appropriate equipment (roulette, magnetic compass, system global positioning–GPS, etc.). The methodology of site placement must ensure the representativeness of the site against the weight of the soil units in the analyzed area, and its fixation in the territory will respect the rule of moving if necessary, namely, 100, 200, 300, 400 m compared with the normal point, in the north, south, east, or west order. The new geographical coordinates will be noted.
- Setting up of the reference plots (sites) in the materialized points with areas of 400 m^2 (20 × 20 m) located in homogeneous, repre-sentative conditions. On the soil profile sheet or on an attached tab are entered the site location and the new geographical coordinates, and at the end of the work, a sketch board is given, on a scale of 1:

25,000 or greater, comprising the position of all sites within each site.

- The placement and execution of a main soil profile is done in a representative place within the site up to the physiologically useful depth (parental material or groundwater); the description of the soil profile will be carried out in accordance with the Methodology, by completing the fields in the Type M file "Field conditions and morphological data," taking into account the Profisol instructions. Greater attention will be paid to the delimitation of the horizons and the thickness of the sub-horizons. In the case of horizons with a thickness greater than 50 cm, they will be divided into sub-horizons. For the 0–50 cm layer, it is necessary to delimit two to three sub-horizons, and for the arable land, the horizon with the plow sole will be delimited.

- On the M–type sheet, all the obligatory, and as far as possible, optional features, including those regarding anthropic modifications, the effect and the status of the improvement works, will be noted. In the "Comments" area of the M–type sheet or in the supplementary sheets, data are entered regarding the position of nearby sources of pollution and the type of pollutants, prevailing winds, etc. The type M file is drawn up in three copies, one of which remains at the executant.

- Collection of the soil samples for chemical and physical analysis is performed from the profile on pedogenetic horizons and sub-horizons in disturbed and undisturbed state, and from the site surface being 0–20 cm (agrochemical sample) in disturbed state. As a general rule for disturbed state, for both chemical and for some physical analysis on samples collected in plastic or paper bags, the following rules are observed: (1) the blade of the harvesting utensil is cleaned every time from the soil with which it comes in contact, (2) the packaging avoids the contact of the label with the soil, (3) the samples do not increase, (4) the samples for chemical analysis from the soil profiles will have 1.5 kg for each horizon or sub–horizon, (5) agrochemical samples are collected in an amount of 1.5 kg from the surface of 400 m^2 around the soil profile, (6) the harvesting depth is 0–20 cm in the case of arable soils from level I sites and 0–10 and 10–20 cm in the wise level I sites, as well as in polluted sites of all uses, (7) soil profile samples for physical analysis will have 0.5 kg, the handling of the samples for physical analysis will be done carefully to avoid deterioration of the structural aggregates, with special attention being paid to very high or very low humidity, (8) at each profile, a micromonolite is collected from samples as far as possible with whole structural aggregates, having noted the iden- tification data on both the lid and the box itself, as well as the name and thickness of the horizons, and (9) after the last subhorizon, a sample of parent material and/or underlying rock is added.

- The samples in the cylinders are collected only from the freshly opened profile (1–3 h) and those from refreshed profiles are not allowed; four cylinders of 100 cm^3 or two to three cylinders of 200 cm^3 are harvested from each horizon or subhorizon, up to 1.5 m deep. In the case of arable soils, the subhorizon with the plow's sole will necessarily be harvested. In soils with shallow groundwater, sampling will be done up to that level, but the description will be up to 1.5 m (Incerti et al., 2007).

1.4.3 THE RISK ASSESSMENT DUE TO THE PRESENCE OF ENVIRONMENTALLY HAZARDOUS SUBSTANCES

The large number of compounds used in the world economy as well as the presence of newcomers every year is arguments in favor of the idea that a significant number of chemicals can be dangerous to humans and the environment.

Starting from the identification of all the dangerous substances for the human being and taking into account that the human being is only part of the environment, the idea that the risk assessment must take into account all the elements of the environment in order to reduce the possibilities of industrial accidents.

The element of danger for the environment is made up of three elements: the source of the danger, the path of transmission and the receiver, and these are interconnected (Sample et al., 2015).

The source of hazard for the environment can be provided, on the one hand, by the quantity of hazardous materials existing, and on the other hand, by processes that provide for the handling or transport of dangerous substances.

A chemical compound or mixture of compounds can be defined and evaluated based on their physical, chemical, and biochemical properties, as well as their ecotoxicological properties or their persistence in the environment. The assessment of the hazard indicator of an industrial unit is made on the basis of the quantity of hazardous materials stored in the territory of the industrial unit.

The hazard indicator is established on the basis of quantitative limits regulated by legislation for each type of material. The physical path through which the released pollutant reaches and propagates into the environment is the path of hazard transmission. If a complete definition of the accident scenario is desired in order to evaluate its consequences, understanding the transport mechanism (s) involved becomes vitally important.

Related to the transmission path, the speed at which the pollutant reaches the source at the receivers is an essential aspect of an accident scenario (Kucher et al., 2018).

The transmission paths are not only limited to the airway (influenced by weather factors) but also include the terrestrial and underground ones through which the pollutant reaches outside the unit at kilometers distance.

A high propagation speed induces a potential for considerable environmental destruction, reducing the possibility of stopping pollutant leakage over long distances, as well as the time required to warn and evacuate pollutant areas.

The thermal accident, given the strong impact on the environment, can significantly affect the source, the transmission paths, and also the receivers in the path.

It should be noted that fires are usually accompanied by emissions of combustion residues which in turn can affect the environment.

The recipients of a chemical or biological accident are directly or indirectly linked to the evaluation of the consequences of that unwanted event. It should be borne in mind that preventive measures can be more easily applied to sources and routes of transmission, while it becomes more difficult to move or modify the receivers.

Due to the different nature of the receptors for standardizing tests, organisms from several classes of species are used, such as fish, algae, crustaceans, molluscs, amphibians, invertebrates, plankton, protozoa, insects (McIlwaine et al., 2014).

The stages of the hazard analysis for an industrial unit, system or process are shown in Figure 1.2 and the stages of the hazard reduction analysis are shown in Figure 1.3.

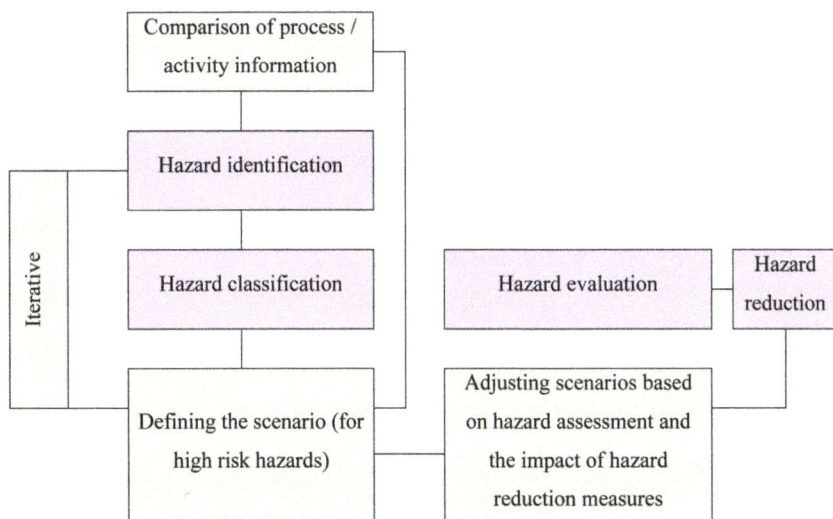

FIGURE 1.2   The stages of hazard assessment.

The analysis of an industrial unit from the point of view of environmental risk assessment not only involves the inventory of hazardous compounds and the quantities held by the respective unit but also involves the assessment of the frequency and consequences of accidents.

Estimating the consequences of accidents can be done based on transport, action, and evolution patterns of hazardous substances escape and includes as appropriate: dilution/dispersion of pollutants in the environment together with the concentration profiles over time in the affected area, dose–toxicity relationships, pollutant and environmental response-related ecotoxicity criteria, accident history, also delayed and long–term effects of pollutant leakage.

Of the total number of chemicals known at this time, about one hundredth of products is being marketed and used in large quantities.

Many of these chemicals are accidentally released and pose a danger to living organisms, human health, and the environment. These "hazardous substances" (pollutants, toxic substances) are characterized and classified on the basis of complex criteria and specified indices (Collins et al., 2016).

For example, taking into account the release mode, "environmental pollutants" (e.g., carbon dioxide, sulfur dioxide, etc.) can be identified as pollutants released typically at low concentrations, but over a long period of time and pollutants released spontaneously and unexpectedly at high concentrations which present an acute hazard (e.g., chemical compounds released in case of industrial accident).

The purpose of hazard reduction measures is to reduce both the likelihood of an accident occurring and to limit the consequences of an accident until they reach an acceptable level (according Fig. 1.3).

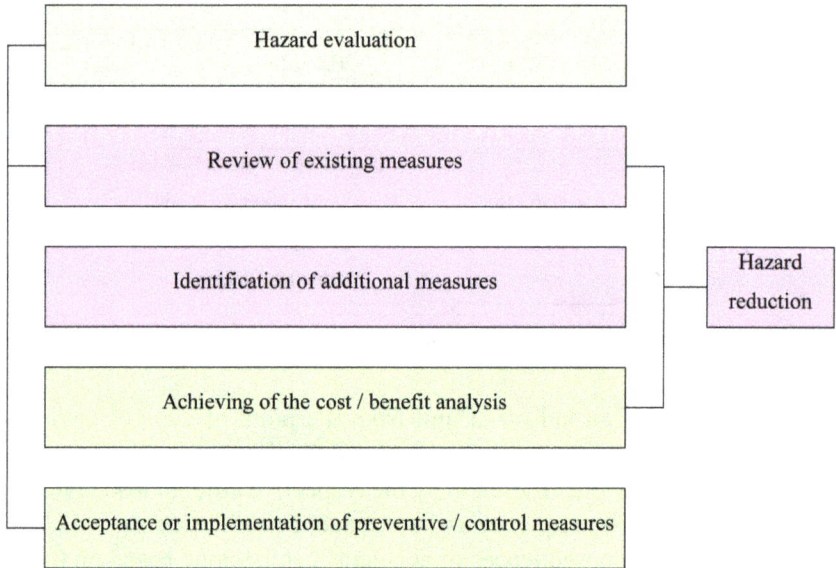

FIGURE 1.3   Stages and measures to reduce the hazard.

The estimation of the degree of danger implies the use of indicators, called indices of the dangerous substance.

In addition to these, there is also a set of environmental quality indicators (soil, air, water) that defines the degree of pollution of an area or classification in the area of acceptability in relation to certain regulations (norms on maximum quantities/concentrations allowed) (Bandow et al., 2018).

Possible effects of toxic substances on health or the environment include intoxication, suffocation, fire, oxidation (ignition initiation), explosion, chemical burns, frostbite, infections, and environmental contamination (soil, air and water).

A possible classification of the substances by the produced effect could be:

• Asphyxiating or suffocating (e.g., methane, carbon dioxide, etc.).

- General toxicants (e.g., carbon monoxide, hydrocyanic acid, etc.).
- Irritants (e.g., strong acids and bases, phenols, etc.).
- Narcotics (e.g., nitrous oxide, hydrocarbons, etc.);
- Various toxicants (metal vapors, volatile metals, etc.).

Hazardous substances to the environment have required the establishment of a separate category, which is included for the purpose of regulating chemicals that present a risk of major accidents (Machado et al., 2018).

Thus, the occurrence in this category of substances very toxic to aquatic organisms, as well as toxic substances that can induce long-term adverse effects in the aquatic environment is justified.

1.5 CONCLUSIONS AND RECOMMENDATIONS

In general, the term "hazard" is associated with the probability of an unwanted event occurring in an industrial installation and is less associated with its consequences. Industrial safety represents a quality of the system, being expressed indirectly through its relation with the notion of hazard. The most common definitions of hazard used in the literature are the following:

- A combination of the likelihood of an unwanted event occurring and its consequences.
- Probability of occurrence of an undesirable event for several possible scenarios of its production.
- The product between the probability of the occurrence of the unwanted event and its quantifiable consequences.

Quantifiable, the hazard can also be defined as a function of the probability that a specific undesirable event will occur in the future with a certain periodicity or in specific circumstances. The hazard can be expressed in terms as:

- Qualitative (high, medium, small, tolerable, intolerable, acceptable).
- Semi-qualitative, when the two components of the hazard (the likelihood of an accident occurring and the consequences of the unwanted event) are qualitatively defined, but then combined to provide pseudoquantitative values of the hazard that allow a hierarchy of the scenarios of unwanted events.

- Quantitative, by numerically evaluating the likelihood function of an unwanted event, as well as indicators that take into account the potential consequences of an accident.

Two major categories of hazard assessment methods for an industrial unit or a chemical process can be highlighted: (1) probabilistic (sampling) methods based on estimating the probability of occurrence of future events and their consequences and (2) analytical methods based on deterministic mathematical models and on the mechanism of the studied phenomena.

KEYWORDS

- **hazards**
- **chemicals**
- **dangerous**
- **soil**
- **threat**
- **contamination**

REFERENCES

Akhigbe, G. E.; Adebiyi, F. M.; Torimiro, N. Analysis and Hazard Assessment of Potentially Toxic Metals in Petroleum Hydrocarbon–Contaminated Soils Around Transformer Installation Areas. *J. Health. Pollut.* **2019,** *9* (24), 191213. doi: 10.5696/ 2156–9614–9.24.191213.

Allen, D. T.; Torres, V. M.; Thomas, J.; Sullivan, D. W.; Harrison, M.; Hendler, A.; Herndon, S. C.; Kolb, C. E.; Fraser, M. P.; Hill, A. D.; Lamb, B. K.; Miskimins, J.; Sawyer, R. F.; Seinfeld, J. H. Measurements of Methane Emissions at Natural Gas Production Sites in the United States. *Proc. Natl. Acad. Sci. USA.* **2013,** *110* (44), 17768–17773.

Alvarenga, P.; Rodrigues, D.; Mourinha, C.; Palma, P.; de Varennes, A.; Cruz, N.; Tarelho, L. A. C. Rodrigues S. Use of Wastes from the Pulp and Paper Industry for the Remediation of Soils Degraded by Mining Activities: Chemical, Biochemical and Ecotoxicological Effects. *Sci. Total Environ.* **2019,** *686,* 1152–1163.

Bandow, N.; Gartiser, S.; Ilvonen, O.; Schoknecht, U. Evaluation of the Impact of Construction Products on the Environment by Leaching of Possibly Hazardous Substances. *Environ. Sci. Eur.* **2018,** *30* (1), 14. doi: 10.1186/s12302–018–0144–2. Review.

Banks, P. J.; Banks, J. C. Relationship between Soil and Groundwater Salinity in the Western Canada Sedimentary Basin. *Environ. Monit. Assess.* **2019,** *191* (12), 761.

Beccaloni, E.; Vanni, F.; Giovannangeli, S.; Beccaloni, M.; Carere, M. Agricultural Soils Potentially Contaminated: Risk Assessment Procedure Case Studies. *Ann. Ist Super Sanita.* **2010,** *46* (3), 303–308.

Bohan, D. A.; Vacher, C.; Tamaddoni–Nezhad, A.; Raybould, A.; Dumbrell, A. J.; Woodward, G. Next–Generation Global Biomonitoring: Large–Scale, Automated Reconstruction of Ecological Networks. *Trends Ecol. Evol.* **2017,** *32* (7), 477–487.

Chen, L. J.; Feng, Q.; Li, C. S.; Song, Y. X.; Liu, W.; Si, J. H.; Zhang, B. G. Spatial Variations of Soil Microbial Activities in Saline Groundwater–Irrigated Soil Ecosystem. *Environ. Manage.* **2016,** *57* (5), 1054–1061.

Cherubin, M. R.; Karlen, D. L.; Cerri, C. E.; Franco, A. L.; Tormena, C. A.; Davies, C. A.; Cerri, C. C. Soil Quality Indexing Strategies for Evaluating Sugarcane Expansion in Brazil. *PLoS One* **2016,** *11* (3), e0150860. doi: 10.1371/journal.pone.0150860.

Chonokhuu, S.; Batbold, C.; Chuluunpurev, B.; Battsengel, E.; Dorjsuren, B.; Byambaa, B. Contamination and Health Risk Assessment of Heavy Metals in the Soil of Major Cities in Mongolia. *Int. J. Environ. Res. Public Health* **2019,** *16* (14). doi: 10.3390/ijerph16142552.

Collins, C. D.; Baddeley, M.; Clare, G.; Murphy, R.; Owens, S.; Rocks, S. Considering Evidence: The Approach Taken by the Hazardous Substances Advisory Committee in the UK. *Environ. Int.* **2016,** *92–93,* 565–568.

De Waele, J.; D'Haene, K.; Salomez, J.; Hofman, G.; De Neve, S. Simulating the Environmental Performance of Post–Harvest Management Measures to Comply with the EU Nitrates Directive. *J. Environ. Manage* **2017,** *187,* 513–526.

Del Buono, D.; Terzano, R.; Panfili, I.; Bartucca, M. L. Phytoremediation and Detoxification of Xenobiotics in Plants: Herbicide–Safeners as a Tool to Improve Plant Efficiency in the Remediation of Polluted Environments. A Mini–Review. *Int. J. Phytoremed.* **2020,** *21,* 1–15.

Di Falco, S.; Zoupanidou, E. Soil Fertility, Crop Biodiversity, and Farmers' Revenues: Evidence from Italy. *Ambio* **2017,** *46* (2), 162–172.

Dijkstra, J. J.; van Zomeren, A.; Brand, E.; Comans, R. N. J. Site–Specific Aftercare Completion Criteria for Sustainable Landfilling in the Netherlands: Geochemical Modelling and Sensitivity Analysis. *Waste Manage.* **2018,** *75,* 407–414.

Eudoxie, G. D.; Mathurin, G.; Lopez, V.; Perminova, O. Assessment of Pesticides in Soil from Obsolete Pesticides Stores: A Caribbean Case Study. *Environ. Monit. Assess.* **2019,** *191* (8), 498.

Gabbert, S.; Hilber, I. Time Matters: A Stock–Pollution Approach to Authorisation Decision–Making for PBT/vPvB Chemicals under REACH. *J. Environ. Manage.* **2016,** *183,* 236–244.

Gopinath, R.; Poopathi, R.; Vasanthavigar, M.; Arun, R.; Mahadevan, M. Stabilized Red Soil–an Efficient Liner System for Landfills Containing Hazardous Materials. *Environ. Monit. Assess.* **2018,** *190* (10), 590.

Green, S. M.; Machin, R.; Cresser, M. S. Effect of Long–Term Changes in Soil Chemistry Induced by Road Salt Applications on N–Transformations in Roadside Soils. *Environ. Pollut.* **2008,** *152* (1), 20–31.

Gulan, L.; Milenkovic, B.; Zeremski, T.; Milic, G.; Vuckovic, B. Persistent Organic Pollutants, Heavy Metals and Radioactivity in the Urban Soil of Priština City, Kosovo and Metohija. *Chemosphere* **2017,** *171,* 415–426.

Incerti, G.; Feoli, E.; Salvati, L.; Brunetti, A.; Giovacchini, A. Analysis of Bioclimatic Time Series and Their Neural Network–Based Classification to Characterise Drought Risk Patterns in South Italy. *Int. J. Biometeorol*. **2007,** *51* (4), 253–263.

Kucher, S.; Dsikowitzky, L.; Ricking, M.; C. H. S.; Schwarzbauer, J. Degree of Phenyl Chlorination of DDT–Related Compounds as Potential Molecular Indicator for Industrial DDT Emissions. *J. Hazard. Mater*. **2018,** *353*, 360–371.

Machado, E. R.; Valle Junior, R. F. D.; Pissarra, T. C. T.; Siqueira, H. E.; Sanches Fernandes, L. F.; Pacheco, F. A. L. Diagnosis on Transport Risk Based on a Combined Assessment of Road Accidents and Watershed Vulnerability to Spills of Hazardous Substances. *Int. J. Environ. Res. Public Health*. **2018,** *15* (9). doi: 10.3390/ijerph15092011.

Malkiewicz, K.; Hansson, S. O.; Rudén, C. Assessment Factors for Extrapolation from Short–Time to Chronic Exposure–Are the REACH Guidelines Adequate? *Toxicol. Lett.* **2009,** *190* (1), 16–22.

Mancini, F. R.; Busani, L.; Tait, S.; La Rocca, C. The Relevance of the Food Production Chain with Regard to the Population Exposure to Chemical Substances and Its Role in Contaminated Sites. *Ann. Ist Super Sanita*. **2016,** *52* (4), 505–510.

McIlwaine, R.; Cox, S. F.; Doherty, R.; Palmer, S.; Ofterdinger, U.; McKinley, J. M. Comparison of Methods Used to Calculate Typical Threshold Values for Potentially Toxic Elements in Soil. *Environ. Geochem. Health*. **2014,** *36* (5), 953–971.

Money, C.; Schnoeder, F.; Noij, D.; Chang, H. Y.; Urbanus, J. ECETOC TRA Version 3: Capturing and Consolidating the Experiences of REACH. *Environ. Sci. Process. Impacts*. **2014,** *16* (5), 970–977.

Mukhopadhyay, S.; Masto, R. E.; Yadav, A.; George, J.; Ram, L. C.; Shukla, S. P. Soil Quality Index for Evaluation of Reclaimed Coal Mine Spoil. *Sci. Total Environ*. **2016,** *542* (Pt A), 540–550.

O'Brien, P. L.; DeSutter, T. M.; Casey, F. X. M.; Wick, A. F.; Khan, E. Wheat Growth in Soils Treated by Ex Situ Thermal Desorption. *J. Environ. Qual*. **2017,** *46* (4), 897–905.

Power, B. A.; Tinholt, M. J.; Hill, R. A.; Fikart, A.; Wilson, R. M.; Stewart, G. G.; Sinnett, G. D.; Runnells, J. L. A Risk–Ranking Methodology for Prioritizing Historic, Potentially Contaminated Mine Sites in British Columbia. *Integr. Environ. Assess Manage*. **2010,** *6* (1), 145–154.

Radziemska, M.; Vaverková, M. D.; Baryła, A. Phytostabilization–Management Strategy for Stabilizing Trace Elements in Contaminated Soils. *Int. J. Environ. Res. Public Health* **2017,** *14* (9). doi: 10.3390/ijerph14090958.

Rajmohan, K. S.; Chandrasekaran, R.; Varjani, S. A Review on Occurrence of Pesticides in Environment and Current Technologies for Their Remediation and Management. *Indian J. Microbiol*. **2020,** *60* (2), 125–138.

Ramón, F.; Lull, C. Legal Measures to Prevent and Manage Soil Contamination and to Increase Food Safety for Consumer Health: The Case of Spain. *Environ. Pollut*. **2019,** *250*, 883–891.

Riccardi, C.; Di Filippo, P.; Pomata, D.; Di Basilio, M.; Spicaglia, S.; Buiarelli, F. Identification of Hydrocarbon Sources in Contaminated Soils of Three Industrial Areas. *Sci. Total Environ*. **2013,** *450–451*, 13–21.

Richardson, P. J.; Lundholm, J. T.; Larson, D. W. Natural Analogues of Degraded Ecosystems Enhance Conservation and Reconstruction in Extreme Environments. *Ecol. Appl*. **2010,** *20* (3), 728–740.

Sample, B. E.; Lowe, J.; Seeley, P.; Markin, M.; McCarthy, C.; Hansen, J.; Aly, A. H. Depth of the Biologically Active Zone in Upland Habitats at the Hanford Site, Washington: Implications for Remediation and Ecological Risk Management. *Integr. Environ. Assess Manage.* **2015,** *11* (1), 150–160.

Saravanan, K.; Anusuya, E.; Kumar, R.; Son, L. H. Real–Time Water Quality Monitoring Using Internet of Things in SCADA. *Environ. Monit. Assess.* **2018,** *190* (9), 556. doi: 10.1007/s10661–018–6914–x.

Saupe, E. E.; Barve, N.; Owens, H. L.; Cooper, J. C.; Hosner, P. A.; Peterson, A. T. Reconstructing Ecological Niche Evolution When Niches Are Incompletely Characterized. *Syst. Biol.* **2018,** *67* (3), 428–438.

Song, B.; Zeng, G.; Gong, J.; Liang, J.; Xu, P.; Liu, Z.; Zhang, Y.; Zhang, C.; Cheng, M.; Liu, Y.; Ye, S.; Yi, H.; Ren, X. Evaluation Methods for Assessing Effectiveness of In Situ Remediation of Soil and Sediment Contaminated with Organic Pollutants and Heavy Metals. *Environ. Int.* **2017,** *105,* 43–55.

Sruthi, S. N.; Shyleshchandran, M. S.; Mathew, S. P.; Ramasamy, E. V. Contamination from Organochlorine Pesticides (OCPs) in Agricultural Soils of Kuttanad Agroecosystem in India and Related Potential Health Risk. *Environ. Sci. Pollut. Res Int.* **2017,** *24* (1), 969–978.

Teng, Y.; Wu, J.; Lu, S.; Wang, Y.; Jiao, X.; Song, L. Soil and Soil Environmental Quality Monitoring in China: A Review. *Environ. Int.* **2014,** *69,* 177–199.

Vanni, F.; Scaini, F.; Beccaloni, E. Agricultural Areas in Potentially Contaminated Sites: Characterization, Risk, Management. *Ann. Ist Super Sanita.* **2016,** *52* (4), 500–504.

Wang, D.; Zhu, S.; Wang, L.; Zhen, Q.; Han, F.; Zhang, X. Distribution, Origins and Hazardous Effects of Polycyclic Aromatic Hydrocarbons in Topsoil Surrounding Oil Fields: A Case Study on the Loess Plateau, China. *Int J Environ Res Public Health.* **2020,** *17* (4), 1390.

Weber, R.; Bell, L.; Watson, A.; Petrlik, J.; Paun, M. C.; Vijgen, J. Assessment of Pops Contaminated Sites and the Need for Stringent Soil Standards for Food Safety for the Protection of Human Health. *Environ. Pollut.* **2019,** *249,* 703–715.

Yang, J.; Zhang, H.; Ren, C.; Nan, Z.; Wei, X.; Li, C. A Cross–Reconstruction Method for Step–Changed Runoff Series to Implement Frequency Analysis under Changing Environment. *Int. J. Environ. Res. Public Health* **2019,** *16* (22). doi: 10.3390/ijerph16224345.

Ye, S.; Zeng, G.; Wu, H.; Zhang, C.; Dai, J.; Liang, J.; Yu, J.; Ren, X.; Yi, H.; Cheng, M.; Zhang, C. Biological Technologies for the Remediation of Co–Contaminated Soil. *Crit. Rev. Biotechnol.* **2017,** *37* (8), 1062–1076.

Yu, J.; Qu, F.; Wu, H.; Meng, L.; Du, S.; Xie, B. Soil Phosphorus Forms and Profile Distributions in the Tidal River Network Region in the Yellow River Delta Estuary. *Sci. World J.* **2014,** 912083. doi: 10.1155/2014/912083.

Zeng, S.; Ma, J.; Yang, Y.; Zhang, S.; Liu, G. J.; Chen, F. Spatial Assessment of Farmland Soil Pollution and Its Potential Human Health Risks in China. *Sci. Total Environ.* **2019,** *687,* 642–653.

Živančev, J. R.; Ji, Y.; Škrbić, B. D.; Buljovčić, M. B. Occurrence of Heavy Elements in Street Dust from Sub/Urban Zone of Tianjin: Pollution Characteristics and Health Risk Assessment. *J. Environ. Sci. Health A Tox. Hazard Subst. Environ. Eng.* **2019,** *54* (10), 999–1010.

CHAPTER 2

Environmental Degradation as a Multifaceted Consequence of Human Development

ZULAYKHA KHURSHID DIJOO[1*], and RIZWANA KHURSHID[2]

[1]*Department of Environmental Sciences/Centre of Research for Development, University of Kashmir, Srinagar, India*

[2]*Department of Management Studies, Iqbal Institute of Technology and Management, Srinagar, India*

Corresponding author. E-mail: zulaykhazehra@gmail.com

ABSTRACT

The subject of environmental degradation is at the vanguard of the green debate. Our environs are waning day by day with practically no portion of the earth been unscathed by it. The principal basis of environmental degradation is human ruckus. All through the preceding numerous decades, industrialization, urbanization, immense agricultural extension are liable for the reckless alteration in the environs. Environmental degradation is fashioned by the amalgamation of a successfully extensive plus escalating population, persistently intensifying financial advances in addition to the function of asset gruelling and poisoning technology. Environmental degradation is an outcome of profuse socioeconomical, technical, and institutional undertakings. To tackle the challenge of environmental degradation, various researches have been piloted worldwide in order to enhance the environmental eminence. Curtailing the exposure to risk factors by improving quality of air, water, sanitation as well as pollution-free energy resources has considerable paybacks. While, at the same time, it can contribute extensively to the accomplishment of the Millennium

Development Goals related to environmental sustainability, health, and development. To survive in these worsening risky settings, there is an urgent requirement for optimal consumption plus management of resources, sustainable development besides implementation of green technologies and participation of people in progressive accomplishments.

2.1   INTRODUCTION

The environment is something with which we are well acquainted with. It is the whole lot of things that creates our surroundings besides influencing our adeptness to exist on this planet. The notion of environmental degradation is as deep-rooted as the human presence on earth. Globally, Environmental degradation is an extremely grave problem covering a multitude of concerns counting with pollution, loss of biodiversity, extinction of flora and fauna, deforestation, desertification, and global warming (Brown et al., 1987; Tian et al., 2004). It is the deterioration of the environment by means of resource depletion (Bourque et al., 2005; Malcolm and Pitelka, 2000) and a modification in the environment accepted to be detrimental and unwanted (Johnson et al., 1997). The United Nations International Strategy for Disaster Reduction explains Environmental Degradation as "The reduction of the capacity of the environment to meet social and ecological objectives, and needs" (ISDR, 2004). Although, world economy has advanced but environmental degradation persistently triggers countless complications globally owing to excessive pressures from increasing population and depleting resources. However, governments and international bodies all over the world have legislated numerous rules and guidelines for safeguarding of our environment, still environmental degradation is among the 10 threats formally signaled by the High Level Threat Panel of the United Nations (ISDR, 2004). Unquestionably, the basis of environmental degradation is human disruption to the balance of nature. The extent of this impact fluctuates with the source of disruption, the habitat under effect and the organisms that inhabit it. Environmental adversities are multidimensional in character varying from local to global issues. Locally, it encompasses land degradation, pollution, soil erosion, etc. while globally, issues like climate change, loss of biodiversity, global warming, and depletion of the ozone layer are the major threats. Environmental degradation also befalls due to superfluous exploitation of natural resources outside their capacity for attaining financial progress. Environmental degradation and its subsequent drawbacks impact all kinds of developmental missions worldwide. Environmental degradation is advancing impetuously affecting mostly developing countries. The impacts of environmental degradation and its extent are currently so immense that their penalties can be partially restricted, not ended. Subject to the impairment, some environments will not recuperate. Hence, better managing of environs is indispensable for global prosperity. It is not a one-time fallaciously declared extravagance for well-off nations apprehensive of aesthetics but a dynamic basis for viable means of support for the underprivileged.

2.2 CAUSES OF ENVIRONMENTAL DEGRADATION

2.2.1 HUMAN ACTIVITIES

Environmental degradation is the consequence of developmental and technological courses of man (Figure 2.1). The intensity of the degradation fluctuates with the source, the locale, as well as the floras and faunas that dwell there.

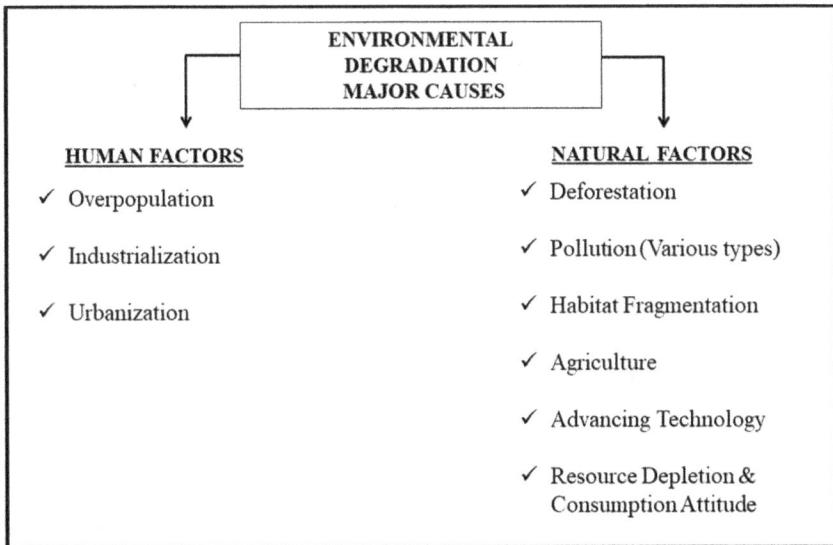

ENVIRONMENTAL DEGRADATION MAJOR CAUSES

HUMAN FACTORS
- ✓ Overpopulation
- ✓ Industrialization
- ✓ Urbanization

NATURAL FACTORS
- ✓ Deforestation
- ✓ Pollution (Various types)
- ✓ Habitat Fragmentation
- ✓ Agriculture
- ✓ Advancing Technology
- ✓ Resource Depletion & Consumption Attitude

FIGURE 2.1 Environmental degradation—causal factors.

This has resulted in drastic environmental modifications destructive to the whole planet. It is a resultant of numerous kinds of pollution, resource exhaustion, growing reliance on energy, and other environmentally destructive technologies. The environmental devastation is credited to exponential population progression as well as expeditious rising industrialization. Man's malicious conduct with environment has speeded up the rate of scientific development besides technological development (Figure 2.2).

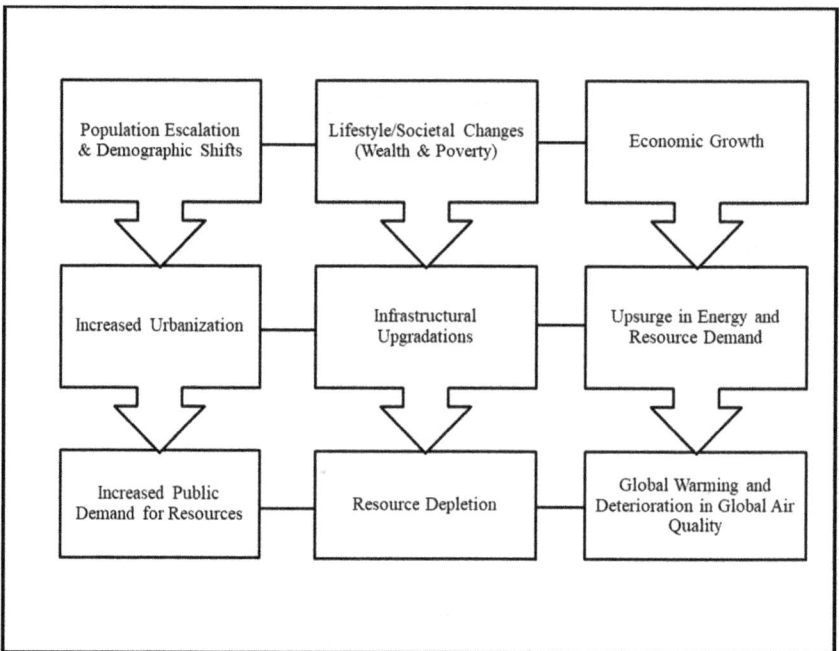

FIGURE 2.2   Flow chart specifying the many facets of developmental changes.

2.2.1.1   OVERPOPULATION

Overpopulation is a towering trouble for our environment. Earth's population is multiplying swiftly and this speedy progress has set stress on natural resources ensuing environmental degradation. The degree of environmental degradation due to overpopulation differs through various continents of the world. Death rates have decreased as a result of improved medical services, hence, ensued increased life expectancy. Additional inhabitants on Earth necessitate surplus need for food, clothes, and shelter etc. An increasing population infers growing consumption levels. More land is required for agricultural activities as well as more homes to adjust increased number of people. Increasing population means soared extractions from the water bodies for domestic, agricultural, and industrial usages. The largest of these will be agriculture as it is the key nonclimatic driver of environmental degradation plus water level decline. The coming years are going to be prospective for populations to demand more amenities. These stresses in environmental pollution increased waste generation rates, deforestation, loss of biodiversity, destruction of the ecosystems thereby leading to environmental degradation.

2.2.1.2   INDUSTRIALIZATION

Industrialization is the social plus economic conversion of human society from agrarian society to industrial society. By the advent of it, many labor-intensive processes were mechanized which resulted in increased and timely production at the cost of environmental degradation. The greatest negative result of industrialization is environmental pollution. Industries all-round the globe are emitting smoke laced with poisonous gases and releasing chemicals that are contaminating the air and water. The increasing concentration of pollutants like chlorofluorocarbons has created an undesirable hole in the ozone layer exposing the populations to deadly ultraviolet radiation placing the earth at great danger. The little accessible water is being affected by increasing pollution levels with climate change is also adding to the woes of the ever growing overall population. While, the water availability is decreasing due to reduced stream flow and groundwater levels, nevertheless, certain areas are experiencing an upsurge in freshwater supply because of disproportionate precipitation patterns.

2.2.1.3   URBANIZATION

Urban areas where majority of the world populations reside are the unhealthiest of places to live. By 2050, two-thirds of the total earth's population will inhabit in urban expanses. Increased urban populations put pressure on the environment through their consumption of food, energy, land, and water. The polluted urban setting recompenses back by distressing the health as well as the quality of life of its occupants.

2.2.2 NATURAL FACTORS

Nature triggers environmental tribulations also. Events, such as avalanches, earthquakes, tornadoes, Tsunamis, wildfires, etc. can completely devastate the life of every being in that particular area to the juncture where it can no longer sustain. Though, humans are not blameworthy for all the natural beginnings of environmental degradation as Earth itself gives rise to some of the challenging environmental problems. Whereas environmental degradation is extremely associated with the actions of man, the certainty is that the environment is unceasingly altering. Some environments naturally cut down to the position where they are powerless to do well to the life which was intended to dwell around (Figure 2.3).

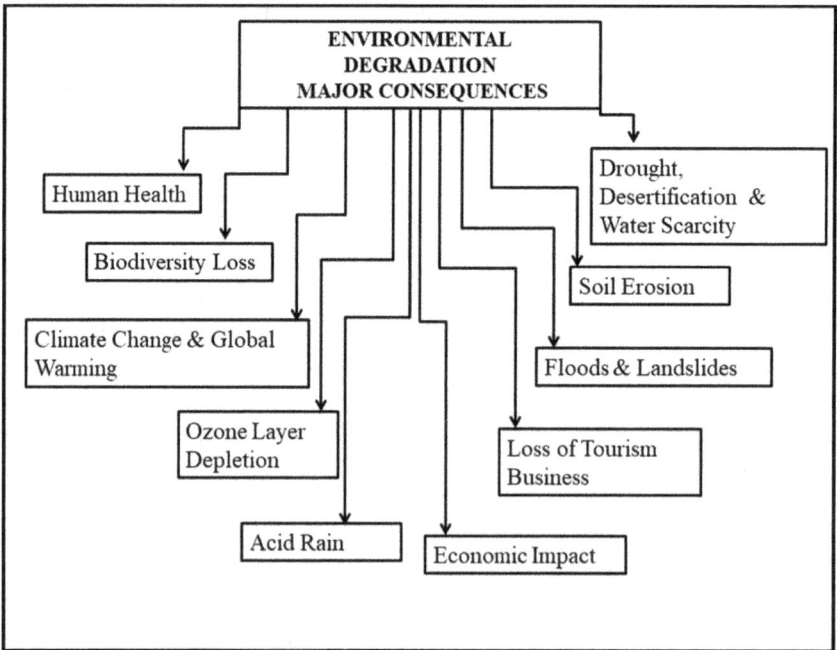

FIGURE 2.3   Major consequences of environmental degradation.

2.2.2.1 DEFORESTATION

Forests are priceless possessions of a country for the reason that they offer raw resources to modern businesses, lumber for construction activities, homes for wild life and microbes. Forest soils are highly fertile as well as highly nutrient loaded containing elevated levels of organic matter extending protection to soils by binding the soil particles with their roots thereby guarding the soils. They promote the infiltration of precipitation, hence, allowing enhanced groundwater recharge and at the same time reduce surface runoff. Hence, frequency, intensity, and dimension of floods are diminished. Overpopulation plus urbanization are among the chief bases of deforestation. Deforestation causes global warming by way of reduced forest size releasing carbon back into the atmosphere. It also gives rise to environmental degradation by augmented speed of soil erosion, upsurge in the sediment load of the water bodies, siltation, multiplies the occurrence of droughts, modifications in the precipitation patterns, strengthening of greenhouse effect, and intensification in the damaging strength of the storms, etc.

2.2.2.2 AIR POLLUTION

Growing levels of air contamination present a grave hazard to human health as well as longevity (Malik and Maurya, 2014; Yadav et al., 2019). The air we inhale is an indispensable component of our sustenance. Tragically, contaminated air is normal all over the globe (EPHA, 2009) specifically in industrialized nations (Kan, 2009). Polluted air is a hazard to the overall health (Health and Energy, 2007). The chief pollutants present in contaminated air consists of particulate matter, PAHs, Pb, tropospheric ozone, heavy metals, gases (SO_x, NO_x, CO) etc. All these contaminants can not be successfully eliminated through Earth's biogeochemical cycles (European Public Health Alliance, 2009). Air pollution at any place results in shorter life expectancy for its inhabitants (Holland et al., 1979). Urbanization combined with industrialization and increased number of motor vehicles is deteriorating the air quality. Other aspects like unconstrained environmental guidelines, less proficient manufacturing techniques, poor upkeep of automobiles intensify the difficulties (Mishra, 2003). A decline in pollution levels is likely to diminish the worldwide liability of sickness like respiratory system issues, heart ailments as well as lung cancer. Abridged pure air quality is the foremost distress equally for industrialized as well as unindustrialized countries.

2.2.2.3   WATER POLLUTION

Water Pollution is a grave issue all around the world specifically in unprivileged nations of the world. In addition to this, groundwater shortage and its contamination is another matter of concern (Karikari and Ansa-Asare, 2006). Water is among the most demandable basic service by man. Water pollution is a serious danger to all forms of life. The WHO declares one-sixth of the planet's population, that is, around 1.1 billion individuals are deprived of access to clean water while as 2.4 billion individuals survive in the absence of basic public health (European Public Health Alliance, 2009). Polluted water comprises of industrial wastes, sewage water, agricultural run-off water and household wastes (Ashraf et al., 2010) triggering harm to human health as well as the environment (European Public Health Alliance, 2009). This polluted water disturbs the quality of soils as well as of plant life (Carter, 1985). Few of its impacts are spotted instantly; however, some impacts remain dormant for extensive periods of time (Ashraf et al., 2010). Water contaminants are potent drivers of ailments, such as cholera, typhoid, arsenicosis, and diarrhea, etc. More than 50 nations comprising of 20 million hectares of global area are treated with contaminated or partly treated contaminated water (Hussain et al., 2007) including parts of all continents (Avdeev and Korchagin, 1994; Carter, 1985; Kan, 2009; Khan, 2011; Krześlak and Korytkowski, 1994; Wu et al., 1999). This polluted water results in various health hazards as well as death of humans and aquatic life alike. This also disrupts the crop productivity (Ashraf et al., 2010; Scipeeps, 2009). Globally, water pollution-related ailments are among the prominent trigger of human death. Besides, water pollution disrupts oceans, lakes, rivers, etc. rendering it an extensive plus global issue (Scipeeps, 2009).

2.2.2.4   SOLID WASTE POLLUTION

Land pollution is drastic environmental devastation humankind is encountering currently (Khan, 2004). Inappropriate managing of solid wastes is counted among the principal sources of environmental degradation with rampant consequences (Kimani, 2007). To dump enormous quantities of wastes, landfills are put to use. These are established inside the city premises by reason of outsized quantity of discarded wastes produced by households, industries, factories, and hospitals. Landfills contaminate the environment by affecting the health of the environment as well as of the local people residing there. It also contaminates the soil by leaching of harmful waste materials like heavy metals. Landfills pollute the environment and destroy the beauty of the city. Heavy metal industries generate wastes that are dumped into landfills short of necessarily required treatment (Lenkova and Vargova, 1994; Spassov, 1994).

2.2.2.5   PLASTIC POLLUTION

The volume of plastic waste manufactured and discarded yearly is massive and subsequently the quantity of waste formed is proportionate. In order to get rid of this mammoth plastic waste, it is burnt which results in the discharge of injurious gases like dioxins and furans into the air. Besides, a gigantic quantity of it culminates in oceans and other water bodies leading to environmental degradation along with the death of many sea animals.

2.2.2.6   HABITAT FRAGMENTATION

Habitat fragmentation brings long-term environmental effects capable of wrecking complete ecological units. Habitats turn out to be fragmented when developmental activities disrupt expanses of land for building of roads and dams cutting over the habitats of various plants and animals. Although it might perhaps appear good superficially, nevertheless, there are severe penalties. The chief detriments are primarily experienced by particular plant and animal populations dedicated to their bioregion or in need of huge extents of territory to maintain a strong genomic heritage.

2.2.2.7   AGRICULTURE

The environmental bearing of agriculture is remarkably severe. It includes range of elements from the soil, to water, the air, animal and soil diversity, plants, as well as the food itself. Agricultural run-off is a lethal fountain of toxins that cause serious environmental degradation. Surface water wash downs the soil along with chemical fertilizers plus pesticides used on the farmlands into water bodies. Other environmental problems linked to farming are climate change, deforestation, genetic engineering, irrigation problems, pollutants, soil degradation, and waste.

2.2.2.8   ADVANCING TECHNOLOGY

The technological applications cause inevitable environmental effects. The principle of technology is to utilize, regulate, or expand processes to profit humanity. Any disruption of these regular procedures by technology results in objectionable environmental consequences. These technologies generate orderliness in the human economy (manifested in buildings, factories, transportation, communication systems etc.) but at the cost of growing disorder in the environs which can be connected to its undesirable environmental consequences.

2.2.2.9   RESOURCE DEPLETION AND CONSUMPTION ATTITUDE

With increasing population, more number of people strives for advanced living conditions. With the purpose often countering the supplies from the people, undue quantities of resources are extracted. Majority of the populations all around the world hinge on natural resources for its living and the residue of the population depend on these resources openly for food, fuel, industrial raw material, and recreation, etc. (Raven et al., 1998). Consumption attitude is another issue for our world and it has reached to a juncture where it is not supportable for our environs. In due course of time, it heads to exhaustion of resources.

2.3   CONSEQUENCES OF ENVIRONMENT DEGRADATION

2.3.1   HUMAN HEALTH

Human well-being is at the receiving end of environmental degradation. Pollution of every type whether air, water, land, or noise is injurious to human life and welfare. Millions of individuals die annually due to secondary results of air contamination (Adakole and Oladimeji, 2006). Regions exposed to poisonous air contaminants cause respiratory complications like pneumonia and asthma (Brauer et al., 2007; Gehring et al., 2002; Jacquemin et al., 2009; Mannino et al., 1998; McConnell et al., 2006; Modig et al., 2006), asthma exacerbations (D'Amato et al., 2005; Heinrich and Wichmann, 2004; Künzli et al., 2000; Nel, 2005), cancer (Ries et al., 1999; European Public Health Alliance, 2009), neurobehavioral ailments (Blaxill, 2004; Landrigan et al., 2002; Mendola et al., 2002; Schettler, 2002;Stein et al., 2002), cardiovascular problems (European Public Health Alliance, 2009). Air contaminants also cause premature death (European Public Health Alliance, 2009), headaches plus dizziness, irritation of eyes, nose, mouth and throat, declined lung working (Colls, 2002; Gauderman et al., 2005). It has been reported that air pollution causes interference in endocrine system (Colls, 2002; Crisp et al., 1998), reproductive as well as immune systems (Colls, 2002; European Public Health Alliance, 2009). Elevated air contaminant concentrations are also associated with infant mortality (Fereidoun et al., 2007). Polluted drinking water gives birth to ailments, such as giardiasis, amoebiasis, hookworm, ascariasis, typhoid, liver and kidney damage, Alzheimer's disease, non-Hodgkin's lymphoma, and multiple sclerosis. It can also cause hormonal disruption affecting growth as well as reproductive processes. In addition to it, contaminated beach water causes stomach aches, encephalitis, hepatitis, diarrhea, vomiting, gastroenteritis, respiratory infections, pink eye, and rashes (Water Pollution Effects, 2006). Photosynthesis in aquatic plants is affected by pollutants, hence upsetting the survival of these plants.

2.3.2 BIODIVERSITY LOSS

Biodiversity is vital in order to retain the equilibrium of the environment by way of tackling pollution, renewing nutrients, safeguarding water resources, and balancing climate. Deforestation, global warming, over-population, and environmental pollution are among the chief sources of biodiversity loss. However, no reliable estimations of the range of biodiversity are accessible. Several species are on the brim of extinction. It is necessary to implement a reliable policy for resources management as well as biodiversity conservation.

2.3.3   CLIMATE CHANGE AND GLOBAL WARMING

Climate change is the most significant of all the environmental issues globally with its increasing effects becoming noticeable especially during previous few decades. The adversative effects of climate change on the environment and human health in addition to the economy has lately increased to the highest of the economic, social, and political policies at global consultations on environment. Most of the climatic alterations are due to human activities, hence promoting and undertaking environment-friendly human activities is a significant factor to reduce the frequency as well as intensity detrimental climate change effects. The primary indicator of climate change is global warming. It is the leading factor causing environmental degradation and catastrophes. Earth's climatic systems are heating in an unprecedented manner as never before in the history. This might disrupt the equilibrium of ecosystems. It will affect the normal climatic systems of Earth by fluctuations in meteorological conditions. It will also increase the intensity and frequency of extreme weather events in result to irregularities in surface temperature and precipitation patterns. It has numerous adversarial effects on human well-being and agricultural production. Heat allied ailments and deaths will also increase as a result of enhanced frequency of diseases like malaria, dengue, yellow fever and encephalitis because of the increase in the number of mosquitoes and other vectors. Therefore, the environmental effects due to climate change and global warming have a destructive consequence on the environment.

2.3.4   OZONE LAYER DEPLETION

Ozone layer protects the planet from dangerous ultraviolet radiations coming from the sun. The presence of chlorofluorocarbons, hydrochlorofluorocarbons in the atmosphere is triggering the ozone layer depletion, due to which more destructive rays comeback to the earth (Buhaug et al., 2010; Raven et al., 1998). There are accounts of ozone depletion above Antarctica letting hazardous ultraviolet radiations to Earth. Overexposure to these radiations causes skin cancer and cataracts.

2.3.5   ACID RAIN

Rainwater having pH of less than or equal to 5.6 is referred to as acid rain and it has been reported in many cities of the world. This acidic rain is harmful for plants, fishes, birds, soil, and water. It also corrodes metals and building materials. The effects of acid rain have been recorded in parts of the world. It can acidify the water or soil to such an extent that life becomes unsustainable. Plants can perish and the animals that are dependent on them vanish. Acid rain can acidify and pollute water bodies and soil render them unsuitable for nutrition and habitation.

2.3.6 DROUGHT, DESERTIFICATION, AND WATER SCARCITY

Drought, desertification, and water scarcity are among the main three outcomes of climate change which may extensively cause climate-induced relocation. All three of these are to be expected to intensify due to global warming. These issues will probably impact almost 33% of the present total world population. Droughts are expected to relocate millions of individuals globally, resulting in food insecurity as well as affecting the human livelihoods. Similarly, large portions of the earth (one-fifth) are threatened with desertification with extreme detrimental effects on social, economic, and political settings. Environmental degradation will also end up causing famines as a reaction to high temperature besides water shortage. Sea level rise will spread causing salinization of groundwater ensuing decline in freshwater accessibility. Furthermore, altering precipitation patterns will burden clean water availability. Temperature upsurge can cut snow season length in winters and escalate the speed of snow melting in spring and summers influencing soil moisture content, flood, and drought possibilities. It can result in reduced water resources during summer.

2.3.7 SOIL EROSION

Soil erosion is also one of the many issues counted under environmental degradation. Soil erosion can be triggered by natural agents, such as wind, water or by human actions, such as deforestation or the building of dams. In both ways, soil erosion can cause grave environmental degradation, as vast stretches of land become no use due to it.

2.3.8 FLOODS AND LANDSLIDES

Attributable to deforestation, floods besides landslides are prospective to become extra widespread and frequent as there will not be sufficient vegetation available for storing rainwater and holding of soil, thus prompting floods and landslides. This may result in damage to lives and belongings with identical harm to the local flora and fauna.

2.3.9 LOSS OF TOURISM BUSINESS

The descent of environs can be a massive impediment aimed at tourism business that depends on travelers for their everyday means of support. Environmental degradation by way of reduction in green cover, biodiversity loss, stinking dumping sites, poisonous air, and water is capable of withdrawing maximum number of visitors from visitor-rich countries.

2.3.10 ECONOMIC IMPACT

The massive price that a country pays by reason of environmental degradation can entail immense economic bearings in terms of restoring vegetation cover, cleansing landfills, reducing pollution, and waste production as well as conservation and protection of species. Environmental degradation and damages to ecosystem facilities will openly impact soil erosion, nutrient diminution, harsh climatic conditions, and decreased availability of water for irrigation. These issues separately may perhaps reduce more than 50% of total crop yield annually.

2.4 CONCLUSION

Environmental degradation is in itself a myriad of numerous environmental concerns. The prime reason of environmental degradation is ascribed to the fast progression of the population plus economic advancement as well as the unjustified exploitation of limited natural resources. Economic development besides varying expenditure patterns have resulted in increasing requirement for energy and transportation undertakings. Emerging nations in pursuit of financial progress besides poverty decline are anticipated to place financial development, energy for all and industrial development at the frontline of their objectives afore environmental concerns. Hence, convincing undeveloped nations to participate in environmental objectives; predominantly cutting back

GHG discharges necessitate extensive financial, scientific, and technological backing from technologically advanced nations. In addition to this, the global society prerequisites reimbursement for the financial deficits linked with cutting environmental pollution. The most effective method to control pollution and depletion of resources besides other environmental issues is by strengthening of legal framework, as there are many loopholes present in modern laws and regulations regarding environment. No doubt, governments have previously sculpted their opinion of monetary and societal approaches to resolve the environmental issues but then its execution continues to be unsatisfactory. This demands more than a few indifferent approaches to be effective. Its execution as well as advancement calls for responsiveness from the highest to the lowest of levels. This will assist the global long-term environmental targets to be achieved ensuing sustainable development.

KEYWORDS

- **environmental degradation**
- **environmental pollution**
- **resource depletion**
- **sustainable development**
- **economic impact**

REFERENCES

Ashraf, M. A.; Maah, M. J.; Yusoff, I.; Mehmood, K. Effects of Polluted Water Irrigation on Environment and Health of People in Jamber, District Kasur, Pakistan. *Int. J. Basic Appl. Sci.* **2010,** *10* (3), 37–57.

Adakole, J. A.; Oladimeji, A. O. The effects of Pollution on Phytoplankton in a Stretch of River Kubanni, Zaria, Nigeria. *Proceedings of the 15th Annual Conference of Fisheries Society of Nigeria (FISON)*, 2006; pp 151–158.

Avdeev, O.; Korchagin, P. Organization and Implementation of Contaminated Waste Neutralization in the Ukraine—National Report II. *Central Eur. J. Public Health* **1994,** *2* (suppl), 51–52.

Brown, B. J.; Hanson, M. E.; Liverman, D.M; Merideth, R. W. Global Sustainability: Toward Definition. *Environ. Manage.* **1987,** *11* (6), 713–719.

Buhaug, H.; Gleditsch, N. P.; Theisen, O. M. Implications of Climate Change for Armed Conflict. Social Dimensions of Climate Change: Equity and Vulnerability in a Warming World, 2010; pp 75–101.

Brauer, M.; Hoek, G.; Smith, H. A.; de Jongste, J. C.; Gerritsen, J.; Postma, D. S. Air Pollution and Development of Asthma, Allergy and Infections in a Birth Cohort. *Eur. Soc. Clin. Respiratory Physiol.* **2007,** *29* (5), 879–888.

Blaxill, M. F. What's Going on? The Question of Time Trends in Autism. *Public Health Rep.* **2004,** *119* (6), 536–551.

Bourque, C. P. A.; Cox, R. M.; Allen, D. J.; Arp, P. A.; Meng, F. R. Spatial Extent of Winter Thaw Events in Eastern North America: Historical Weather Records in Relation to Yellow Birch Decline. *Global Change Biol.* **2005,** *11* (9), 1477–1492.

Carter, F. W. Pollution Problems in Post-War Czechoslovakia. *Trans. Inst. Br. Geo.* **1985,** *10* (1), 17–44.

Colls, J. *Air Pollution,* 2nd ed.; Spon Press: New York, 2002.

Crisp, T. M.; Clegg, E. D.; Cooper, R. L.; Wood, W. P.; Anderson, D. G.; Baetcke, K. P.; Hoffmann, J. L.; Morrow, M. S.; Rodier, D. J.; Schaeffer, J. E.; Touart, L. E.; Zeeman, M. G.; Patel, Y. M. Environmental Endocrine Disruption: An Effects Assessment and Analysis. *Environ. Health Perspect.* **1998,** *106* (1), 11–56.

D'Amato, G.; Liccardi, G.; D'Amato, M.; Holgate. S. Environmental Risk Factors and Allergic Bronchial Asthma. *Clin. Exp. Allergy* **2005,** *35* (9), 1113–1124.

European Public Health Alliance, EPHA Air, Water Pollution and Health Effects. 2009. www.epha.org

Gauderman, W. J.; Avol, E.; Gilliland, F.; Vora, H.; Thomas, D.; Berhane, K.; McConnell. R.; Kuenzli, N.; Lurmann, F.; Rappaport, E.; Margolis, H.; Bates, D.; Peters, J. The Effect of Air Pollution on Lung Development from 10 to 18 Years of Age. *Nw Engl. J. Med.* **2005,** *352* (12), 1276.

Fereidoun, H.; Nourddin, M. S.; Rreza, N. A.; Mohsen, A.; Ahmad, R.; Pouria, H. The Effect of Long-Term Exposure to Particulate Pollution on the Lung Function of Teheranian and Zanjanian Students. *Pak. J. Physiol.* **2007,** *3* (2), 1–5.

Gehring, U.; Cyrys, J.; Sedlmeir, G.; Brunekreef, B.; Bellander, T.; Fischer, P. Traffic Related Air Pollution and Respiratory Health During the First 2 Years of Life. *Eur. Respiratory J.* **2002,** *19* (4), 690–698.

Health and Energy, Air Pollution Health Effects. 2007. www.healthandenergy.com

Heinrich, J.; Wichmann, H. E. Traffic Related Pollutants in Europe and Their Effect on Allergic Disease. *Curr. Opin. Allergy Clin. Immunol.* **2004,** *4* (5), 341–348.

Holland, W. W.; Bennett, A. E.; Cameron, I. R.; Florey, C. V.; Leeder, S. R.; Shilling, R. S. F.; Swan, A. V.; Waller, R. E. Health Effects of Particulate Pollution: Reappraising the Evidence. *Am. J. Epidemiol,* **1979,** *110* (5), 525–659.

Hussain, I.; Raschid, L.; Hanjra, M. A.; Marikar, F.; Van der Hoek, W. A Framework for Analyzing Socioeconomic, Health and Environmental Impacts of Wastewater Use in Agriculture in Developing Countries, 2007, IWMI.

Jacquemin, B.; Sunyer, J.; Forsberg, B.; Aguilera, I.; Briggs, D.; Garcia-Esteban, R. Home Outdoor NO2 and New Onset of Self-Reported Asthma in Adults. *Epidemiology* **2009,** *20* (1), 119–126.

ISDR: Terminology. The International Strategy for Disaster Reduction, 2004.

Johnson, P.; Sau, L. M.; Winter-Nelson, A. E. Meanings of Environmental Terms. *J. Environ. Qual.* **1997**, *26*, 581–589.

Kan, H. Environment and Health in China: Challenges and Opportunities. *Environ. Health Perspect.* **2009**, *117* (12), A530–A531.

Karikari, A. Y.; Ansa-Asare, O. D. Physico-Chemical and Microbial Water Quality Assessment of Densu River of Gha-na. *West Afr. J. Appl. Ecol.* **2006**, *10* (1), 1–10.

Khan, A. Air Pollution in Lahore. *The Dawn II* (2), **2011**, 283. the-editor/air-pollution-in-lahore-070

Khan, S. I. Dumping of Solid Waste: A Threat to Environment. *The Dawn*, **2004**. http://66.219.30.210/weekly/science/archive/040214/science13.htm

Kimani, N. G. Environmental Pollution and Impacts on Public Health: Implications of the Dandora Dumping Site Municipal in Nairobi, Kenya, United Nations Environment Programme, 2007; pp 1–31.

Krześlak, A.; Korytkowski, J. Hazardous Wastes in Poland-National Report. *Central Eur. J. Public Health* **1994**, 44–40.

Künzli N.; Kaiser, R.; Medina, S.; Studnicka, M.; Chanel, O.; Filliger, P.;Herry, M.; Horak, F. Jr.; Puybonnieux-Texier, V.; Quénel, P.; Schneider, J.; Seethaler, R.; Vergnaud, J-C.; Sommer, H. Public Health Impact of Outdoor and Traffic Related Air Pollution: A European Assessment. *Lancet* **2000**, *356* (9232), 795–801.

Landrigan, P. J.; Schechter, C. B.; Lipton, J. M.; Fahs, M. C.; Schwartz, J. Environmental Pollutants and Disease in American Children: Estimates of Morbidity, Mortality, and Costs for Lead Poisoning, Asthma, Cancer, and Developmental Disabilities. *Environ. Health Perspect.* **2002**, *110* (7), 721–728.

Nel, A. Air Pollution Related Illness: Effects of Particles. *Science* **2005**, *308* (5723), 804–806.

Lenkova, K.; Vargova, M. Hazardous Wastes in the Slovak Republic-National Report. *Central Eur. J. Public Health* **1994**, *2*, 43–48.

Mannino, D. M.; Homa, D. M.; Pertowski, C. A.; Ashizawa, A.; Nixon, L. L.; Johnson, C. A.; Ball, L. B.; Jack, E.; Kang, D. S. Surveillance for Asthma-United States, 1960–1995. *MMWR CDC Surveillance Summaries* **1998**, *47* (1), 1–27.

Mendola, P.; Selevan, S. G.; Gutter, S.; Rice, D. Environmental Factors Associated with a Spectrum of Neuro Developmental Deficits. *Mental Retard. Dev. Disab. Res. Rev.* **2002**, *8* (3), 188–197.

Mishra, V. Health Effects of Air Pollution, Background Paper for Population- Environment Research Network (PERN), Cyber Seminar, **2003**, December 1–15.

Malcolm, J. R.; Pitelka, L. *Ecosystems & Global Climate Change: A Review of Potential Impacts on US Terrestrial Ecosystems and Biodiversity*; Pew Center on Global Climate Change: Arlington, 2000.

McConnell, R.; Berhane, K.; Yao, L.; Jerrett, M.; Lurmann, F.; Gilliland, F. Traffic, Susceptibility, and Childhood Asthma. *Environ. Health Perspect.* **2006**, *114* (5), 766–772.

Malik, D. S.; Maurya, P. K. Heavy Metal Concentration in Water, Sediment, and Tissues of Fish Species (Heteropneustis Fossilis and Puntius Ticto) from Kali River, India. *Toxicol. Environ. Chem.* **2014**, *96* (8), 1195–1206.

Modig, L.; Jarvholm, B.; Ronnmark, E.; Nystrom, L.; Lundback, B. Andersson, C.; Forsberg, B. Vehicle Exhaust Exposure in an Incident Case Control Study of Adult Asthma. *Eur. Respirat. J.*, **2006**, *28* (1), 75–81.

Raven, P. H.; Berg. L. R.; Johnson, B. G. *Environment*, 2nd ed.; Saunders College Publishing, Harcourt Brace College Publishers, Saunders College Publishing: New York, 1998.

Ries, L. A. G.; Smith, M. A.; Gurney, J. G.; Linet, M.; Tamra, T.; Young, J. L.; Bunin, G. R. (Eds.). *Cancer Incidence and Survival among Children and Adolescents: United States SEER Program 1975–1995*; National Cancer Institute, SEER Program: Bethesda, MD, 1999.

Spassov, A. Identification of Problem Related to Solid Wastes in Bulgaria-National Report. *Central Eur. J. Public Health* **1994**, *2*, 21–23.

Schettler, T. *Changing Patterns of Disease: Human Health and the Environment*; San Francisco Medicine, **2002**. http://www.sfms.org

Scipeeps, Effects of Water Pollution. **2009**. http://scipeeps.com/effects-ofwater- pollution/

Stein, J.; Schettler, T.; Wallinga, D.; Valenti, M. In Harm's Way: Toxic Threats to Child Development. *J. Dev. & Behav. Pediatr.* **2002**, *23*, S13–S22.

Tian, Q.; Zhou, Z. Q.; Jiang, S. R.; Ren, L. P.; Qiu, J. Research Progress in Butachlor Degradation in the Environment. *Pesticides-Shenyang* **2004**, *43* (5), 205–208.

Yadav, K. K.; Kumar, S.; Pham, Q. B.; Gupta, N.; Rezania, S.; Kamyab, H.; Talaiekhozani, A. Fluoride Contamination, Health Problems and Remediation Methods in Asian Groundwater: A Comprehensive Review. *Ecotoxicol. Environ. Safety* **2019**, *182*, 109362.

Water Pollution Effects, in Grinning Planet, Saving the Planet One Joke at a Time, 2006. http://www.grinningplanet.com

Wu, C.; Maurer, C.; Wang, Y.; Xue, S.; Davis, D. L. Water Pollution and Human Health in China. *Environ. Health Perspect.* **1999**, *107* (4), 251–256.

CHAPTER 3

Persistent Organic Pollutants: Sources, Impacts, and Their Remediation by Microalgae

DIG VIJAY SINGH, ATUL KUMAR UPADHYAY, RANJAN SINGH, and D. P. SINGH*

Department of Environmental Science, Babasahib Bhimrao Ambedkar University, Lucknow, India

Corresponding author. E-mail: dpsingh_lko@yahoo.com

ABSTRACT

Persistent organic pollutants (POPs) are hazardous chemicals, which are commonly introduced into the environment by several anthropogenic activities. The unplanned discharge of chemicals has led to widespread contamination and accumulation in diverse organisms in the ecosystem. Several persistent pollutants are also lacking the regulatory standards due to the dearth of knowledge about their impacts on the environment. Techniques used for their remediation involves huge capital investment and are also inefficient in the removal of POPs from the contaminated ecosystem. Microalgae have shown potential to eliminate POPs from the environment by employing several mechanisms. Among the various mechanisms, biodegradation is one of the most effective mechanisms for the complete removal of POPs from the environment. Thus, using the microalgae to remove POPs not only conserves resource but also provides viable alternative for effective remediation of wastewater and contaminated environment.

3.1   INTRODUCTION

Persistent organic pollutants (POPs) involve a wide range of toxic chemicals, which are intentionally or unintentionally released into the environment (Xu et al., 2013). Several human activities are responsible for generation of the POPs (Pariatamby and Kee, 2016). Two major sectors, agriculture and industries, are responsible for release of the majority of POPs into the environment (Tombesi et al., 2014; Quinn et al., 2009). The emission pattern of POPs varies due to the urbanization/industrialization process (Hung, et al., 2013). Rapid production in agriculture with the use of agrochemicals (Carvalho, 2017) releases toxic pollutants like aldrin, dieldrin, toxapene, and polycyclic aromatic hydrocarbons (PAHs) into the environment (Guo et al., 2019). Because of liphophilicity (Lee et al., 2010), toxicity, and the long persistence of POPs, it is one of the most growing concerns for the world (Zacharia, 2019; Adeola, 2004). The concerns regarding the deleterious impacts of persistent pollutants on the ecosystem health (Sweetman et al., 2005) are escalating with population and industrial growth (Rhind, 2009). Long-distance transport (Beyer et al., 2000), persistence (Jacob, 2013; Rasheed et al., 2019), and ability to bio-accumulate (Jacob, 2013), as well as biomagnify along food chain (Alava et al., 2018) are some of the unique characteristics of POPs (Zacharia, 2019). Thus, chemicals having such properties and toxicology are referred to as POPs. POPs have several negative impacts on the ecosystem (Islam et al., 2018) as their rapid accumulation in different tropic levels can disturb the functioning of the food chain (Neamtu et al., 2007) and ultimately interrupt the stability of the biosphere (Langenbach, 2013). POPs, being recalcitrant (Panda, 2018), are transported to very distant places from the place of origin (Bacaloni et al., 2011), and thus, causes threats to the survival of the plants and animals in the receiving environment (Ashraf, 2017; Lushchak et al., 2018; Jamieson et al., 2017). The half-life of POPs is long and is very resistant to degradation (Ashraf, 2017). The slow degradation rate and semi-volatility (Jayaraj et al., 2016) is also responsible for their build up in food chain (Bartrons Vilamala et al., 2012; Franzaring and van der Eerden, 2000). POPs move easily through aerial or water medium and eventually reach the organism's body (Ashraf, 2017). The widespread transport of POPs to the regions in which such compounds have not been used or banned from many years can destruct the stability of the environment (World Health Organization, 2003). The perpetual exposure to persistent pollutants (González-Mille et al., 2010) can ultimately inhibit the growth of diverse organisms (Wu et al., 2008) and disturb the balance of the aquatic ecosystem (González-Mille et al., 2010; Varsha et al., 2011; Mushtaq et al., 2020; Singh et al., 2020).

Wastewater discharged with varied toxic pollutants lead to more accumulation of pollutants into the aquatic environment (Rasalingam et al., 2014). Wastewater of different origin (Mushtaq et al., 2020) has numerous contaminants (Mondal et al., 2019), which, even in small quantity, can disturb the ecosystem functioning (Tang, 2013; UNEP, 2005; Mushtaq et al., 2020). The existing treatment facilities are not capable of removing the POPs from contaminated environment thereby leading to their accumulation in diverse ecosystems (Abdel-Razek et al., 2019). Several technologies for the removal of persistent pollutants have been examined, but, due to huge investment and ineffectiveness in removal of pollutants, such technologies do not seem to be proficient (Pariatamby and Kee, 2016). Current technologies for wastewater treatment are still lacking the information about the diverse pollutants, which even in minute quantity can lead to severe health complications (Ashraf, 2017). Several persistent pollutants are also lacking the regulatory standards due to the dearth of knowledge about their impacts on the environment. Wastewater treatment techniques are also incapable of removing such toxic pollutants (Singh et al., 2020), so designing and operation of treatment plants should be done in such a way so that toxic impacts posed by these pollutants in the coming years can be minimized (Samer, 2015). The concentration of these pollutants is escalating day-by-day but effective as well as environment-friendly techniques are required to decrease the damage posed by such toxic pollutants (Pariatamby and Kee, 2016). Numerous health and ecological impacts are posed by such pollutants in diverse habitats (Alharbi et al., 2018), thus their effective removal is required for the protection of the environment (Sutherland and Ralph, 2019). Exploiting microalgae for the elimination of POPs is the feasible approach for effective remediation of wastewater and contaminated environment (Kumar et al., 2015). Remediation by microalgae (phycoremediation) is considered a reliable alternative (Kang, 2014) for the removal of pollutants (Kurashvili et al., 2018; Kumar et al., 2018) due to their cosmopolitan nature, ability to grow in extreme conditions, huge diversity (Mostafa, 2012), and efficient mechanisms for the elimination of toxic pollutants from contaminated environment (Rodas et al., 2009). Thus, using microalgae to eliminate

POPs can provide the opportunity to remediate contaminated environment in a cost-effective and environmentally sound manner.

3.2 POPS ACCUMULATION AT GLOBAL LEVEL

POPs are ubiquitous (Wu et al., 2008), persistent, toxic, and recalcitrant in nature (Batt et al., 2017). In 2001, 92 countries signed a convention in which almost 12 POPs (aldrin, dieldrin, endrin, chlordane, heptachlor, hexachloro benzene (HCB), mirex, toxaphene, dichloro diphenyl trichloroethane (DDT), polychlorinated biphenyls (PCBs), polychlorinated dibenzo-p- dioxins (PCDDs), and polychlorinated dibenzofuranes (PCDFs)) (Kim et al., 2019), known as dirty dozen (Pariatamby and Kee, 2016), were pledged to diminish their discharge into the environment. In 2009, nine chemicals (chlordecone, lindane, α-HCH, β-HCH, γ-HCH, hexabromobiphenyl, polybrominated diphenylether, and pentachlorobenzene) were added and in 2011, one more chemical (Endosulfan) was added to the POPs list (Xu et al., 2013). The Stockholm Convention for POPs is a well-known international convention that helps in preventing the environmental contamination by these pollutants globally (Lallas, 2001). Due to international regulations, the ban on the use, as well as production of POPs, has led to the decrease in their concentration (Teran et al., 2012) in the environment (Ibrahim, 2007).

POPs are semi-volatile, which travel distant places through ocean, as well as wind current, and get trapped in cold regions (Burkow and Kallenborn, 2000). POPs accumulate far from the place of production and contaminates the distant regions (Polar region) having no industries and also low population (Hung, et al., 2013). Deposition in alpine region led to the exposure of highland inhabitants to different types of POPs as compared to lowland inhabitants (Hung et al., 2013). POPs accumulation in the Arctic region (Kirby, 2010) came into light when pesticides were found in polar bear in 1970s (Ashraf, 2017; El-Shahawi et al., 2010). The Arctic is a very vulnerable region due to continuous accumulation of POPs in the region (Pariatamby and Kee, 2016) as well as in the organisms (Bacaloni et al., 2011). The presence of some POPs in the Arctic region is very astonishing as residues of such persistent pollutants have been observed which were banned in USA and Canada (Ashraf, 2017). The local communities feeding on organisms residing in the Arctic region are at high risk due to the consumption of food contaminated with POPs (Gibson et al., 2016). POPs in the atmosphere of the Arctic and Antarctic regions are in low concentration but after entering into the food chain can reach to top organisms thus becoming the common contamination source for the inhabitants (Hung et al., 2013). The potential toxicity associated with POPs and their predominance in diverse ecosystems are grave concerns for the society (Link, 2006).

3.3 POPs IN INDIA

The use of pesticides in India started in 1948 (Gupta, 2004) and, by 1978, approximately 44,000 metric tonnes of pesticides were produced in India in which Lindane and DDT alone account for 76% of the total pesticide production. The dramatic change in the use of pesticides in India and at global level has been observed, which is one of the major concerns of the world (Devi et al., 2017). Pesticides, mainly from the organochlorine group, consist of majority of chemicals (aldrin, chlordane, DDT, dieldrin, endrin, heptachlor, mirex, toxaphene, and hexachlorobenzene) that are enlisted in the Stockholm convention as POPs (Tsai, 2010). In India, there is a tremendous shift in the production as well as in the use of pesticide in agricultural sector (Aktar et al., 2009). Pesticides are not only used in enormous quantity to increase grain production (Pariatamby and Kee, 2016) but are also responsible for contamination of all environmental components (Carvalho, 2017; Gushit et al., 2013; Gill and Garg, 2014). Pesticides from organochlorine group are more frequently used in India (Subramaniam and Solomon, 2006) due to their easy availability and simple way of their application in different sectors (Dhananjayan et al., 2012). Agrochemicals are extensively used in agriculture (Zhang et al., 2018) but high application rate and mobility (Pal et al., 2014) has led to leaching of these pollutants (Sharma et al., 2019) into the diverse ecosystems (Tam et al., 2012).

In India and several Asian as well as African countries, despite banning the use of POPs in agricultural as well as industrial sector, DDT are used illegitimately in these countries (UNEP, 2008). Industrial processes also generate huge quantity of the POPs, which upon disposing into the environment can disturb the functioning of the environment (World Health Organization, 2009). The swift increase in agriculture and industries production is due to the enormous use of pesticides mainly DDT and HCHs (Hung et al., 2013). DDT and HCH concentration in river water of India was found higher than the permissible limits (Anon., 1998). In India, environment is contaminated by mainly two POPs (DDTs and HCHs) and data on several other pollutants are lacking thus making it difficult to assess the exposure of POPs at national level (Hung et al., 2013). These synthetic chemical has severe environmental impacts as these chemical persist for the several years and have potential of bio-accumulation in the diverse organisms (Langenbach, 2013). Rivers are important medium to scrutinize POPs about their transport and distribution in the environment (Hung et al., 2013). POPs presence in water can contaminate humans (Sharma et al., 2014) as water is used for drinking (Mushtaq et al., 2020) and crop production (Edokpayi et al., 2017). Higher POPs concentration in the surface water can be due to the extensive use of antifouling paints in boats (Amaraneni and Pillala, 2000; Wezenbeek, et al., 2018, Lin et al., 2009), and dry and wet deposition from the atmosphere (O'Driscoll et al., 2013). Groundwater is also contaminated by POPs (Prakash et al., 2004), which can be due their widespread exploitation in agricultural sector (Sharma et al., 2014). POPs can also contaminate soil either by direct use of such chemicals in agricultural field (Namiki et al., 2013), disease control (Wang et al., 2012), and industrial production (Weber et al., 2018) or indirect deposition from the atmosphere (Hanedar et al., 2019). Soil acts as the long-term storehouse of POPs due to their affinity for organic matter (Mechlińska et al., 2009; Pereira et al., 2010). POPs accumulation in the soil can enhance exposure of smaller organism and eventually via diet and respiratory pathways leading to the contamination of higher organisms (Miglioranza et al., 1999). Continuous POPs accumulation in the soil can have severe impact on the water quality (Kolpin et al., 1996), organisms (Jacob, 2013), and human health (Kammenga et al., 2000). Several countries has banned the POPs use but poor implementation of laws (Link, 2006), cheapness of these pesticides, and high effectiveness (Gudorf and Huchingson, 2010) are promoting their use in an illegal manner (Agnihotri et al., 1996; Imphal Free Press, 2008). POPs levels decrease below threshold level in developed countries due to strict ban on these compounds but in developing countries, these are still major environmental degraders due to poor monitoring and implementation of international legislations.

3.4 SOURCES OF POPS IN THE ENVIRONMENT

POPs are frequently released by several human activities in the different components of the environment (Guo et al., 2019). Majority of POPs are released from the pesticide plants and industries (Figure 3.1). Several chemicals were used after World War II (Gay, 2012) in order to increase the production in agricultural sectors (Oerke, 2005). Those chemicals proved very beneficial in increasing grain production, pest (Popp et al., 2013), and diseases control in agricultural crops (Aktar et al., 2009). Pesticides during their production, storage, transport, and ultimately their use in agriculture can lead to the release of certain chemicals, which are very hazardous for the environment (Ozkara et al, 2016). POPs from pesticide industries (Tieyu et al., 2005) are halogenated chemicals having strong bond between carbon and halogen groups and thereby are more resistant to degradation (Zacharia, 2019). POPs devoid of halogen group are also persistent because of their chemical structure, which is very stable (Guo et al., 2019). Several persistent pollutants like Dioxins, Furan, and PAHs are released during industrial processes like combustion of waste, coke production, and also during the synthesis of chlorinated substances (Zacharia, 2019). PCBs have several industrial uses, which upon releasing into the environment, can cause severe damage to environment health (Olatunji, 2019). The harmful effects related to persistent chemicals on human health and environment led to the banning of PCBs in 1970s (Hens and Hens, 2018; Ruiz-Fernández et al., 2014) but the older electrical equipments can release these compounds into the environment (Della-Valle et al., 2013). Both PCB and PBDE (polybrominated diphenylether) have not only identical structure but also have identical physicochemical properties (Luthe et al., 2008). PCBs were used in industries as lubri-cants, wood preservative, cooling fluid in transformers, cable insulation, heat exchange fluid, and plasticizers in paints (Erickson and Kaley, 2011) while PBDE were used in the electrical equipments, insulation wire (Tam et al., 2012) adhesives, mobile phones, and computers (Aleksa et al., 2012). Electronic waste generated at national and global level has relation with huge release of PCBs and PBDEs that led to contamination of the environment (Eguchi et al., 2012). Countries like the USA and Europe has banned the use of PBDE (Olisah et al., 2018) but products manufactured using PBDE is major source of contamination in such countries (Siddiqi et al., 2003). Both persistent pollutants are observed in the countries where such chemicals have never been produced (Guo et al., 2019) but due to the import of equipments, such harmful compounds are now encountered in these countries. PBDEs are used in the production of numerous products (Siddiqi et al., 2003) but after releasing into the environment pose serious damage to several organisms (Tam et al., 2012) and also disturb the balance of diverse ecosystems (Siddiqi et al., 2003; O'Driscoll et al., 2016).

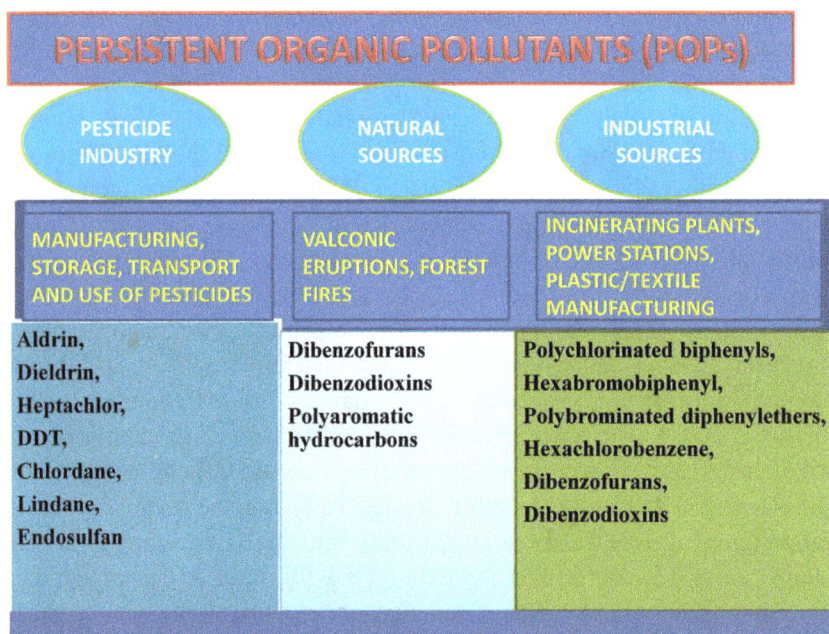

FIGURE 3.1   Sources of persistent organic pollutants.

PAHs are the group of chemicals having two or more than two benzene rings and can be found in every ecosystem (Lawal, 2017). Natural as well as human activities (Pariatamby and Kee, 2016) are also responsible for release of PAHs into the environment (Brown and Wania, 2008). The leading source for production of PAHs is fire including natural as well as human induced burning (Lee, 2010). PAHs are still released into the atmosphere from vehicles upon combustion of fuels (Lee, 2010). Industries, power plants, and heating stations are also the source of PAHs in the environment

(Lawal, 2017). PAHs are produced mainly due to incomplete combustion of wood, coal, and fuel (Abdel-Shafy and Mansour, 2016). Food material gets contaminated by PAHs due to several processes like direct putting of food material in fire, roasting, and frying (Zelinkova and Wenzl, 2015). Through smoking and by consumption of contaminated food, PAHs enter into the human body (Ramesh et al., 2004). Consumption of contaminated diets is responsible for more than 70% of exposure in nonsmoker persons (Martorell et al., 2010). Incineration of waste material is responsible for release of dioxin and furan into the environment (Lali, 2018). The high concentration of these pollutants in vegetation and animals are found in the adjoining place of incineration (Lopes et al., 2015). The toughest task for the scientific fraternity is to assess the risks related to POPs contamination in dietary supplement.

DDT, an important pesticide, has beneficial uses in the agriculture but have detrimental impacts on environmental health (Nicolopoulou-Stamati et al., 2016). DDT was used extensively after "World War II" (1945–1972) in the USA for agriculture crops mostly cotton (Stockwell, 2008) and was also used to provide protection to soldiers against insect borne diseases (Kitchen et al., 2009). DDT as insecticide is banned but is still used for control of malaria in India (Sharma et al., 2014). Despite banning of DDT in several countries (Guimarães et al., 2007), the compound remain in the ecosystem for long time owing to its stability (Brown and Wania, 2008) and reluctance to degradation (van den Berg, 2009). Being hydrophobic, DDT can be easily transferred through food chain and finally reach to human body via digestive tract or skin (Bernardes et al., 2015; Snedeker, 2001). Extensive use of DDT not only results in environmental contamination (Ozkara et al., 2016) but also rapid bio-accumulation in the organisms (Sweetman et al., 2005; Jayaraj et al., 2016), phenomena observed by Rachel Carson in 1962 and brought to public attention by mentioning in her book *"Silent Spring"* (Bernardes et al., 2015). The harmful effects of DDT on living organism have propelled several countries to stop their use of DDT in agriculture and health sectors.

3.5   IMPACTS OF POPS ON ENVIRONMENT AND HUMAN HEALTH

POPs were used in different fields ranging from agriculture to various industrial processes (Ukalska-Jaruga et al., 2020) and their presence in different ecosystem is now a global issue (Link, 2006). POPs reach to environment through different ways such as leaching from agricultural fields, direct discharge of effluent from industries, and deposition from atmosphere (Blanchard et al., 2004; Geissen et al., 2015). Concerns related to POPs presence in different environmental components (Jacob, 2013) are increasing due to their enormous impacts on humans (Figure 3.2) and environment (Qing et al., 2006). Persistent pollutants like pesticides are toxic to living organism in every ecosystem (Damalas and Eleftherohorinos, 2011; Bernardes et al., 2015) and are known to affect vital biological processes from cellular up to organizational level of ecosystem (Schäfer et al., 2011). POPs once released into the environment can reach food crops; livestock and aquatic organisms thus contaminate them and become the sources of contamination for humans (Thompson and Darwish, 2019). These POPs affect the environment by two processes: transport of pollutants to distant places from the point of origin and their bio-accumulation in the organism (Matthies et al., 2016), which increase with each tropic level in the food chain (Alava and Gobas, 2012). The hazards caused by POPs are numerous and their ability to biomagnify along food chain can further enhance the degradation of environmental health (Carvalho, 2017). The accumulation of persistent pollutants at each top tropic level can have deleterious impacts on the human and animal health (Alharbi et al., 2018). Exposure of humans to persistent pollutants is more than 90% due to contaminated dietary intake of especially animal origin (Cok et al., 2009; Polder et al., 2010). Humans are directly affected by these chemicals in the environment and reach their body through skin, inhalation, and consumption of contaminated food (World Health Organization, 2009). The enormous release of these persistent pollutants can lead to their accumulation in fatty tissue of fish and chickens thereby leading to the contamination of food sources (Filazi et al., 2017). Son et al. (2012) studied different food materials and observed that fish have PCB concentration four times higher as compared to rice. Fishes are mostly contaminated by POPs and become the source of contamination for humans due to regular intake of fish in the food (Fair et al., 2018). Although vegetables and rice have low concentration of PCB but can become the major source of this contamination in the body as regular and higher amount of food and vegetables are consumed (Guo et al., 2019). The other major global concern of POPs is their widespread presence in blood and milk of humans (Massart et al., 2005). Persistent pollutants are responsible for birth defects; cancer; and disturbed functioning of the reproductive system (Sweeney et al., 2015; Gregoraszczuk and Ptak, 2013), immune system

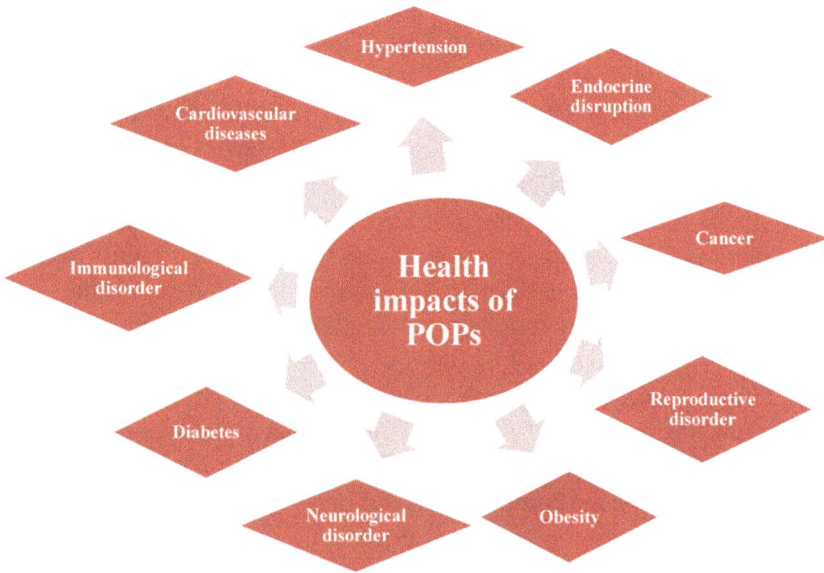

FIGURE 3.2 Health impacts of persistent organic pollutants.

(Qing Li et al., 2006), brain, and endocrine system (Geyer et al., 2000; Schug et al., 2015) in animals as well as humans (Ruzzin, 2012; Mitra et al., 2011). The other deleterious impacts of pesticides on humans are liver damage; breast cancer, disturb hormonal activity, and disruption of genetic material (Rodgers et al., 2018). Children's are more susceptible compared to adults to the risks posed by persistent pollutants as their cell are in developing phase and are more sensitive to the attack of contaminants (Ferguson and Solo-Gabriele, 2016). Exposure of POPs during developmental phase of new-born children can have detrimental impacts on functioning of the body and even permanent undesirable changes in growth and development of organisms (National Research Council, 2004). POPs are also known to cause several carcinogenic as well as mutagenic effects on humans (Ashraf et al., 2015). Persistent pollutants are now more frequently encountered in drinking water creating more hurdles in protecting humans from their severe impacts (Xu et al., 2013). Pesticides causes severe toxicity to living organisms (Abdel-Razek et al., 2019) by either altering their functional roles in the ecosystem or by causing death of living creatures (Lushchak et al., 2018). Only few POPs have permissible limits based on their toxicity in the environment but still several persistent pollutants are emerging without having any permissible limits. Thus, it also becomes necessary to know the pathways of their entry into the food and protect human health from deterioration.

Keeping in view the detrimental impacts of POPs on the environment, the Stockholm Convention framed guidelines, which not only encouraged destroying the stockpiles of POPs but also restricted the production and use of these compounds (Link, 2006). As per the Stockholm Convention, the chemicals should be managed properly in the ecofriendly manner (UNEP, 2005) and before discharging should be converted into the form nonhazardous to the environment (Kovner, 2009). POPs concentration is declining (Kong et al., 2014; Alharbi et al., 2018), but some developing are still using these persistent pollutants (Stephenson et al., 1995), which further impede the steps taken by the international organization to combat their harmful environmental effects. Strict regulations are required in developing countries to combat the use of persistent pollutant. Systematic and detailed studies about their toxic effects on the environment and focus on development of effective techniques are required, which can reduce the menace of persistent pollutants effectively (Guo et al., 2019).

3.6   RESISTANCE OF POPS TO DEGRADATION

POPs are the polyhalogenated compounds, which, because of lipid solubility, accumulate in the fatty tissues of animals (Zacharia, 2019). The continuous accumulation of persistent pollutant in the environment is due to their reluctance to degradation by natural agents (Pariatamby and Kee, 2016). POPs, the carbon-based compounds, are very reluctant to degradation (Ashraf, 2017). Such compounds are also very stable and nonreactive toward hydrolysis (Zacharia, 2019) and degradation (Bharagava, 2017). In carbon-halogen bond, halogens are electronegative and carbon is poor in electrons (Cavallo et al., 2016). The charge on the elements (C, Cl) results in electrostatic attractions thereby contributing to the remarkable bond strength of carbon–chlorine bond, which is known to be one of the strongest bonds in organic chemistry. More addition of halogens to the same carbon atom of chemical compound further strengthens the bond between the carbon and chlorine. The halogen content in the persistent pollutants also determines their stability as well as solubility in the fatty tissues (Zacharia, 2019).

3.7   MICROALGAE AS PROMISING ORGANISM FOR POPs REMEDIATION

Microalgae are microscopic organism predominant in the aquatic ecosystem (Randrianarison and Ashraf, 2017) that serves as the source of food for diverse organisms. Microalgae are the group of organisms that often results in the formation of complex communities. The natural ability of the microalgae to produce oxygen (Chapman, 2013) can eliminate the use of mechanical aerators for degradation of pollutants (Praveen and Loh, 2015). Utilization of microalgae for the removal of persistent pollutants (Table 3.1) from contaminated sites is the wonderful approach toward environmental protection (Ke et al., 2012). Microalgae have shown immense potential of removing heavy metals (Salama et al., 2019) and are also gaining attention for exclusion of persistent pollutants in a cost-effective manner. The discharge of wastewater into water bodies can disturb the structure as well as functioning of receiving ecosystems (Edokpayi et al., 2017) and the functioning of several ecosystems has been disturbed by the continuous discharge of toxic pollutants into the environment (Varsha et al., 2011). Microalgae are at the bottom of food chain (Torres et al., 2008) and any damage to the microalgae can have significant effects on the food web (Tam et al., 2012) but some microalgae are capable of inducing certain pathways, which can overcome the damages caused by toxic pollutants. Several microalgae show enhanced growth, photosynthesis, and metabolic function when they acclimatize in varied environmental conditions (Choi and Lee, 2015). Microalgae have potential to tolerate toxic pollutants and also have ability to remove them from wastewater (Molazadeh et al., 2019). It was also observed in several studies that removal rate was enhanced after the microalgae were pre-exposed to pollutants for acclimatization (De-Bashan and Bashan, 2010). Microalgae like *Chlorella* and *Scenedesmus* have shown immense capacity to remove persistent pollutants (bisphenol-A, endosulfan) from wastewater (Subashchandrabose et al., 2013; Dosnon-Olette et al., 2010). The effects of POPs on the microalgae has been studied extensively and was observed inhibited (Ramakrishnan et al., 2010) or profuse growth of the microalgae in POPs-contaminated environment, which indicates that the tolerance is species specific (Tam et al., 2012). The effective mitigation is possible by selecting and screening such organisms that adjust in contaminated sites and grow profusely in the contaminated sites (Delrue et al., 2016). Thus, the selection of the microalgae for wastewater and contaminated-environment remediation should be done on the basis of their adaptability and robust growth in contaminated environment. The diverse application and popularization of the microalgae at global level as the remediator of wastewater (Rawat et al., 2016) with simultaneous protection of environment have made them potential organism for achieving the goals of sustainable development (Avagyan, 2008).

TABLE 3.1   Removal of Persistent Pollutants from Contaminated Ecosystem by Microalgae Species.

Persistent pollutants	Microalgae species	References
Endosulfan	*Chlorococcum* sp. *Anabaena* sp.	Sethunathan et al., 2004; Shivaramaiah, 2000
DDT	*Arthrospira platensis*	Tabagari et al., 2019
	Chlorococcum sp., *Anabaena* sp. and *Nostoc* sp.	Megharaj et al., 2000)
PAHs	*Scenedesmus obliquus*	Ibrahim and Gamila, 2004
	Skeletonema costatum and *Nitzschia sp*	Hong et al., 2008
Lindane	*Anabaena azotica*	Zhang et al., 2012
	Selenastrum capricornutum	Friesen-Pankratz et al., 2003
Dibenzo-*p*-dioxin	*Scenedesmus* sp.	Todd et al., 2002
PCBs	*Chlamydomonas reinhardtii*	Jabusch and Swackhamer, 2004)
Aldrin	*Anabaena* sp.	Park and Lee, 2010.
17b-estradiol (E2) and diethylstilbestrol	*Raphidocelis subcapitata.*	Liu et al., 2018
17α-ethynylestradiol (EE2)	*Selenastrum capricornutum* and *Chlamydomonas reinhardtii*	Hom-Diaz et al., 2015
	Scenedesmus	Wang et al., 2017
	Chlorella sp.	Sole and Matamoros, 2016
	Navicula incerta	Liu et al., 2010

3.8   MECHANISMS OF POPS REMOVAL FROM THE CONTAMINATED ENVIRONMENT

Microalgae are one of the strong candidates known to adapt in diverse contaminated conditions (Ma et al., 2018; Metsoviti et al., 2020). Several technologies related to microalgae for the removal of persistent pollutants have been exploited and positive results signify their potential role in the removal of persistent pollutants at a wide scale. The main mechanisms adopted by the microalgae for remediation of persistent pollutants from contaminated environment are adsorption, pollutant uptake, and biodegradation (Wang et al., 2019). Bio-adsorption is the process in which pollutants get binded to the cell surface (Shen et al., 2017); bio-uptake is transport of pollutants inside the cell, which occur by the utilization of energy; while biodegradation involves the degradation of complex pollutants into simple one that can be used by organisms for growth and development (Sutherland and Ralph, 2019).

Microalgae removes persistent pollutants by adsorption, which is a passive process involving binding of pollutants to cell wall (Bilal et al., 2018) or extracellular polysaccharides (Abdi and Kazemi, 2015). It is because of electrostatic interactions that the cationic pollutants get attracted to negatively charged cell wall, thus enhancing the removal of pollutants from contaminated sites (Ayangbenro and Babalola, 2017; Bilal et al., 2018). Several processes like ion exchange, chelates, and complex formation are involved in the adsorption process (Bilal et al., 2018). Some studies related to the microalgae shows low removal of pollutants, which can be due to hydrophilic nature of these pollutants while the few microalgae (*Scenedesmus obliquus* and *Chlorella pyrenoidosa)* having efficient capacity to remediate the lipophilic-persistent pollutants from contaminated environment (de Wilt et al., 2016; Peng et al., 2014a). Adsorption of pollutants by the microalgae varies as lipophillic compounds are adsorbed more (>70%) compared to hydrophilic compounds (0%–20%) from industries (Gojkovic et al., 2019). Adsorption of persistent pollutants on microalgal cell wall occurs on both living and dead biomass as the receptors on cell wall remains viable for long time even after the death of the microalgae (Choi and Lee, 2015). Adsorption does not involve the use of metabolic energy for decontamination and thus represent dead biomass as an efficient alternative (Dixit and Singh, 2014) for pollutant removal from contaminated sites (Baghour, 2019). Thus, bio-adsorption utilizing living or dead biomass is an effective treatment option for pollutant removal from contaminated ecosystem (Derco and Vrana, 2018).

Wastewater is abundant in diverse compounds (Mushtaq et al., 2020) and due to this, the binding sites in the microalgae can become saturated with nontarget compounds thus decreasing the efficiency of the process.

Multiple contaminants in wastewater can elevate competition for binding sites and also disturb the interaction between contaminants and microalgae. Apart from decreasing binding sites at the surface of cell wall, presence of multiple components in wastewater can also decrease the adsorption rate of specific pollutants. Thus, screening of hyper-adsorbent microalgae species is necessary to enhance the efficiency of the remediation process in environmental friendly manner (Sutherland and Ralph, 2019). Several microalgae species from *Anabaena, Chlorella,* and *Scenedesmus* genera are known as hyper-accumulators (Bwapwa et al., 2017; Kumar et al., 2016) and need is to exploit the diverse microalgae species for the effective treatment of contaminated sites (Brar et al., 2017).

The process of bio-adsorption through microalgae can also be enhanced by either physical or chemical treatment, which can become feasible methods for enhancing remediation potential of the microalgae (Ali et al., 2018; Bilal et al., 2018). Modification of cell surface can enhance the interaction between cell surface and different pollutants through deprotonation, which is responsible for increasing negative charges on the cell surface of the microalgae (Bilal et al., 2018). Temperature also affects the process of bio-adsorption as, in endothermic process, with the decrease in temperature, the process of adsorption also decreases and vice versa (Zeraatkar et al., 2016). The process of adsorption is also affected by the proper growth of microalgae. Optimized growth increases the number of microalgae cell, which eventually increase the functional groups for the binding of pollutants to cell surface (Xia et al., 2017).

The uptake of metal in microalgae occurs by using metabolic energy and binds with proteins inside cell leading to their elimination from the wastewater (Kanamarlapudi et al., 2018). Bio-uptake of pollutants by microalgae is another mechanism for the removal of lipophillic pollutants (Bai and Acharya, 2016). *Desmodesmus subspicatus* (Maes et al., 2014) and *Nannochloris* sp. (Bai and Acharya, 2016) account for 23% and 42% removal of 17á-ethinylestadiol and triclosan by bio-uptake mechanism in both microalgae. Enhanced uptake of persistent pollutants is possible by selecting the species, which can acclimatize in polluted environment and have high uptake rate. The other most-promising technique for remediation of pollutants from contaminated sites is by bio-degradation process (Baghour, 2019; Delrue et al., 2016). Apart from adsorption or bio-uptake mechanism, biodegradation is a promising mechanism for the effective removal of persistent pollutants from wastewater (Sutherland and

Ralph, 2019). In biodegradation process, pollutants are degraded by using them as carbon source or acceptor/donor of electrons and the other way to degrade these pollutants is by secretion of enzymes (Chekroun et al., 2014) that attack compounds and lead to their removal from contaminated sites (Karigar and Rao, 2011; Tiwari et al., 2017; Sutherland and Ralph, 2019). Degradation of pollutants can occur outside or inside the cell or both inside and outside. The initial degradation can occur outside the cell and after transporting inside the cell, the further degradation process occurs inside the cell (Tiwari et al., 2017). Degradation of contaminants by microalgae occurs in two phases (Pflugmacher and Sandermann, 1998). In the first phase, enzymes makes the contaminant water loving by exposing or adding OH (hydroxyl) group through different reactions, like hydrolysis and redox reaction, while in the second phase, enzymes such as gluthathione-S-transferases enhance the reaction of glutathione with diverse compounds devoid of electrons leading to the opening of the epoxide ring in the cell in order to provide protection against oxidative damage (Xiong et al., 2018a). The wide range of enzymes in the microalgae plays a pivotal role in the degradation of diverse contaminants and also protection against cellular damage induced by these pollutants (Wang et al., 2019). Microalgae species from *Chlamydomonas, Chlorella*, and *Scenedesmus* genus have been studied widely and have shown immense potential to remove pollutants from contaminated sites (Sutherland and Ralph, 2019). Thus, phycoremediation is a promising mechanism for the complete elimination of toxic pollutants from wastewater.

3.9 CONCLUSION

POPs, because of unique features, poses serious risk in the protection of the environment. High mobility, liphophilicity, slow degradation, and bio-accumulation potential elevates the damage caused by POPs to the environmental and human health. Human exposure owing to the consumption of POPs-contaminated food causes severe health complications and may eventually lead to death of organisms. To ensure the protection of human health, it became vital to foster the removal of POPs from the ecosystem and totally ban their production as well use in industrial and agriculture sector. The existing technologies are inefficient and expensive in removing the persistent pollutants from diverse ecosystems. The need is to rely on techniques that are environmentally sustainable and feasible. Microalgae have shown immense potential to remove persistent pollutants by different mechanisms thereby acting as a promising organism for POPs removal from contaminated ecosystems.

KEYWORDS

- **persistent organic pollutants**
- **microalgae**
- **wastewater**
- **contamination**
- **remediation**
- **biodegradation**

REFERENCES

Abdel-Razek, M. A.; Abozeid, A. M.; Eltholth, M. M.; Abouelenien, F. A.; El-Midany, S. A.; Moustafa, N. Y.; Mohamed, R. A. Bioremediation of a Pesticide and Selected Heavy Metals in Wastewater from Various Sources Using a Consortium of Microalgae and Cyanobacteria. *Sloven. Vet. Res.* **2019,** *56,* (22 Suppl).

Abdel-Shafy, H. I.; Mansour, M. S. A Review on Polycyclic Aromatic Hydrocarbons: Source, Environmental Impact, Effect on Human Health and Remediation. *Egypt. J. Petrol.* **2016,** *25,* 107–123.

Abdi, O.; Kazemi, M. A Review Study of Biosorption of Heavy Metals and Comparison between Different Biosorbents. *J. Mater. Environ. Sci.* **2015,** *6,* 1386–1399.

Adeola, F. O. Boon or Bane? The Environmental and Health Impacts of Persistent Organic Pollutants (POPs). *Human Ecol. Rev.* **2004,** 27–35.

Agnihotri, N. P.; Kulshreshtha, G.; Gajbhiye, V. T.; Mohapatra, S. P.; Singh, S. B. Organochlorine Insecticide Residues in Agricultural Soils of the Indo-Gangetic Plain. *Environ. Monit. Assess.* **1996,** *40,* 279–288.

Aktar, W.; Sengupta, D.; Chowdhury, A. Impact of Pesticides Use in Agriculture: Their Benefits and Hazards. *Interdiscipl. Toxicol.* **2009,** *2,* 1–12.

Alava, J. J.; Cisneros-Montemayor, A. M.; Sumaila, U. R.; Cheung, W. W. Projected Amplification of Food Web Bioaccumulation of MeHg and PCBs Under Climate Change in the Northeastern Pacific. *Sci. Rep.* **2018,** *8,* 1–12.

Alava, J. J.; Gobas, F. A. P. C. Assessing biomagnification and Trophic Transport of Persistent Organic Pollutants in the Food Chain of the Galapagos Sea Lion (*Zalophus wollebaeki*): Conservation and Management Implications. New Approaches to the Study of Marine Mammals, 2012; pp 77–108.

Aleksa, K.; Carnevale, A.; Goodyer, C.; Koren, G. Detection of Polybrominated Biphenyl Ethers (PBDEs) in Pediatric Hair as a Tool for Determining in Utero Exposure. *Foren. Sci. Int.* **2012**, *218*, 37–43.

Alharbi, O. M. L.; Basheer, A. A.; Khattab, R. A.; Ali, I. Health and Environmental Effects of Persistent Organic Pollutants. *J. Mol. Liq.* **2018**, *263*, 442–453.

Ali, M. E.; El-Aty, A. M. A.; Badawy, M. I.; Ali, R. K. Removal of Pharmaceutical Pollutants from Synthetic Wastewater Using Chemically Modified Biomass of Green Alga *Scenedesmus obliquus*. *Ecotoxicol. Environ. Saf.* **2018**, *151*, 144–152.

Amaraneni, S. R.; Pillala, R. R. Kolleru Lake Water Pollution by Pesticides. *Indian J. Environ. Health* **2000**, *42*, 169–175.

Anon. Council Directive 98/83/EC of 3 November 1998 on the Quality of Water Intended for Human Consumption. *Off. J. Eur. Communities* **1998**, *330*, 32–53.

Ashraf, M. A. Persistent Organic Pollutants (POPs): A Global Issue, a Global Challenge, 2017.

Avagyan, A. B. A Contribution to Global Sustainable Development: Inclusion of Microalgae and Their Biomass in Production and Bio Cycles. *Clean Technol. Environm. Policy* **2008**, *10*, 313–317.

Ayangbenro, A. S.; Babalola, O. O. A New Strategy for Heavy Metal Polluted Environments: a Review of Microbial Biosorbents. *Int. J. Environ. Res. Public Health* **2017**, *14*, 94.

Bacaloni, A.; Insogna, S.; Zoccolillo, L. Remote Zones Air Quality-Persistent Organic Pollutants: Sources, Sampling and Analysis. *Air Qual. Monitor., Assess. Manage.* **2011**, 223.

Baghour, M. Algal Degradation of Organic Pollutants. In: Martínez, L., Kharissova, O., Kharisov, B. (Eds.) *Handbook of Ecomaterials*; Springer: Cham, 2019.

Bai, X.; Acharya, K. Removal of Trimethoprim, Sulfamethoxazole, and Triclosan by the Green Alga *Nannochloris* sp. *J. Hazard. Mater.* **2016**, *315*, 70–75.

Bartrons Vilamala, M.; Grimalt Obrador, J.; Catalan, J. Food Web Bioaccumulation of Organohalogenated Compounds in High Mountain Lakes. *Limnetica* **2012**, *31*, 0155–0164.

Batt, A. L.; Wathen, J. B.; Lazorchak, J. M.; Olsen, A. R.; Kincaid, T. M. Statistical Survey of Persistent Organic Pollutants: Risk Estimations to Humans and Wildlife Through Consumption of Fish from US Rivers. *Environ. Sci. Technol.* **2017**, *51*, 3021–3031.

Bernardes, M. F. F.; Pazin, M.; Pereira, L. C.; Dorta, D. J. Impact of Pesticides on Environmental and Human Health. *Toxicol. Stud.-Cells, Drugs Environ.* **2015**, 195–233.

Beyer, A.; Mackay, D.; Matthies, M.; Wania, F.; Webster, E. Assessing Long-Range Transport Potential of Persistent Organic Pollutants. *Environ. Sci. Technol.* **2000**, *34*, 699–703.

Bharagava, R. N. *Environmental Pollutants and Their Bioremediation Approaches*; CRC Press, 2017.

Bilal, M.; Rasheed, T.; Sosa-Hernández, J. E.; Raza, A.; Nabeel, F.; Iqbal, H. Biosorption: An Interplay between Marine Algae and Potentially Toxic Elements—a Review. *Marine Drugs* **2018**, *16* (2), 65.

Blanchard, M.; Teil, M.; Ollivon, D.; Legenti, L.; Chevreuil, M. Polycyclic Aromatic Hydrocarbons and Polychlorobiphenyls in Wastewaters and Sewage Sludges from the Paris Area (France). *Environ. Res.* **2004**, *95*, 184–197.

Brar, A.; Kumar, M.; Vivekanand, V.; Pareek, N. Photoautotrophic Microorganisms and Bioremediation of Industrial Effluents: Current Status and Future Prospects. *3 Biotech* **2017**, *7*, 18.

Brown, T. N.; Wania, F. Screening Chemicals for the Potential to be Persistent Organic Pollutants: A Case Study of Arctic Contaminants. *Environ. Sci. Technol.* **2008**, *42*, 5202–5209.

Burkow, I. C.; Kallenborn, R. Sources and Transport of Persistent Pollutants to the Arctic. *Toxicol. Lett.* **2000**, *112*, 87–92.

Bwapwa, J. K.; Jaiyeola, A. T.; Chetty, R. Bioremediation of Acid Mine Drainage Using Algae Strains: A Review. *S. Afr. J. Chem. Eng.* **2017**, *24*, 62–70.

Carvalho, F. P. Pesticides, Environment, and Food Safety. *Food Energy Sec.* **2017**, *6*, 48–60.

Cavallo, G.; Metrangolo, P.; Milani, R.; Pilati, T.; Priimagi, A.; Resnati, G.; Terraneo, G. The Halogen Bond. *Chem. Rev.* **2016**, *116* (4), 2478–2601.

Chapman, R. L. Algae: The World's Most Important "Plants"—an Introduction. *Mitigation Adapt. Strateg. Global Change* **2013**, *18*, 5–12.

Chekroun, K. B.; Sánchez, E.; Baghour, M. The Role of Algae in Bioremediation of Organic Pollutants. *Int. Res. J. Public Environ. Health* **2014**, *1*, 19–32.

Choi, H. J.; Lee, S. M. Heavy Metal Removal from Acid Mine Drainage by Calcined Eggshell and Microalgae Hybrid System. *Environ. Sci. Pollut. Control Ser.* **2015**, *22*, 13404–13411.

Cok, I.; Donmez, M. K.; Uner, M.; Demirkaya, E.; Henkelmann, B.; Shen, H.; Kotalik, J.; Schramm, K. W. Polychlorinated Dibenzo-P-Dioxins, Dibenzofurans and Polychlorinated Biphenyls Levels in Human Breast Milk from Different Regions of Turkey. *Chemosphere* **2009**, *76*, 1563–1571.

Damalas, C. A.; Eleftherohorinos, I. G. Pesticide Exposure, Safety Issues, and Risk Assessment Indicators. *Int. J. Environ. Res. Public Health* **2011**, *8*, 1402–1419.

De-Bashan, L. E.; Bashan, Y. Immobilized Microalgae for Removing Pollutants: Review of Practical Aspects. *Bioresour. Technol.* **2010**, *101*, 1611–1627.

DellaValle, C. T.; Wheeler, D. C.; Deziel, N. C.; De Roos, A. J.; Cerhan, J. R.; Cozen, W.; Severson, R. K.; Flory, A. R.; Locke, S. J.; Colt, J. S.; Hartge, P. Environmental Determinants of Polychlorinated Biphenyl Concentrations in Residential Carpet Dust. *Environ. Sci. Technol.* **2013**, *47*, 10405–10414.

Delrue, F.; Álvarez-Díaz, P. D.; Fon-Sing, S.; Fleury, G.; Sassi, J. F. The Environmental Biorefinery: Using Microalgae to Remediate Wastewater, a Win-Win Paradigm. *Energies* **2016**, *9*, 132.

Derco, J.; Vrana, B. Introductory Chapter: Biosorption. *Biosorption* **2018**, 1.

Devi, P. I.; Thomas, J.; Raju, R. K. Pesticide Consumption in India: A Spatiotemporal Analysis. *Agric. Eco. Res. Rev.* **2017**, *30*, 163–172.

de Wilt, A.; Butkovskyi, A.; Tuantet, K.; Leal, L. H.; Fernandes, T. V.; Langenhoff, A.; Zeeman, G. Micropollutant Removal in an Algal Treatment System Fed with Source Separated Wastewater Streams. *J. Hazard. Mater.* **2016**, *304*, 84–92.

Dhananjayan, V.; Ravichandran, B.; Rajmohan, H. R. Organochlorine Pesticide Residues in Blood Samples of Agriculture and Sheep Wool Workers in Bangalore (Rural), India. *Bull. Environ. Contam. Toxicol.* **2012**, *88*, 497–500.

Dixit, S.; Singh, D. P. An Evaluation of Phycoremediation Potential of Cyanobacterium *Nostoc muscorum*: Characterization of Heavy Metal Removal Efficiency. *J. Appl. Phycol.* **2014**, *26*, 1331–1342.

Dosnon-Olette, R.; Trotel-Aziz, P.; Couderchet, M.; Eullaffroy, P. Fungicides and Herbicide Removal in *Scenedesmus* Cell Suspensions. *Chemosphere* **2010**, *79*, 117–123.

Edokpayi, J. N.; Odiyo, J. O.; Durowoju, O. S. Impact of Wastewater on Surface Water Quality in Developing Countries: A Case Study of South Africa. *Water Qual.* **2017,** 401–416.

Eguchi, A.; Nomiyama, K.; Devanathan, G.; Subramanian, A.; Bulbule, K. A.; Parthasarathy, P. et al. Different Profile of Anthropogenic and Naturally Produced Organohalogen Compounds in Serum From Residents Living Near a Coastal Area and E-Waste Recycling Workers in India. *Environ. Int.* **2012,** *47,* 8–16.

El-Shahawi, M. S.; Hamza, A.; Bashammakh, A. S.; Al-Saggaf, W. T. An Overview on the Accumulation, Distribution, Transformations, Toxicity and Analytical Methods for the Monitoring of Persistent Organic Pollutants. *Talanta* **2010,** *80,* 1587–1597.

Erickson, M. D.; Kaley, R. G. Applications of Polychlorinated Biphenyls. *Environ. Sci. Pollut. Res.* **2011,** *2,* 135–151.

Fair, P. A.; White, N. D.; Wolf, B.; Arnott, S. A.; Kannan, K.; Karthikraj, R.; Vena, J. E. Persistent Organic Pollutants in Fish from Charleston Harbor and Tributaries, South Carolina, United States: A Risk Assessment. *Environ. Res.* **2018,** *167,* 598–613.

Ferguson, A.; Solo-Gabriele, H. Children's Exposure to Environmental Contaminants: An Editorial Reflection of Articles in the IJERPH Special Issue Entitled. Children's Exposure to Environmental Contaminants, 2016.

Filazi, A.; Yurdakok-Dikmen, B.; Kuzukiran, O.; Sireli, U. T. Chemical Contaminants in Poultry Meat and Products. *Poult. Sci.* **2017,** *15,* 171.

Franzaring, J.; van der Eerden, L. J. Accumulation of Airborne Persistent Organic Pollutants (POPs) in Plants. *Basic Appl. Ecol.* **2000,** *1* (1), 25–30.

Friesen-Pankratz, B.; Doebel, C.; Farenhorst, A.; Goldsborough, L. G. Interactions between Algae (*Selenastrum capricornutum*) and Pesticides: Implications for Managing Constructed Wetlands for Pesticide Removal. *J. Environ. Sci. Health* **2003,** *2,* 147–155.

Gay, H. Before and after Silent Spring: From Chemical Pesticides to Biological Control and Integrated Pest Management—Britain, 1945–1980. *Ambix* **2012,** *59* (2), 88–108.

Geissen, V.; Mol, H.; Klumpp, E.; Umlauf, G.; Nadal, M.; van der Ploeg, M.; van de Zee, S. E. A. T. M.; Ritsema, C. J. Emerging Pollutants in the Environment: A Challenge for Water Resource Management. *Int. Soil Water Conserv. Res.* **2015,** *3* (1), 57–65.

Geyer, H. J.; Rimkus, G. G.; Scheunert, I.; Kaune, A.; Schramm, K. W.; Kettrup, A.; Zeeman, M.; Muir, D. C.; Hansen, L. G.; Mackay, D. Bioaccumulation and Occurrence of Endocrine-Disrupting Chemicals (EDCs), Persistent Organic Pollutants (POPs), and Other Organic Compounds in Fish and Other Organisms Including Humans. In *Bioaccumulation–New Aspects and Developments*; Springer: Berlin, Heidelberg, 2000; pp 1–166.

Gibson, J.; Adlard, B.; Olafsdottir, K.; Sandanger, T. M.; Odland, J. O. Levels and Trends of Contaminants in Humans of the Arctic. *Int. J. Circumpolar Health* **2016,** *75* (1), 33804.

Gill, H. K.; Garg, H. Pesticide: Environmental Impacts and Management Strategies. *Pesticides-Toxic Aspects* **2014,** *8,* 187.

Gojkovic, Z.; Lindberg, R. H.; Tysklind, M.; Funk, C. Northern Green Algae Have the Capacity to Remove Active Pharmaceutical Ingredients. *Ecotoxicol. Environ. Safety* **2019,** *170,* 644–656.

González-Mille, D. J.; Ilizaliturri-Hernández, C. A.; Espinosa-Reyes, G.; Costilla-Salazar, R.; Díaz-Barriga, F.; Ize-Lema, I.; Mejía-Saavedra, J. Exposure to Persistent Organic Pollutants (POPs) and DNA Damage as an Indicator of Environmental Stress in Fish of

Different Feeding Habits of Coatzacoalcos, Veracruz, Mexico. *Ecotoxicology* **2010**, *19* (7), 1238–1248.

Gregoraszczuk, E. L.; Ptak, A. Endocrine-Disrupting Chemicals: Some Actions of POPs on Female Reproduction. *Int. J. Endocrinol.* **2013**.

Gudorf, C. E.; Huchingson, J. E. *Boundaries: A Casebook in Environmental Ethics*; Georgetown University Press, 2010.

Guimarães, R. M.; Asmus, C. I. R. F.; Meyer, A. DDT Reintroduction for Malaria Control: The Cost-Benefit Debate for Public Health. *Cadernos de Saúde Pública* **2007**, *23*, 2835–2844.

Guo, W.; Pan, B.; Sakkiah, S.; Yavas, G.; Ge, W.; Zou, W.; Tong, W.; Hong, H. Persistent Organic Pollutants in Food: Contamination Sources, Health Effects and Detection Methods. *Int. J. Environ. Res. Public Health* **2019**, *16* (22), 4361.

Gupta, P. K. Pesticide Exposure—Indian Scene. *Toxicology* **2004**, *198* (1–3), 83–90.

Gushit, J. S.; Ekanem, E. O.; Adamu, H. M.; Chindo, I. Y. Analysis of Herbicide Residues and Organic Priority Pollutants in Selectedroot and Leafy Vegetable Crops in Plateau State, Nigeria. *World J. Analyt. Chem.* **2013**, *1* (2), 23–28.

Hanedar, A.; Güneş, E.; Kaykioglu, G.; Çelik, S. O.; Cabi, E. Presence and Distributions of POPS in Soil, Atmospheric Deposition, and Bioindicator Samples in an Industrial-Agricultural Area in Turkey. *Environ. Monitor. Assess.* **2019**, *191* (1), 42.

Hens, B.; Hens, L. Persistent Threats by Persistent Pollutants: Chemical Nature, Concerns and Future Policy Regarding PCBs—What Are We Heading for. *Toxics* **2018**, *6* (1), 1.

Hom-Diaz, A.; Llorca, M.; Rodríguez-Mozaz, S.; Vicent, T.; Barceló, D.; Blánquez, P. Microalgae Cultivation on Wastewater Digestate: β-estradiol and 17α-Ethynylestradiol Degradation and Transformation Products Identification. *J. Environ. Manage.* **2015**, *155*, 106–113.

Hong, Y. W.; Yuan, D. X.; Lin, Q. M.; Yang, T. L. Accumulation and Biodegradation of Phenanthrene and Fluoranthene by the Algae Enriched from a Mangrove Aquatic Ecosystem. *Marine Pollut. Bull.* **2008**, *56* (8), 1400–1405.

Hung, H.; MacLeod, M.; Guardans, R.; Scheringer, M.; Barra, R.; Harner, T.; Zhang, G. Toward the Next Generation of Air Quality Monitoring: Persistent Organic Pollutants. *Atmospheric Environ.* **2013**, *80*, 591–598.

Ibrahim, M. B. M.; Gamila, H. A. Algal Bioassay for Evaluating the Role of Algae in Bioremediation of Crude Oil: II. Freshwater Phytoplankton Assemblages. *Bull. Environ. Contamination Toxicol.* **2004**, *73* (6), 971–978.

Ibrahim, M. S. Chapter 14 Persistent Organic Pollutants in Malaysia. In An Li, S. T. G. J. J. P. G., Paul, K. S. L. (Eds.), *Dev. Environ. Sci.* **2007**, *7*, 629–655.

Imphal Free Press. Concern Over Excessive DDT Use in Jiriban Fields, 2008. http://www. kanglaonline.com/index.php?template=headline&newsid=42015&typeid=1

Islam, R.; Kumar, S.; Karmoker, J.; Kamruzzaman, M.; Rahman, M. A.; Biswas, N.; Tran, T. K. A.; Rahman, M. M. Bioaccumulation and Adverse Effects of Persistent Organic Pollutants (POPs) on Ecosystems and Human Exposure: A Review Study on Bangladesh Perspectives. *Environ. Technol. Innov.* **2018**, *12*, 115–131.

Jabusch, T. W.; Swackhamer, D. L. Subcellular Accumulation of Polychlorinated Biphenyls in the Green Alga *Chlamydomonas reinhardtii*. *Environ. Toxicol. Chem. Int. J.* **2004**, *23* (12), 2823–2830.

Jacob, J. A Review of the Accumulation and Distribution of Persistent Organic Pollutants in the Environment. *Int. J. Biosci. Biochem. Bioinfo.* **2013**, *3* (6), 657.

Jamieson, A. J.; Malkocs, T.; Piertney, S. B.; Fujii, T.; Zhang, Z. Bioaccumulation of Persistent Organic Pollutants in the Deepest Ocean Fauna. *Nat. Ecol. Evol.* **2017,** *1* (3), 1–4.

Jayaraj, R.; Megha, P.; Sreedev, P. Organochlorine Pesticides, Their Toxic Effects on Living Organisms and Their Fate in the Environment. *Interdiscipl. Toxicol.* **2016,** *9* (3–4), 90–100.

Kammenga, J. E., Dallinga, R.; Donker, M. H.; Kohler, H. R.; Simonsen, V.; Triebskorn, R. et al. Biomarkers in Terrestrial Invertebrates for Ecotoxicological Soil Risk Assessment. *Rev. Environ. Contam. Toxicol.* **2000,** *164,* 93–147.

Kanamarlapudi, S. L. R. K.; Chintalpudi, V. K.; Muddada, S. Application of Biosorption for Removal of Heavy Metals from Wastewater. *Biosorption* **2018,** *18,* 69.

Kang, J. W. Removing Environmental Organic Pollutants with Bioremediation and Phytoremediation. *Biotechnol. Lett.* **2014,** *36* (6), 1129–1139.

Karigar, C. S.; Rao, S. S. Role of Microbial Enzymes in the Bioremediation of Pollutants: A Review. *Enzyme Res.* **2011.**

Ke, L.; Wong, Y. S.; Tam, N. F. Toxicity and Removal of Organic Pollutants by Microalgae: A Review. In *Microalgae: Biotechnology, Microbiology, and Energy*; Nova Science Publishers, Inc., 2012; pp 101–140.

Kim, Y. A.; Park, J. B.; Woo, M. S.; Lee, S. Y.; Kim, H. Y.; Yoo, Y. H. Persistent Organic Pollutant-Mediated Insulin Resistance. *Int. J. Environm. Res. Public Health* **2019,** *16* (3), 448.

Kirby, R. T. Persistent Organic Pollutant Accumulation in the Arctic. *Sustain. Dev. Law Policy* **2010,** *8* (3), 13.

Kitchen, L. W.; Lawrence, K. L.; Coleman, R. E. The Role of the United States Military in the Development of Vector Control Products, Including Insect Repellents, Insecticides, and Bed Nets. *J. Vector Ecol.* **2009,** *34* (1), 50–61.

Kolpin, D. W.; Thurman, E. M.; Goolsby, D. A. Occurrence of Selected Pesticides and Their Metabolites in Near-Surface Aquifers of the Mid-Western United States. *Environ. Sci. Technol.* **1996,** *30,* 335–340.

Kong, D.; MacLeod, M.; Hung, H.; Cousins, I. T. Statistical Analysis of Long-Term Monitoring Data for Persistent Organic Pollutants in the Atmosphere at 20 Monitoring Stations Broadly Indicates Declining Concentrations. *Environ. Sci. Technol.* **2014,** *48,* 12492–12499.

Kovner, K. Persistent Organic Pollutants: A Global Issue, a Global Response. US Environmental Protection Agency (EPA), December 2009.

Kumar, D.; Pandey, L. K.; Gaur, J. P. Metal Sorption by Algal Biomass: From Batch to Continuous System. *Algal Res.* **2016,** *18,* 95–109.

Kumar, K. S.; Dahms, H. U.; Won, E. J.; Lee, J. S.; Shin, K. H. Microalgae–a Promising Tool for Heavy Metal Remediation. *Ecotoxicology and Environmental Safety* **2015,** *113,* 329–352.

Kumar, P. K.; Krishna, S. V.; Verma, K.; Pooja, K.; Bhagawan, D.; Himabindu, V. Phycoremediation of Sewage Wastewater and Industrial Flue Gases for Biomass Generation from Microalgae. *SA J. Chem. Eng.* **2018,** *25,* 133–146.

Kurashvili, M.; Varazi, T.; Khatisashvili, G.; Gigolashvili, G.; Adamia, G.; Pruidze, M.; Gordeziani, M.; Chokheli, L.; Japharashvili, S.; Khuskivadze, N. Blue-Green Alga Spirulina as a Tool Against Water Pollution by 1, 1'-(2, 2, 2-trichloroethane-1, 1-diyl) bis (4-chlorobenzene)(DDT). *Annals of Agrarian Science* **2018,** *16* (4), 405–409.

Lali, Z. Release of Dioxins from Solid Waste Burning and its Impacts on Urban Human Population-A Review. *J. Pollut. Eff. Cont.* **2018**, *215* (6). doi: 10.4172/2375-4397.1000215

Lallas, P. L. The Stockholm Convention on Persistent Organic Pollutants. *Am. J. Int. L.* **2001**, *95* (3), 692–708.

Langenbach, T. Persistence and Bioaccumulation of Persistent Organic Pollutants (POPs). In *Applied Bioremediation-Active and Passive Approaches Rijeka: InTech*, 2013; p 307.

Lawal, A. T. Polycyclic Aromatic Hydrocarbons. A Review. *Cogent Environ. Sci.* **2017**, *3* (1), 1339841.

Link, T. *Country Situation on Persistent Organic Pollutants (POPs) in India*; India: New Delhi, 2006.

Liu, W.; Chen, Q.; He, N.; Sun, K.; Sun, D.; Wu, X.; Duan, S. Removal and Biodegradation of 17b-estradiol and Diethylstilbestrol by the Freshwater Microalgae *Raphidocelis subcapitata. Int. J. Environ. Res. Public Health* **2018**, *15*, 452.

Lee, B. K. Sources, Distribution and Toxicity of Polyaromatic Hydrocarbons (PAHs) in Particulate Matter. In *Air Pollution*; IntechOpen, 2010.

Lee, D. H.; Steffes, M. W.; Sjodin, A.; Jones, R. S.; Needham, L. L.; Jacobs, D. R. Jr. Low Dose of Some Persistent Organic Pollutants Predicts Type 2 Diabetes: A Nested Case-Control Study. *Environ. Health Perspect.* **2010**, *118*, 1235–1242.

Lin, T.; Hu, Z.; Zhang, G.; Li, X.; Xu, W.; Tang, J. Levels and Mass Burden of DDTs in Sediments from Fishing Harbors: The Importance of DDT-Containing Antifouling Paint to the Coastal Environment of China. *Environ. Sci. Technol.* **2009**, *43*, 8033–8038.

Liu, Y.; Guan, Y.; Gao, Q.; Tam, N. F. Y.; Zhu, W. Cellular Responses, Biodegradation and Bioaccumulation of Endocrine Disrupting Chemicals in Marine Diatom Navicula Incerta. *Chemosphere* **2010**, *80* (5), 592–599.

Lopes, E. J.; Okamura, L. A.; Yamamoto, C. I. Formation of Dioxins and Furans during Municipal Solid Waste Gasification. *Braz. J. Chem. Eng.* **2015**, *32* (1), 87–97.

Lushchak, V. I.; Matviishyn, T. M.; Husak, V. V.; Storey, J. M.; Storey, K. B. Pesticide Toxicity: A Mechanistic Approach. *EXCLI J.* **2018**, *17*, 1101.

Luthe, G.; Jacobus, J. A.; Robertson, L. W.; Receptor Interactions by Polybrominated Diphenyl Ethers versus Polychlorinated Biphenyls: A Theoretical Structure–Activity Assessment. *Environ. Toxicol. Pharmacol.* **2008**, *25* (2), 202–210.

Ma, M.; Gong, Y.; Hu, Q. Identification and Feeding Characteristics of the Mixotrophic Flagellate *Poterioochromonas malhamensis,* a Microalgal Predator Isolated from Outdoor Massive *Chlorella* Culture. *Algal Res.* **2018**, *29*, 142–153.

Martorell, I.; Perelló, G.; Martí-Cid, R.; Castell, V.; Llobet, J. M.; Domingo, J. L. Polycyclic Aromatic Hydrocarbons (PAH) in Foods and Estimated PAH Intake by the Population of Catalonia, Spain: Temporal Trend. *Environ. Int.* **2010**, *36*, 424–432.

Massart, F.; Harrell, J. C.; Federico, G.; Saggese, G. Human Breast Milk and Xenoestrogen Exposure: A Possible Impact on Human Health. *J. Perinatol.* **2005**, *25*, 282–288.

Matthies, M.; Solomon, K.; Vighi, M.; Gilman, A.; Tarazona, J. V. The Origin and Evolution of Assessment Criteria for Persistent, Bioaccumulative and Toxic (PBT) Chemicals and Persistent Organic Pollutants (POPs). *Environ. Sci.: Process. Impacts* **2016**, *18* (9), 1114–1128.

Mechlińska, A.; Gdaniec-Pietryka, M.; Wolska, L.; Namieśnik, J. Evolution of Models for Sorption of PAHs and PCBs on Geosorbents. *Trends Anal. Chem.* **2009**, *28*, 466–482.

Megharaj, M.; Kantachote, D.; Singleton, I.; Naidu, R. Effects of Long-Term Contamination of DDT on Soil Microflora with Special Reference to Soil Algae and Algal Transformation of DDT. *Environ. Pollut.* **2000,** *109* (1), 35–42.

Metsoviti, M. N.; Papapolymerou, G.; Karapanagiotidis, I. T.; Katsoulas, N. Effect of Light Intensity and Quality on Growth Rate and Composition of *Chlorella vulgaris. Plants* **2020,** *9* (1), 31.

Miglioranza, K. S. B; Moreno, J. E. A.; Moreno, V. J.; Osterrieth, M. L., Escalante, A. H. Fate of Organochlorine Pesticides in Soils and Terrestrial Biotal of "Los Padres" Pond Watershed, Argentina. *Environ. Pollut.* **1999,** *105*, 91–99.

Mitra, A.; Chatterjee, C.; Mandal, F. B. Synthetic Chemical Pesticides and Their Effects on Birds. *Res. J. Environ. Toxicol.* **2011,** *5* (2), 81–96.

Molazadeh, M.; Ahmadzadeh, H.; Pourianfar, H. R.; Lyon, S.; Rampelotto, P. H. The Use of Microalgae for Coupling Wastewater Treatment with CO_2 Biofixation. *Front. Bioeng. Biotechnol.* **2019,** 7.

Mondal, M.; Halder, G.; Oinam, G.; Indrama, T; Tiwari, O. N. Bioremediation of Organic and Inorganic Pollutants Using Microalgae. In *New and Future Developments in Microbial Biotechnology and Bioengineering*; Elsevier, 2019; pp 223–235.

Mostafa, S. S. Microalgal Biotechnology: Prospects and Applications. *Plant Science* **2012,** *12*, 276–314.

Mushtaq, N.; Singh, D. V.; Bhat, R. A.; Dervash, M. A; bin Hameed, O. Freshwater Contamination: Sources and Hazards to Aquatic Biota. In *Fresh Water Pollution Dynamics and Remediation*; Springer: Singapore, 2020; pp 27–50.

Namiki, S.; Otani, T.; Seike, N. Fate and Plant Uptake of Persistent Organic Pollutants in Soil. *Soil Sci. Plant Nutr.* **2013,** *59* (4), 669–679.

National Research Council. *Children's Health, the Nation's Wealth: Assessing and Improving Child Health*; National Academies Press, 2004.

Neamtu, S.; Bors, A. M.; Stefan, S. Risk Assessment of Some Persistent Organic Pollutants on Environment and Health. *Studies* **2007,** *13*, 15.

Nicolopoulou-Stamati, P.; Maipas, S.; Kotampasi, C.; Stamatis, P.; Hens, L. Chemical Pesticides and Human Health: The Urgent Need for a New Concept in Agriculture. *Front. Public Health* **2016,** *4*, 148.

O'Driscoll, K.; Robinson, J.; Chiang, W. S.; Chen, Y. Y.; Kao, R. C.; Doherty, R. The Environmental Fate of Polybrominated Diphenyl Ethers (PBDEs) in Western Taiwan and Coastal Waters: Evaluation with a Fugacity-Based Model. *Environ. Sci. Pollut. Res.* **2016,** *23* (13), 13222–13234.

O'Driscoll, K.; Mayer, B.; Ilyina, T.; Pohlmann, T. Modelling the Cycling of Persistent Organic Pollutants (POPs) in the North Sea System: Fluxes, Loading, Seasonality, Trends. *J. Marine Syst.* **2013,** *111*, 69–82.

Oerke, E. C. Crop Losses to Pests. *J. Agr. Sci.* **2005,** *144*, 31–43.

Olatunji, O. S. Evaluation of Selected Polychlorinated Biphenyls (PCBs) Congeners and Dichlorodiphenyltrichloroethane (DDT) in Fresh Root and Leafy Vegetables Using GC-MS. *Sci. Rep.* **2019,** *9* (1), 1–10.

Olisah, C.; Okoh, O. O.; Okoh, A. I. A Bibliometric Analysis of Investigations of Polybrominated Diphenyl Ethers (PBDEs) in Biological and Environmental Matrices from 1992–2018. *Heliyon* **2018,** *4* (11), 00964.

Ozkara, A.; Akyıl, D.; Konuk, M. Pesticides, Environmental Pollution, and Health. In *Environmental Health Risk-Hazardous Factors to Living Species*; IntechOpen, 2016.

Pal, A.; He, Y. L.; Jekel, M.; Reinhard, M.; Gin, K. Y. H. Emerging Contaminants of Public Health Significance as Water Quality Indicator Compounds in the Urban Water Cycle. *Environ. Int.* **2014,** *71*, 46–62.

Panda, D. Cavitation Based Wastewater Treatment of Recalcitrant Organic Pollutants (Dicofol, BDE-209, HBCD, PFOS, PFOA) (Doctoral Dissertation, University of Nottingham), 2018.

Pariatamby, A.; Kee, Y. L. Persistent Organic Pollutants Management and Remediation. *Procedia Environ. Sci.* **2016,** *31*, 842–848.

Park, B. S.; Lee, S. E. Biotransformation of Aldrin and Chlorpyrifos-Methyl by Anabaena sp. PCC 7120. *Korean J. Environ. Agric.* **2010,** *29* (2), 184–188.

Peng, F. Q.; Ying, G. G.; Yang, B.; Liu, S.; Lai, H. J.; Liu, Y. S.; Chen, Z. F.; Zhou, G. J. Biotransformation of Progesterone and Norgestrel by Two Freshwater Microalgae (*Scenedesmus obliquus* and *Chlorella pyrenoidosa*): Transformation Kinetics and Products Identification. *Chemosphere* **2014a,** *95*, 581–588.

Pereira, R. C.; Martinez, M. C. M.; Cortizas, A. M.; Macias, F. Analysis of Composition, Distribution and Origin of Hexachlorocyclohexane Residues in Agricultural Soils from NW Spain. *Sci. Total Environ.* **2010,** *408*, 5583–5591.

Pflugmacher, S.; Sandermann, H. Cytochrome P450 Monooxygenases for Fatty Acids and Xenobiotics in Marine Macroalgae. *Plant Physiol.* **1998,** *117* (1), 123e128.

Polder, A.; Savinova, T. N.; Tkachev, A.; Løken, K. B.; Odland, J. O.; Skaare, J. U. Levels and Patterns of Persistent Organic Pollutants (POPS) in Selected Food Items from Northwest Russia (1998–2002) and Implications for Dietary Exposure. *Sci. Total Environ.* **2010,** *408*, 5352–5361.

Popp, J.; Pető, K.; Nagy, J. Pesticide Productivity and Food Security. A Review. *Agron. Sustain. Dev.* **2013,** *33* (1), 243–255.

Prakash, O.; Mrutyanjay, S.; Vishaha, R.; Charu, D.; Rinha, P.; Rup, L. Residues of Hexachlorohexane Isomers in Soil and Water Samples from Delhi and Adjoining Area. *Curr. Sci.* **2004,** *87*, 73–77.

Praveen, P.; Loh, K. C. Photosynthetic Aeration in Biological Wastewater Treatment Using Immobilized Microalgae-Bacteria Symbiosis. *Appl. Microbiol. Biotechnol.* **2015,** *99* (23), 10345–10354.

Qing Li, Q.; Loganath, A.; Seng Chong, Y.; Tan, J.; Philip Obbard, J. Persistent Organic Pollutants and Adverse Health Effects in Humans. *J. Toxicol. Environ. Health, Part A* **2006,** *69* (21), 1987–2005.

Quinn, L.; Pieters, R.; Nieuwoudt, C.; Borgen, A. R.; Kylin, H.; Bouwman, H. Distribution Profiles of Selected Organic Pollutants in Soils and Sediments of Industrial, Residential and Agricultural Areas of South Africa. *J. Environ. Monitor.* **2009,** *11* (9), 1647–1657.

Ramakrishnan, B.; Megharaj, M.; Venkateswarlu, K.; Naidu, R.; Sethunathan, N. The Impacts of Environmental Pollutants on Microalgae and Cyanobacteria. *Crit. Rev. Environ. Sci. Technol.* **2010,** *40* (8), 699–821.

Ramesh, A.; Walker, S. A.; Hood, D. B.; Guillén, M. D.; Schneider, K.; Weyand, E. H. Bioavailability and Risk Assessment of Orally Ingested Polycyclic Aromatic Hydrocarbons. *Int. J. Toxicol.* **2004,** *23* (5), 301–333.

Randrianarison, G.; Ashraf, M. A. Microalgae: A Potential Plant for Energy Production. *Geol. Ecol. Landscapes* **2017**, *1* (2), 104–120.

Rasalingam, S.; Peng, R.; Koodali, R. T. Removal of Hazardous Pollutants from Wastewaters: Applications of TiO_2-SiO_2 Mixed Oxide Materials. *J. Nanomater.* 2014.

Rasheed, T.; Bilal, M.; Nabeel, F.; Adeel, M.; Iqbal, H. M. Environmentally-Related Contaminants of High Concern: Potential Sources and Analytical Modalities for Detection, Quantification, and Treatment. *Environ. Int.* **2019**, *122*, 52–66.

Rawat, I.; Gupta, S. K.; Shriwastav, A.; Singh, P.; Kumari, S.; Bux, F. Microalgae Applications in Wastewater Treatment. In *Algae Biotechnology*; Springer: Cham, 2016; pp 249–268.

Rhind, S. M. Anthropogenic Pollutants: A Threat to Ecosystem Sustainability? *Philos. Trans. R. Soc. B Biol. Sci.* **2009**, *364* (1534), 3391–3401.

Rodas, V. L.; Martínez, D. C.; Salgado, E.; Sanz, A. M. A Facinating Example of Microalgal Adaptation to Extreme Crude Oil Contamination in a Natural Spill in Arroyo Minero, Río Negro, Argentina. *In Anales de la Real Academia Nacional de Farmacia Real Academia Nacional de Farmacia* **2009**, *4*, 883–900.

Rodgers, K. M.; Udesky, J. O.; Rudel, R. A.; Brody, J. G. Environmental Chemicals and Breast Cancer: An Updated Review of Epidemiological Literature Informed by Biological Mechanisms. *Environ. Res.* **2018**, *160*, 152–182.

Ruiz-Fernández, A. C.; Ontiveros-Cuadras, J. F.; Sericano, J. L.; Sanchez-Cabeza, J.-A.; Liong Wee Kwong, L.; Dunbar, R. B.; Mucciarone, D. A.; Pérez-Bernal, L. H.; Páez-Osuna, F. Long-Range Atmospheric Transport of Persistent Organic Pollutants to Remote Lacustrine Environments. *Sci. Total Environ.* **2014**, *493*, 505–520.

Ruzzin, J. Public Health Concern Behind the Exposure to Persistent Organic Pollutants and the Risk of Metabolic Diseases. *BMC Public Health* **2012**, *12* (1), 298.

Salama, E. S.; Roh, H. S.; Dev, S.; Khan, M. A.; Abou-Shanab, R. A.; Chang, S. W.; Jeon, B. H. Algae as a Green Technology for Heavy Metals Removal from Various Wastewater. *World J. Microbiol. Biotechnol.* **2019**, *35* (5), 75.

Samer, M. Biological and Chemical Wastewater Treatment Processes. *Wastewater Treat. Eng.* **2015**, 1–50.

Schäfer, R. B.; van den Brink, P. J.; Liess, M. Impacts of Pesticides on Freshwater Ecosystems. *Ecol. Impacts Toxic Chem.* **2011**, 111–137.

Schug, T. T.; Blawas, A. M.; Gray, K.; Heindel, J. J.; Lawler, C. P. Elucidating the Links between Endocrine Disruptors and Neurodevelopment. *Endocrinol.* **2015**, *156* (6), 1941–1951.

Sethunathan, N.; Megharaj, M.; Chen, Z. L.; Williams, W. D.; Lewis, G.; Naidu, R. Algal Degradation of a Known Endocrine Disrupting Insecticide, α-Endosulfan, and Its Metabolite, Endosulfan Sulfate, in Liquid Medium and Soil. *J. Agric. Food Chem.* **2004**, *52*, 3030–3035.

Sharma, A.; Kumar, V.; Shahzad, B.; Tanveer, M.; Sidhu, G. P. S.; Handa, N.; Kohli, S. K.; Yadav, P.; Bali, A. S.; Parihar, R. D.; Dar, O. I. Worldwide Pesticide Usage and Its Impacts on Ecosystem. *SN Appl. Sci.* **2019**, *1* (11), 1446.

Sharma, B. M.; Bharat, G. K.; Tayal, S.; Nizzetto, L.; Čupr, P.; Larssen, T. Environment and Human Exposure to Persistent Organic Pollutants (POPs) in India: A Systematic Review of Recent and Historical Data. *Environ. Int.* **2014**, *66*, 48–64.

Shen, N.; Birungi, Z. S.; Chirwa, E. Selective Biosorption of Precious Metals by Cell-Surface Engineered Microalgae, 2017.

Shivaramaiah, H. M. Organochlorine Pesticides in Agroecosystems: Monitoring Residues with Suitable Strategies for Their Management and Remediation. Ph.D. thesis, University of Sydney, Sydney, Australia, 2000.

Siddiqi, M. A.; Laessig, R. H.; Reed, K. D. Polybrominated Diphenyl Ethers (PBDEs): New Pollutants–Old Diseases. *Clin. Med. Res.* **2003,** *1* (4), 281–290.

Singh, D. V.; Bhat, R. A.; Dervash, M. A.; Qadri, H.; Mehmood, M. A.; Dar, G. H.; Hameed, M.; Rashid, N. Wonders of Nanotechnology for Remediation of Polluted Aquatic Environs. In *Fresh Water Pollution Dynamics and Remediation*; Springer: Singapore, 2020; pp 319–339.

Snedeker, S. Pesticides and Breast Cancer Risk: A Review of DDT, DDE and Dieldrin. *Environ. Health Perspect.* **2001,** *109*, 35–47.

Solé, A., Matamoros, V. Removal of Endocrine Disrupting Compounds from Wastewater by Microalgae Co-Immobilized in Alginate Beads. *Chemosphere* **2016,** *164*, 516–523.

Son, M. H.; Kim, J. T.; Park, H.; Kim, M.; Paek, O. J.; Chang, Y. S. Assessment of the Daily Intake of 62 Polychlorinated Biphenyls from Dietary Exposure in South Korea. *Chemosphere* **2012,** *89* (8), 957–963.

Stephenson, M. D.; Martin, M.; Tjeerdema, R. S. Long-Term Trends in DDT, Polychlorinated Biphenyls, and Chlordane in California Mussels. *Arch. Environ. Contam. Toxicol.* **1995,** *28*, 443–450.

Stockwell, R. J. The Family Farm in the Post-World War II Era: Industrialization, the Cold War and Political Symbol (Doctoral Dissertation, University of Missouri–Columbia), 2008.

Subashchandrabose, S. R.; Ramakrishnan, B.; Megharaj, M.; Venkateswarlu, K.; Naidu, R. Mixotrophic Cyanobacteria and Microalgae as Distinctive Biological Agents for Organic Pollutant Degradation. *Environ. Int.* **2013,** *51*, 59–72.

Subramaniam, K.; Solomon, J. Organochlorine Pesticides BHC and DDE in Human Blood in and Around Madurai, India. *Indian J. Clin. Biochem.* **2006,** *21* (2), 169.

Sutherland, D. L.; Ralph, P. J. Microalgal Bioremediation of Emerging Contaminants-Opportunities and Challenges. *Water Res.* **2019,** 114921.

Sweeney, M. F.; Hasan, N.; Soto, A. M.; Sonnenschein, C. Environmental Endocrine Disruptors: Effects on the Human Male Reproductive System. *Rev. Endocr. Metabol. Disorders* **2015,** *16* (4), 341–357.

Sweetman, M.; Vall, K.; Predouros, K.; Tones. The Role of Soil Organic Carbon in the Global Cycling of Persistent Organic Pollutants (POPs): Interpreting and Modelling Field Data. *Chemosphere* **2005,** *60*, 959–970.

Tabagari, I.; Kurashvili, M.; Varazi, T.; Adamia, G.; Gigolashvili, G.; Pruidze, M.; Chokheli, L.; Khatisashvili, G.; von Fragstein und Niemsdorff, P. Application of *Arthrospira* (*Spirulina*) *platensis* against Chemical Pollution of Water. *Water* **2019,** *11* (9), 1759.

Tam, N. F. Y.; Wang, P.; Gao, Q. T.; Wong, Y. S. Toxicity of Water-Borne Persistent Organic Pollutants on Green Microalgae. In *5th International Scientific Conference on Water, Climate and Environment* (BALWOIS), 2012.

Tang, H. P. Recent Development in Analysis of Persistent Organic Pollutants under the Stockholm Convention. *Trends Analyt. Chem.* **2013,** *45*, 48–66.

Teran, T.; Lamon, L.; Marcomini, A. Climate Change Effects on POPs' Environmental Behaviour: A Scientific Perspective for Future Regulatory Actions. *Atmospheric Pollut. Res.* **2012**, *3* (4), 466–476.

Thompson, L. A.; Darwish, W. S. Environmental Chemical Contaminants in Food: Review of a Global Problem. *J. Toxicol.* **2019**.

Tieyu, W.; Yonglong, L.; Hong, Z.; Yajuan, S. Contamination of Persistent Organic Pollutants (POPs) and Relevant Management in China. *Environ. Int.* **2005**, *31* (6), 813–821.

Tiwari, B.; Sellamuthu, B.; Ouarda, Y.; Drogui, P.; Tyagi, R. D.; Buelna, G. Review on Fate and Mechanism of Removal of Pharmaceutical Pollutants from Wastewater Using Biological Approach. *Bioresour. Technol.* **2017**, *224*, 1–12.

Todd, S. J.; Cain, R. B.; Schmidt, S. Biotransformation of Naphthalene and Diaryl Ethers by Green Microalgae. *Biodegradation* **2002**, *13* (4), 229–238.

Tombesi, N.; Pozo, K.; Harner, T. Persistent Organic Pollutants (POPs) in the Atmosphere of Agricultural and Urban Areas in the Province of Buenos Aires in Argentina Using PUF Disk Passive Air Samplers. *Atmospheric Pollut. Res.* **2014**, *5* (2), 170–178.

Torres, M. A.; Barros, M. P.; Campos, S. C.; Pinto, E.; Rajamani, S.; Sayre, R. T.; Colepicolo, P. Biochemical Biomarkers in Algae and Marine Pollution: A Review. *Ecotoxicol. Environ. Safety* **2008**, *71* (1), 1–15.

Tsai, W. T. Current Status and Regulatory Aspects of Pesticides Considered to be Persistent Organic Pollutants (POPs) in Taiwan. *Int. J. Environ. Res. Public Health* **2010**, *7* (10), 3615–3627.

Ukalska-Jaruga, A.; Lewińska, K.; Mammadov, E.; Karczewska, A.; Smreczak, B.; Medyńska-Juraszek, A. Residues of Persistent Organic Pollutants (POPs) in Agricultural Soils Adjacent to Historical Sources of Their Storage and Distribution—The Case Study of Azerbaijan. *Molecules* **2020**, *25* (8), 1815.

UNEP. General Technical Guidelines for the Environmentally Sound Management of Wastes Consisting of, Containing or Contaminated with Persistent Organic Pollutants (POPs). Basel Convention Series. SBC Nr.2005/1, 2005a.

UNEP. Ridding the World of POPs: A Guide to the Stockholm Convention on Persistent Organic Pollutants, 2005b.

UNEP. United Nations Environment Programme. Persistent Organic Pollutants. Workshop Proceeding. United Nations Environment Programme, 2008. http://www.chem.unep.ch/Pops/POPs_Inc/proceedings/coverpgs/procovers.htm (accessed Jan 2013).

van den Berg, H. Global Status of DDT and Its Alternatives for Use in Vector Control to Prevent Disease. *Environ. Health Perspect.* **2009**, *117* (11), 1656–1663.

Varsha, Y.; Naga Deepthi, C. H.; Chenna, S. An Emphasis on Xenobiotic Degradation in Environmental Clean Up. *J. Bioremed. Biodegrad.* **2011**, *11*, 1–10.

Wang, C.; Dong, D.; Zhang, L.; Song, Z.; Hua, X.; Guo, Z. Response of Freshwater Biofilms to Antibiotic Florfenicol and Ofloxacin Stress: Role of Extracellular Polymeric Substances. *Int. J. Environ. Res. Public Health* **2019**, *16* (5), 715.

Wang, L.; Xiao, H.; He, N.; Sun, D.; Duan, S. Biosorption and Biodegradation of the Environmental Hormone Nonylphenol by Four Marine Microalgae. *Sci. Rep.* **2019**, *9* (1), 1–11.

Wang, P.; Wong, Y. S.; Tam, N. F. Y. Green Microalgae in Removal and Biotransformation of Estradiol and Ethinylestradiol. *J. Appl. Phycol.* **2017**, *29* (1), 263–273.

Weber, R.; Herold, C.; Hollert, H.; Kamphues, J.; Ungemach, L.; Blepp, M.; Ballschmiter, K. Life Cycle of PCBs and Contamination of the Environment and of Food Products from Animal Origin. *Environ. Sci. Pollut. Res.* **2018,** *25* (17), 16325–16343.

Wang, T. Y.; Tan, B.; Lu, Y. L.; HCHs and DDTs in Soils around Guanting Reservoir in Beijing, China: Spatial-Temporal Variation and Countermeasures. *Sci. World J.* **2012.**

Wezenbeek, J. M.; Moermond, C. T. A.; Smit, C. E. Antifouling Systems for Pleasure Boats: Overview of Current Systems and Exploration of Safer Alternatives, 2018.

World Health Organization. Children's Health and the Environment. WHO Training Package for the Health Sector-World Health Organization, 2009. http://www. who. int/ceh

World Health Organization. *Health Risks of Persistent Organic Pollutants from Long-Range Transboundary Air Pollution* (No. EUR/03/5042687); WHO Regional Office for Europe: Copenhagen, 2003.

Wu, R. S.; Chan, A. K.; Richardson, B. J.; Au, D. W.; Fang, J. K.; Lam, P. K.; Giesy, J. P. Measuring and Monitoring Persistent Organic Pollutants in the Context of Risk Assessment. *Marine Pollut. Bull.* **2008,** *57* (6–12), 236–244.

Xia, L.; Huang, R.; Li, Y.; Song, S.; The Effect of Growth Phase on the Surface Properties of Three Oleaginous Microalgae (*Botryococcus* sp. FACGB-762, *Chlorella* sp. XJ-445 and *Desmodesmus bijugatus* XJ-231). *PloS One* **2017,** *12* (10).

Xiong, J. Q.; Kurade, M. B.; Jeon, B. H. Can Microalgae Remove Pharmaceutical Contaminants from Water? *Trends Biotechnol.* **2018a,** *36* (1), 30–44.

Xu, F. L.; Jorgensen, S. E.; Shimizu, Y.; Silow, E. Persistent Organic Pollutants in Fresh Water Ecosystems. *Sci. World J.,* **2013.**

Xu, W., Wang, X. and Cai, Z. Analytical Chemistry of the Persistent Organic Pollutants Identified in the Stockholm Convention: A Review. *Analytica Chimica Acta* **2013,** *790,* 1–13.

Younes, M.; Galal-Gorchev, H. Pesticides in Drinking Water-A Case Study. *Food Chem. Toxicol.* **2000,** *38,* S87–S90.

Zacharia, J. T. Degradation Pathways of Persistent Organic Pollutants (POPs) in the Environment. In *Persistent Organic Pollutants*; IntechOpen, 2019.

Zelinkova, Z.; Wenzl, T. The Occurrence of 16 EPA PAHs in Food–A Review. Polycyclic Aromatic Compounds **2015,** *35* (2–4), 248–284.

Zeraatkar, A. K.; Ahmadzadeh, H.; Talebi, A. F.; Moheimani, N. R.; McHenry, M. P. Potential Use of Algae for Heavy Metal Bioremediation, a Critical Review. *J. Environ. Manage.* **2016,** *181,* 817–831.

Zhang, L.; Yan, C.; Guo, Q.; Zhang, J.; Ruiz-Menjivar, J. The Impact of Agricultural Chemical Inputs on Environment: Global Evidence from Informetrics Analysis and Visualization. *Int. J. Low-Carbon Technol.* **2018,** *13* (4), 338–352.

Zhang, S.; Qiu, C. B.; Zhou, Y.; Jin, Z. P.; Yang, H. Bioaccumulation and Degradation of Pesticide Fluroxypyr Are Associated with Toxic Tolerance in Green Alga *Chlamydomonas reinhardtii*. *Ecotoxicology* **2012,** *20* (2), 337–347.

CHAPTER 4

Microbial Intervention for Degradation of Agricultural Wastes

TAWSEEF AHMAD MIR[1*], MUATASIM JAN[1], and MIR SAJAD RABANI[2]

[1]*Centre of Research for Ethnobotany, Govt. Model Science College, Jiwaji University, Gwalior, India*

[2]*Microbiology Research Lab., School of Studies in Botany, Jiwaji University Gwalior, Madhya Pradesh, India*

Corresponding author. E-mail: mirtawseef787@gmail.com

ABSTRACT

Agricultural wastes are unproductive yields of preparation and handling of agricultural productions that may contain some material beneficial for man. Estimates of agricultural wastes generated are limited, but they are believed of contributing a considerable amount to the total waste materials in the developed world. Agricultural wastes are usually produced from a number of sources such as cultivation, aquaculture, and livestock. Agricultural waste biomass is mostly composed of cellulose, hemicellulose, lignin, and other inorganic substances. In the process of degradation and recycling of agricultural wastes, microorganisms play an important role. The compost generated by the conversion of agro-residues provides numerous benefits (enhanced soil fertility and soil health) that can increase improved soil biodiversity, agricultural productivity, and reduced ecological hazards. Microorganisms including bacteria and fungi have confirmed to improve the degradation process based on the previous studies. The individuality of microorganisms and their unique functions have made them possible applicants for degradation and decomposition of agricultural waste products.

4.1   INTRODUCTION

Human inhabitations generate an enormous quantity of solid agricultural wastes ranging from domestic to agricultural and industrial waste materials. Besides all other wastes, agricultural wastes are of concern because it directly affects our agricultural system. A number of countries, whether developed or developing, choose burning and landfills for degradation of agricultural wastes. All these methods of waste clearance are not so much sophisticated and are indeed problematic. If not managed exactly, the wastes can lead to consequences of several ecological and health problems (Ansari, 2011).

The agricultural wastes can be well described as the remains of the growing and handling of crude agricultural materials including vegetables, fruits, poultry, meat, crops, and dairy products. These are nonproductive yields of various agricultural products that can benefit man but whose commercial value is less than the expenditure of collection, transportation, and handling. The content of agricultural wastes depends on the way and category of agricultural actions. Agricultural wastes can be in different forms (slurries, liquids, or solids). Agricultural wastes, sometimes referred to as agro-wastes, are formed of animal wastes (manure and carcases), food handling wastes (it is believed that 20% of canned maize becomes beneficial whether 80% becomes waste), food processing waste (only 20% of maize is canned for beneficial use and 80% is waste), crop wastes (crop stems, sugarcane residues, vegetables, pruning wastes, and fruit wastes), and toxic agricultural wastes (insecticides, pesticides, and herbicides being used for better growth and disease free crops). Agricultural waste estimations are limited, but they are believed to contribute a better proportion to the total waste materials produced in the developed countries. Increasing agricultural yield has indeed lead to increased quantity of agro-wastes, livestock, and other industrial wastes. Probably, there is most likely a significant escalation in the agro-wastes worldwide if underdeveloped countries continue to strengthen agricultural systems. It has been assessed that around 998 million tonnes of agro-wastes are produced per year (Agamuthu, 2009). Eighty percent of the total wastes produced in the farming process are of organic nature that can produce a sum of 5.27 kg/day/1000 kg manure on wet weight basis (Overcash, 1973).

Since the very end of the 20th century to present, the overall output of world-wide grain production has improved from 500 to 700 million tonnes

(FAO-UN, 2018). Among these, only cereals make 80% of human food consumption (Solà et al., 2018). During its natural growth and storage, food is attacked by pests. China, for example, is a huge agricultural hub, but about 8.8% of the total crop output is lost every year due to attack by numerous pests (Pimentel et al., 2001). On the contrary, India produces about 250 million tonnes of grains every year, but 11–15% (27.5–37.5 million tonnes per year) of the total output is lost to pest attacks (Walter, 2016). To prevent such type of loses to pests, pesticides are enormously used to regulate agricultural pests (Singh and Walker, 2006). On one hand, food loss has been reduced to a greater extent by the use of different pesticides but, on the other hand, pesticides are getting enormously distributed in soil, air, water, and agricultural materials. Hence, the extensive use of pesticides creates a possible risk to environment (Fenner et al., 2013). Pesticides not only contaminate the crops and soil, but they also contaminate the underground water table and also marine water ecosystems, which are indeed hazardous for environment and human health (Nayak et al., 2018).

Crop residues (an agricultural waste) are produced in a huge quantity and comprise of a good but under-utilized renewable biomass source. In India, around 620 million tonnes of crop residues (Fig. 4.1) are estimated to be present (Pandey et al., 2009). It is estimated that about half of the quantity of agro-wastes being produced are used for roofing material, feeding animals, fuel, and packing material, while another half is discarded by burning. Burning of these residues in farm fields is a low cost and simple type of waste degradation. However, this process of burning of agro-residues disturbs the composition air as well as increases chances of soil erosion (Walia et al., 1999). Besides its impact to environment, it also affects human health and causes a number of diseases and also visibility is decreased due to increased fog intensity (Lalchandani, 2012). In addition, extra labor cost, irrigation, and tillage are also increased (Sidhu et al., 1998). Addition of agricultural residues increases gaseous emissions (e.g., methane) particularly from watered soils, which in turn leads to increased global temperature and causes global warming (Conrad, 2002).

A huge amount of waste, either beneficial or harmful to environment, is also generated by livestock and poultry industries. Poultry or livestock excreta, feed losses, beddings etc., can be included in these types of agro-wastes (Bouwman and Booij, 1998). For the maintenance of crop production and soil fertility, animal wastes are indeed valuable sources of nutrients. It has been studied that phosphorus and nitrogen portion of animal feed gets excreted in urine and faeces (Tamminga et al., 2000). This portion is usually used as organic fertilizer. However, inconsiderate disposal of these wastes on farm-lands and releasing directly to water channels leads to percolation of these wastes to underground water through fissures and cracks poses a possible threat risk to human health because there is presence of various myriads of pathogens in livestock waste (Davies, 1997). If not managed properly, animal wastes are usually associated with health risk. Research is needed to find out the potential strategies and techniques for degradation or conversion of livestock agricultural wastes for the development of better livestock production.

In addition, food processing wastes are also produced as solid and liquid wastes. Initial material that cannot be exploited in generation of products is solid waste (pips, skins, and fruit fibers removed in juice production). Food wastes generated worldwide annually can be used to get a number of valuable components such as phenols, antioxidants, and other pectin substances. A huge portion of these wastes, however, is lignocellulosic, which mostly remains unutilized and needs to be degraded (Van Dyk et al., 2013).

Microorganisms, as stated by a number of studies, play a vital role in degradation and conversion of agro-wastes. Biodegradation of agro-residues produces compost that is beneficial in different ways such as it improves biodiversity of soil and reduces ecological hazards and soil fertility that in turn increases agro yield. (Singhand Nain, 2014).

It was in the 1940s when studies on degradation of pesticide residues by the action of microorganisms started, but the research related to the process and mechanism of degradation is being studied deeply (Audus, 1949). Microorganisms like bacteria in natural conditions can degrade pesticide remains at low cost and prevent environmental pollution to a greater extent. But one thing to be noted here is that microbes have slow efficacy for pesticide degradation and also the natural environment is changeable that can be a possible factor to affect the feasibility of microbes toward pesticides. But a number of researches have been carried out to find the mechanism of degradation of pesticide and other agricultural residues. Keeping this in view, a great sum of bacteria have been isolated that have the ability to degrade pesticide residues (Akbar and Sultan, 2016).

Using possible decomposing microbes for the conversion or degradation of agro residues to compost and using bacterial or fungal species for fortification of organic residues are having greater significance (Dhananjaya et al., 2019). The objective of this chapter is to elaborate the significance of microbes in the process of agricultural waste degradation.

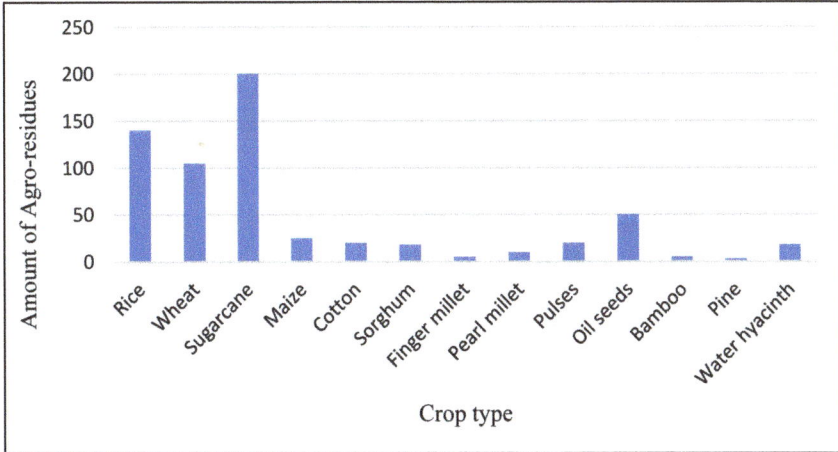

FIGURE 4.1   Agro residues generated from different crops annually.

4.2   AGRICULTURAL WASTE GENERATION

As mentioned earlier, agricultural production is generally associated with residues from inadequate faming systems and rigorous use of chemical fertilizers in farm fields affecting global as well as local environment. Agro-wastes generated is directly proportional to the agricultural activity type (Obi et al., 2016).

4.2.1   WASTES FROM CULTIVATION ACTIVITIES

Agricultural practices are also associated with insect and weed development, which in turn increases pesticide demand to get rid of pests so as to protect crops from diseases. Although chemical fertilizers play a vital role in improving crop yield and quality. However, enormous amount of fertilizes is used by farmers regardless of plant need (Hai et al., 2010). Misuse of fertilizers is carried out to improve the agricultural productivity, but these fertilizers are absorbed to a different degree based on the types of plants, characteristics of land, and application of fertilizer used (Thao, 2003). Almost all of the fertilizers being used in fields is absorbed but a little part remains in the soil. This remaining portion of fertilizers moves into water bodies such as lakes, ponds, or rivers due to irrigation or runoff, which in turn results in the pollution of water bodies and sometimes denitrated part, causes pollution in air (Obi et al., 2016).

Most of the Indian farmers are practicing residue burning of rice straw residues in farm fields after being harvested to clear the field in rice–wheat cropping system for timely sowing of the next crop in the same year. The reason behind this is low nutritive value and more cost of labor to clear the field for the following crop. However, crop-residue burning causes air pollution by releasing harmful gases like CO_2, CH_4, N_2O, H_2S, O_3, and smog. These gases mostly affect public life and disturb soil's physical, biological, and chemical properties by destroying beneficial soil microorganism. The employment of effective agricultural waste management can mitigate the air pollution and also provide better inputs to crop (Patel et al., 2020).

4.2.2   LIVESTOCK PRODUCTION WASTES

Solid wastes including organic materials, manure, urine, water coming from animal bathing, CH_4, and H_2S from animals and wastes coming from the process of maintaining hygiene at livestock locations are all livestock wastes. As most of the livestock residing places are built adjoining to residential areas so the pollution created by production of livestock is a grave concern. Odors originating from livestock production due to process of digestion of livestock residues, breakdown of organic manure, processing of urine substances, and digestion of foods causes air pollution. Smell for these processes are dependent on density of animals, ventilation facilities, humidity, and temperature in livestock places. Amount of NH_3, H_2S, and CH_4 are not the same along all the stages of process of digestion and it is dependent on food components, animal status, microbes, and organic matter. Due to unprocessed source of livestock residues, greenhouse gasses are produced. Besides the production of greenhouse gases, it is having adverse impacts on soil fertility and water pollution can also be caused. Water in livestock wastes is about 75–95% of total concentration, while inorganic materials, organic materials, eggs of parasites, and pathogens that can cause a number of diseases to man and can also affect environment (Hai et al., 2010).

Forage waste is a residue from livestock production system and is also an agricultural waste. Forage waste includes silage, silage effluent, and hay. Waste incorporated with manure and straw that becomes waste during feeding process is referred to as hay. Silage system can also be capable of producing possible waste. If silage waste runoff is not managed properly, it enters water bodies and contaminates them. If waste forage is degraded by burning process, it causes bad odors and, in addition, creates air pollution (Lane and Associates, 1993).

4.2.3 AQUACULTURE WASTES

Increased production of aqua species has resulted in the escalation of use of feeds. The concentration and quality of feeds being used in aquaculture plays a vital role in the generation of wastes from the production system (Miller and Semmens, 2002). Waste from metabolic actions of aqua species is one of the primary sources of wastes from aquaculture production system. Approximately 30% of the feeds used turn out to be waste in a properly managed production system. Ambient temperature plays a vital role in feeding rates in aquaculture. As the temperature in the system is increased, it eventually increases feeding rate and that in turn is associated with increase in waste generation. In production units, water flow should be maintained properly, as properly flowing water minimizes the disintegration of aqua species faeces and is good for fast settling of settle-able wastes. This is indeed good because when faeces are not disintegrated, it can be collected easily and completely that in turn decreases organic wastes dissolved in water (Mathieu and Timmons, 1995).

4.3 AGRICULTURAL WASTE BIOMASS

The biomass of agricultural waste is cheaply composed of Cellulose-$(C6H10O5)n$, hemicellulose-$(C_5H_8O_4)m$, lignin $\{(C_9H_{10}O_3)(OCH_3)0.9–1.7\}$ x, and a number of other organic molecules (Sjostrom, 1993). Cellulose content in biomass comprises of 30–50%, hemicellulose-15–35%, and lignin-10–20% (Petterson, 1984; Mielenz, 2001). Cellulose and hemicel-lulose, these two are firmly associated with lignin portion of biomass by hydrogen and covalent bonds. This tight bonding makes it extremely tough and also resilient to any treatment of degradation (Mielenz, 2001; Knauf and Moniruzzaman, 2004). All of the components of agricultural waste biomass are described later.

4.3.1   CELLULOSE

Most abundant organic material found on earth is cellulose, it is a linear chain of some hundreds to tens of thousands of repeated D-glucose components and these are linked together by $\beta 1(1 \rightarrow 4)$ glycoside bonds (Zhang and Lynd, 2004). Cellulose is an abundant material and is having 100,000 molecular weight. It is a biologically abundant form of fixed carbon, which is responsible for the main component of structural system of cell wall in algae and green plants. However, lignocellulosic matter has both amorphous and crystalline systems of cellulose. These forms are bounded tightly in arranged parallel bundles of cellulose chains, which are formed due to strong hydrogen bonding between chains, this forms a crystalline structure and the other less arranged and easily noticeable designates amorphous region.

4.3.2   HEMICELLULOSE

As cellulose is the first most-abundant organic material on earth, hemi-cellulose takes a second number in this regard (Hendriks and Zeeman, 2009). It is a heterogeneous polymer and comprises of methyl pentoses, pentoses, carboxylic acids, and hexoses. These parts of the said polymer can be used for the preparation process in production of ethanol and several other valuable products in addition for human use. Taking the example of hardwood, hemicelluloses are mainly found in the form of xylan and, in contrast, in softwood, it is found in the form of glucomannan (Singh et al., 2008). Having a haphazard branched and amorphous structure with a very low resistance for hydrolytic processes, hemicelluloses are hydrolyzed very easily to their monomer units by acids (Mod et al., 1981). Being the most abundant form of polysaccharidein hemicellulose family, xylan comprises of D-xylopyranose, which are linked together by β-1,4 linkage which is having upto 30,000 molecular weight and 200 degrees of polymerization.

4.3.3   LIGNIN

The third-largest existing polymer on earth is lignin. It comprises of a phenyl propane component (p-coumaryl alcohol, sinapyl alcohol, and coniferyl alcohol), linked by ester bonds forming a possible complex with that of hemicellulose for encapsulating cellulose to make it resilient to hydrolysis of chemicals and enzymes (Hendricks and Zeeman, 2009; Agbor et al., 2011). Biodelignification intensity is based on the ratio of guaicyl to syringyl units that are present in various types of wood (Ruiz-Dueñas and Martínez, 2009).The molecular weight of lignin is generally less than 20,000. It is mostly found in soft wood plants such as balsam, spruce, pine, and fir than in other hardwood plants such as birch, walnut, oak, and popular. Biomass consisting of lignin is not easily depolymerized of holocellulose (hemicellulose and cellulose) for the production of sugars. Also, lignin may form furan (compounds that have the capability to prevent fermentation) like compounds during the process of degradation (Zaldivar et al., 1999).

4.3.4   PECTIN

A heteropolysaccharide, pectin is present in cell walls of plants containing α-1,4 linked D-galacturonic acid backbone, that may be methyl-esterified or substituted pectin, which generally contains D-galacturonic acids. In contrast, the hairy portion of pectin has D-galcturonic acid residues backbone that can be disturbed by α-1,2-linked L-rhamnose residues (de Souza, 2013). The residues of rhamnose in the hairy portion of pectin have generally L-arabinose and D-galactose residues long-side chains (de Vries, 2003). Pectin backbone degradation mainly needs polysaccharide lyase. Fungal glycoside hydrolases that are responsible for degradation of pectin backbone actually belong to GH family 28 (Martens-Uzunova and Schaap, 2009). Endo and exo-polygalacturonases of GH28 family usually break the α-1,4-glycoside bonds between α-galacturonic acids. Besides this, for the process of hydrolysis of backbone of pectin GH family, enzymes are required, which may include α-rhamnosidases of GH78, unsaturated glucuronyl hydrolysis of GH88, and also unsaturated rhmnogalacturonan hydrolases of GH105 (Brink and de Vries, 2011). For the degradation of cellulose (Martinez et al., 2008), *Trichoderma* spp. are famous; whereas,

Aspergillus spp. have the capability to produce a number of enzymes to degrade pectin (Martens-Uzunova and Schaap, 2009).

4.4 MICROBIAL INTERVENTION IN DEGRADATION OF AGRICULTURAL WASTES

Because of their valuable functions, microorganisms are highly beneficial for the development of mankind in terms of food, health, food quality and safety, increasing crop yield, and for other purposes like disease free plants. Farming systems have been advanced due to their use. All the agricultural wastes contain lignocellulose that comprises of cellulose, hemicellulose, and lignin that makes a large component in their structural system (Pothiraj et al., 2006). A large amount of agro-wastes are subsidize to total production of lignocellulose worldwide due to increasing agricultural practices (Loow et al., 2015).

4.4.1 BIOLOGICAL PRETREATMENT

Soft rot, brown rot, and white rot fungi are generally involved in the biological treatment of agricultural wastes so as to release cellulose from lignin and hemicellulose. Lignin and hemicellulose is degraded by the said treatment but cellulose remains unaffected at the end (Shi et al., 2008). This treatment works in slight and low-energy conditions. Biological pretreatment has a slow rate; that is why, the said treatment needs more time to get completed (Sun and Cheng, 2002). Cellulose, hemicellulose, and lignin biomass degradation have been summarized in Table 4.1.

4.4.1.1 WHITE-ROT FUNGI

White rot, biological pretreatment process in wood is generally carried out by a group of different fungi called lignolytic basidiomycetes, generally referred to as white-rot fungi. For the process of degradation of lignin, white-rot fungi are being used over many years (Vicuna, 2000). Species such as *Pycnoporus cinnabarinus, Pycnoporus sanguineus*, and *Phanerochaete chrysosporium* have been reported to degrade lignin to a greater extent, similarly *Ceriporiopsis subvermispora, phlebia floridensis, p. radiate*, and

TABLE 4.1 Lignocellulosic Biomass and the Effect of Biological Treatment on Them.

Microbe	Biomass	Cellulose loss%	Hemicellulose loss%	Lignin loss%	References
Fungi					
Dichomitus squalens	Woodchips	8.0	–	21.7	Itoh et al. (2003)
Pleurotuso streatus	Woodchips	4.1	–	10.8	Itoh et al. (2003)
Coriolus versicolor	Rice straw	17	48	41	Taniguchi et al. (2005)
Trametes versicolor	Corn straw	47.5	–	56.6	Yu et al. (2010a)
Echinodontiumtaxodii	Woodchips	16.8	29	31	Ferraz et al. (2001)
	Wood	27	51	45	Yu et al. (2009)
	Corn straw	15.8		42.2	Yu et al. (2010b)
Poriamedula-panis	Woodchips	10	15	18	Ferrazet al. (2001)
Ceriporia lacerata	Woodchips	–	8	13	Lee et al. (2007)
Ceriporiopsis subvermispora	Woodchips	5.0	–	6.7	Itoh et al. (2003)
	Sugarcane bagasse	9.1	21.5	38.4	Sasaki et al. (2005)
	Corn stover	<5	27	39.2	Wan and Li (2010)
Stereum hirsutum YJ9	Woodchips	–	7.8	14	Lee et al. (2007)
Gloeophyllum trabeum KU-41	Corn stover	17	43	–	Gao et al. (2012)
Trametes hirusta	Corn stover	34.06	77.84	71.99	Sun et al. (2011)
Gloephyllum trabeum	Woodchips	–	31	–	Monrroyet al. (2011)
Laetoporeus sulphureus	Woodchips	–	6	–	Monrroyet al. (2011)
	Woodchips	21	47	3	Fissoreet al. (2010)
Wolfiporia cocos	Woodchips	6	23	2.6	Ferraz et al. (2001)

ABLE 4.1 *(Continued)*

Microbe	Biomass	Cellulose loss%	Hemicellulose loss%	Lignin loss%	References
Coniophora puteana	Wood blocks	17	-	59.7	Irbeet al. (2011)
Bacterial species					
Bacillus spp.	Rice straw	3.2	16	20	Chang et al. (2014)
Cellulomonas cartea	Sugarcane trash	25.4	–	5.5	Singh et al. (2008)
Cellomonasuda	Sugarcane trash	21.8	–	5.5	Singh et al. (2008)
Bacillus macerans	Sugarcane trash	30.4	–	5.5	Singh et al. (2008)
Zymonas mobilis	Sugarcane trash	26.8	–	8	Singh et al. (2008)
Paenibacillus spp. AY952466	Kraft lignin	–	–	43	Chandra et al. (2007)
Aneurinibacillus aneurinilyticus	Kraft lignin	–	–	56	Chandra et al. (2007)
Bacillus spp. AY952465	Kraft lignin	–	–	37	Chandra et al. (2007)

Trichoderma reesei have shown the capability to degrade carbohydrate content with a decrease of 40–60% in lignin content (Lu et al., 2010). When before chemical treatment bio-delignification was used, it showed lignin degradation up to 80% (Yu et al., 2010a, 2010b).

4.4.1.2   BROWN-ROT FUNGI

As compared to white-rot fungi, brown-rot fungi is responsible for the degradation of cellulose and hemicellulose rather than lignin, which results in the formation of brown-rotted wood because of partial lignin degradation. As already described, cellulose and hemicellulose portions are degraded by brown-rot fungi (Hastrup et al., 2012). *Coniophoraputeana, Laetoporeus sulphurous, Gleophyllumtrabeum*, and *Meruliporia incrassate* are some of the species being used in a number of degradation studies (Rasmussen et al., 2010).

4.4.1.3   SOFT-ROT FUNGI

Daldinia concentric, a type of soft-rot fungi, degrades hardwood and is believed to reduce the weight of birch wood to 53% within a time period of two months (Nilsson et al., 1989). In addition, *Cadophora* and *Paecilomyces* spp. are also found to degrade lignin biomass (Chandel et al., 2013). *Xylariaceous ascomycetes,* which belongs to genera Xylaria, have been classified in white-rot fungi but now these have been kept under soft-rot fungi, because they are believed to cause rots in woods.

4.4.1.4   BACTERIA AND ACTINOMYCETES

When it comes to the biodelignificationof lignocellulosic compounds, bacteria are not so much capable in this regard because either they lack or have capability to produce less amount of lignolytic enzymes. Species of bacteria like *Cellulomonas, Zymomonas*, and *Sphingomonas paucimobilis* are believed to degrade lignocellulosic biomass (Singh et al., 2008). The bacterial delignification process by bacteria can degrade lignin up to 50%, which is somewhat same as that of delignification mediated by fungi. Delignification process by bacteria is generally carried out by some extracellular enzymes called xylanases (Bezalel et al., 1993). On the other hand, the increase in degradation of cellulose might increase lignin elimination as reported for *Cellulomonas cartae* and *Bacillus macerans* species of bacteria (Singh et al., 2008).

4.5 DEGRADATION OF AGRICULTURAL WASTES

Microorganisms including bacteria and fungi have confirmed to improve the degradation process based on the previous studies. Individuality and unique functions of microorganisms have made them possible species to degrade and convert agro-wastes into other value-added products (Kumar and Sai Gopal, 2015). Microbes have the capability to effect litter decomposition and nutrient pool size in soils. Microbes arrest nutrients present in microbial biomass and give out nutrients after the process of decomposition from microbial pool (Sahu et al., 2018). Agro-wastes that comprise of 20–30% cellulose, hemicellulose, and lignin can be degraded or they can also be converted to get a number of value-added products such as biofuel, which can be used in various industrial processes (Sorek et al., 2014). If lignocellulosic biomass is not managed properly or is burnt in farm fields, it creates a lot of environmental problems such as pollution.

4.5.1 CELLULOSE DEGRADATION

Cellulose, a polysaccharide, is made up of linear β-1,4-linked D-glucopyranose chains. Cellulose needs enzyme cellulases for degradation. A number of microbes such as *Trametes versicolor, T. hirsute*, and *Ganoderma lucidam,* have the capability to synthesize cellulose enzymes for the degradation of lignocellulosic component called cellulose (Lee, 2001). Enzyme cellulose can be divided into three different types, which are exoglucanase, endoglucanase, and β-glucosidase. These enzymes are known for their function of degrading cellulose (Bht, 2000). Cellulases can be found free in microorganisms (aerobic microorganisms), or can also be found to be in groups forming enzyme complex (cellulosome) (Lynd et al., 2002). These (cellulases) are the enzymes generated by bacteria, fungi, and various protozoans, which have the capability to covert cellulose into fermentable sugars by catalytic process of conversion. Cellulose can be used as a renewable source of energy and can be converted into a number of valuable compounds

(Shahzadi et al., 2014). Due to various agricultural and industrial processes, a huge concentration of cellulosic waste is accumulated on earth (Kim et al., 2003); so, there is dire need to develop techniques for the utilization of these cellulosic compounds as sources of energy. Cellulosic compounds are present in stems, leaves, corn fibers, crops, woods, sugarcane residues, and litter present in forests. Moreover, other sources of cellulose are citrus peel, coconut biomass, paper pulp, and other wastes coming from agricultural and industrial processes (Sadhu et al., 2013). From last years, researchers are working day and night to develop novel techniques and other novel strains of microbes for the production of cellulase enzyme and to reduce its cost considerably for the degradation or conversion of cellulose for industrial use. These enzymes can be used to get a number of benefits of lignocellulosic compounds (Wen et al., 2005). Production of cellulase enzyme generally depends on various factors such a temperature, pH, medium additives, aeration, and size of inoculum (Robson and Chambliss, 1984). A number of countries particularly India, China, and South Korea are the countries who have a number of manufacturing research centers of cellulase enzyme and can play a vital role in the production of cellulases for industrial use (Acharya and Chaudhary, 2012).

4.5.2 HEMICELLULOSE DEGRADATION

Made up of diverse residues of xyloglucan, mannan, and xylan-like components, hemicellulose is a complex matrix compound. Hemicellulose complex structure requires a joint action of exo and endo-enzymes. Endo-enzymes are meant for internal cleavage of main chain, exo-enzymes are meant for release of monomeric sugar units, and accessory enzymes are responsible for the splitting of side chains of related polymers and other oligosaccharides, and they result in the release of mono, disaccharides, and acetic acid units (de Souza, 2013). The primary component of hemicellulose that is xylan is possibly decomposed by β-1,4-endoxylanase, which disintegrates xylan into its oligosaccharide units, which is then disintegrated to xylose by β-1,4-xylosidase. β-1,4-endoxylanases from fungi are being classified to GH10 and GH11, which are different from each other in terms of specificity of substrate (Biely et al., 1997). GH10 family of fungal endoxylanases are more specific for their substrate than family GH11 of endoxylanases (Brink and de Vries et al., 2011). GH10 endoxylanases have generally the function of degrading xylo-oligosaccharides, xylan backbones, and chains. Therefore, GH10 endoxylanases are required to function for the degradation of xylans (Pollet et al., 2010). β-xylosidases are responsible for the complete disintegration of xylan (de Souza, 2013). Moreover, for the synthesis of oligosaccharides, some β-xylosidases having transxylosylation activity have been documented showing an application for these enzymes (Shinoyama et al., 1991). Galactomannans, generally called as mannans, are comprised of β-1,4-linked D-glucose and D-mannose having side chains of D-galactose (de Souza, 2013). B-mannosidases and β-endomannases, which are generally expressed by *Aspergillus* species, is responsible for the degradation of galactomannans type of hemicellulose (de Vries, 2001). Moreover, there is need of accessory enzymes for the very degradation of hemicellulose, which includes p-coumaricesterases, α-1-arbinofuranosidases, and xylanesterases that have the capability to degrade xylans and mannans of wood (Perez et al., 2002). *Ceriporiopsis subvermispora, Coriolus versicolor, Trametes hirsute*, and *Laetoporeus sulphurous* are some of the notable microorganisms used in the degradation of hemicellulose.

4.5.3   LIGNIN DEGRADATION

Lignin is one of the utmost insoluble polymer complex present in cell wall. It is well known to improve the power and also the resistance of cell wall of plant. The utmost well-specified enzymes responsible for degradation of lignin being produced by microbes are laccase, manganese peroxidase, lignin peroxidase, and a few more. Enzyme lignin peroxidases are involved in catalyzing a number of reactions by the use of H_2O_2, which acts equally as a cofactor for the degradation of lignin or other compounds of lignin (Hattaka, 2001). A little number of fungi has been recognized to generate lignin peroxidases, which may include *Bjerkhandera* spp., *T. versicolor, T. cervina*, and *P. chrysosporium* (Ghosh and Ghosh, 2003). Besides lignin peroxidases, manganese peroxidases also require hydrogen peroxide (H_2O_2) in Mn-dependent catalyzing reaction as an oxidant (Narayanswamy et al., 2013). In addition, in the processes of lignin depolymerization and demethylation, manganese peroxidases play a vital role for oxidizing phenolic and sulphuric compounds and other insaturated fatty acids. Manganese peroxidases are produced by *Pleurotus ostreatus, P. chrysosporium, Trametes* spp., and by other species of family Coriolaceae, Merulaceae, and polyporaceae (Isroi et al., 2011).

Laccase enzyme contains copper in its crystalline structure and associated with family oxidase. Laccase is generally present in fungi, bacteria, and a number of plants. *Pycnoporous cinnabarinus* species is found to produce a better quality of laccase, which can degrade lignin to a greater extent (Geng and Li, 2002). Almost every type of white-rot fungi has the capability of producing good amount lac, which can show improved functions in presence of copper. Induction of lac was shown in presence of some aromatic materials such as VA and two to five xylidine (Isroi et al., 2011). *Streptomyces lavendulae* and *S. cinnamensis* are some of the bacterial species that have the capability to lignin peroxidases, which can degrade lignin like substances (Jing, 2010). Versatile peroxidase, another type of lignin degrading enzyme, has as many activities like that of lignin peroxidases, these are confirmed in a number of *Pleurotus* spp. and also in *Bjerkandera adusta* (Ayala Aceves et al., 2001).

4.5.4   PECTIN DEGRADATION

It is a natural polymer, which exists in cell wall and middle lamella of plants. Pectin is a complex family of a number of polysaccharides and is generally comprised of 17 different monomers, which contain 20 different glycosidic linkages (Ridley et al., 2001). In the degradation of pectin backbone, the fungal glycoside hydrolases involved belong to GH28 family (Martens-Uzunova and Schaap, 2009). In addition, enzymes from some other GH families are also involved in the degradation of pectin. These enzymes (α-rhamnosidases (GH78), unsaturated rhamnogalacturonan hydrolases (GH105), and unsaturated glucuronyl hydrolases (GH88)) are capable of hydrolysing pectin (Brink and de Vries, 2011). Also many enzymes are produced by *Aspergillus* spp., which can degrade pectin (Martens-Uzunova and Schaap, 2009). Chemical hydrolysis is not more advantageous than enzymatic hydrolysis when it comes to pectin degradation, because enzymes have the capability to hydrolyse specific linkages (Schols et al., 1996).

4.5.5   MICROBIAL DEGRADATION OF PESTICIDE RESIDUES

Nowadays, farmers are making enormous use of pesticides to increase their crop yield by killing pests and insects, but misuse of these pesticides is harmful to environment because once used, pesticides making their place in environment are thereby included in agricultural wastes. Pesticides like organochlorine are not degraded easily in natural environment and enter human body either through food or water. Acetanilides, benzoylphenyl ureas, aldrin, chlordimeform, and carbaryland bayleton are some of the mostly used pesticides for increasing agricultural yield (Ye et al., 2018). During the past few years, researchers have been isolating and culturing a number of microbes belonging to bacteria, fungi, algae, and actinomycetes from various sources for the degradation of pesticides (Kafilzadeh et al., 2015).

Pesticide degradation happens either in usual environment or with the help of several microbes having the activity to degrade the same. A number of microbes such as *Flavobacterium, Rhodococous, Pseudomonas, Gliocladium, Pencillium*, and *Trichoderma* are believed to make use of pesticides as a source of carbon (Aislabie and Lloyd-jones, 1995). Kato et al. (2001) have described the degradation of hydrocarbons using *Alcaligenes* and *Burkholderia.*

Perclich and Lockwood (1978) detected the occurrence of pesticide by using *Micrococcus, Bacillus, Psedomonas*, and *vibrio* in irrigational channels and other water samples. It has been detected that pesticides are usually disintegrated with the help of various bacterial species such as *Bacillus* and *Pseudomonas,* this functionality is thought to be because of extensive range of enzymes (Walker et al., 1993). Besides several bacterial species, numerous fungal species (*Coriolus versicolor, Flammulina velupites, Dichomitus squalens, Stereum hirsutum*, and *Hypholoma fasciculare*) are also involved in the degradation process of pesticides such as triazine, dicarboximide, phenylurea, and other organophosphorus pesticides (Bending et al., 2002). While many pesticide compounds are degraded by several microbes, they can be resilient to degradation process because of their insolubility, complex structure, and temperature stability (Godheja et al., 2016).

4.6 CONCLUSION

Agricultural wastes are unproductive yields of preparation and handling of agricultural production that may contain some material beneficial for man. These type of by-products are produced from several agro processes, including cultivation, livestock production, aquaculture, and their biomass. These wastes if not managed properly can be harmful to our environment and to human health as well. Pesticide and chemical fertilizer use are important in the present world so as to save crops from pesticide attack and improve yield. But a lot of problems related to present-day agro system, mostly those that are somehow linked to soil, plants, and human health can be minimized by reduced use of pesticides and other chemicals. Agro-residues that are left behind in farm fields after harvest season create a lot of sanitation problems because of improper and unmanaged degradation process. Due to use of chemical fertilizers in farmlands, organic carbon in soils has reduced to a greater extent. Lack of organic carbon in soil make soil nonresponsive for giving life to vegetation and that indeed makes it to lose various bio functions. Usually farmers burn postharvest residues in farmlands, which disrupt soil microbiota and pollute the air nearby making it difficult for nearby vegetation to carry out biological functions properly. It has been observed that microbes play a great part in the degradation process of agricultural wastes. With the help of microbial interventions, these agricultural wastes and crop residues can be degraded and also be transformed to other beneficial products. Agrowastes can also be transformed to microbe rich compost, which may be helpful for crop production. Microbial degradation of agriculture wastes will help to decrease greenhouse gas emission, environmental pollution, and serve as a sustainable solid waste management strategy.

KEYWORDS

- microbes
- organic waste
- livestock
- biotreatment
- degradation

REFERENCES

Acharya, S.; Chaudhary, A. Bioprospecting Thermophiles for Cellulose Production: A Review. *Braz. J. Microbiol.* **2012**, *47*, 844–856.

Agamuthu, P. *Challenges and Opportunities in Agro-Waste Management: An Asian Perspective.* Inaugural Meeting of First Regional 3R Forum in Asia: Tokyo, Japan, 11–12 Nov 2009.

Agbor, V. B.; Cicek, N.; Sparling, R.; Berlin, A.; Levin, D. B. Biomass Pretreatment: Fundamentals toward Application. *Biotechnol. Adv.* **2011**, *29*, 675–685.

Aislabie, J.; Lloyd-jones, G. A Review of Bacterial Degradation of Pesticides. *Austr. J. Soil Res.* **1995**, *33*, 925–942.

Akbar, S.; Sultan, S. Soil Bacteria Showing a Potential of Chlorpyrifos Degradation and Plant Growth Enhancement. *Braz. J. Microbiol.* **2016**, *47*, 563–570.

Ansari, A. A. Worm Powered Environmental Biotechnology in Organic Waste Management. *Int. J. Soil Sci.* **2011**, *6* (1), 25–30.

Audus, L. J. The Biological Detoxication of 2: 4-Dichlorophenoxyacetic Acid in Soil. *Plant Soil.* **1949**, *2*, 31–36.

Ayala Aceves, M.; Baratto, M. C.; Basosi, R.; Vazquez-Duhalt, R.; Pogni, R. Spectroscopic Characterization of a Manganese-Lignin Peroxidase Hybrid Isozyme Produced by *Bjerkanderaadusta* in the Absence of Manganese: Evidence of a Protein Centred Radical by Hydrogen Peroxide. *J. Mol. Cataly. B Enzym.* **2001**, *16*, 159–167.

Bending, G. D.; Friloux, M.; Walker, A. Degradation of Contrasting Pesticides by White Rot Fungi and Its Relationship with Ligninolytic Potential. *FEMS Microbiol. Lett.* **2002**, *212*, 59–63.

Bezalel, L.; Shoham, Y.; Rosenberg, E. Characterization and Delignification Activity of a Thermostable α-L-arabinofuranosidase from *Bacillus stearothermophilus. Appl. Microbiol. Biotechnol.* **1993**, *40*, 57–62.

Bhat, M. K. Cellulases and Related Enzymes in Biotechnology. *Biotechnol. Adv.* **2000**, *18*, 355–383.

Biely, P.; Vrsanska, M.; Tenkanen, M. Endo-beta-1,4-xylanase Families: Differences in Catalytic Properties. *J. Biotechnol.* **1997**, *57*, 151–166.

Bouwman, A. F.; Booij, H. Global Use and Trade of Foodstuffs and Consequences for the Nitrogen Cycle. *Nutr. Cycl. Agroecosys.* **1998**, *52*, 261–267.

Brink, J. V. D.; de Vries, R. P. Fungal Enzyme Sets for Plant Polysaccharide Degradation. *Appl. Microbiol. Biotechnol.* **2011**, *91* (6), 1477–1492.

Chandel, A. K.; Goncalves, B. C. M.; Strap, J. L.; da Silva, S. Biodelignification of Lignocellulose Substrates: An Intrinsic and Sustainable Pretreatment Strategy for Clean Energy Production. *Crit. Rev. Biotechnol.* **2013**, 1–13.

Chandra, R.; Raj, A.; Purohit, H. J.; Kapley, A. Characterisation and Optimisation of Three Potential Aerobic Bacterial Strains for Kraft Lignin Degradation from Pulp Paper Waste. *Chemosphere.* **2007**, *67*, 839–846.

Chang, Y. C.; Choi, D.; Takamizawa, K.; Kikuchi, S. Isolation of Bacillus sp. Strains Capable of Decomposing Alkali Lignin and Their Application in Combination with Lactic Acid Bacteria for Enhancing Cellulase Performance. *Bioresour. Technol.* **2014**, *152*, 429–436.

Conrad, R. Control of Microbial Methane Production in Wetland Rice Fields. *Nutr. Cycl. Agroecosyst.* **2002**, *64*, 59–69.

Davies, R. H. A Two Year Study of Salmonella Typhimurium DT104 Infection and Contamination on Cattle Farms. *Cattle Practice* **1997**, *5*, 189–194.

de Souza, W. R. Microbial Degradation of Lignocellulosic Biomass. Sustainable Degradation of Lignocellulosic Biomass-Techniques, Applications and Commercialization. *Intech.* **2013**, 207–247.

de Vries, R. P. Regulation of *Aspergillus* Genes Encoding Plant Cell Wall Polysaccharide-Degrading Enzymes; Relevance for Industrial Production. *Appl. Microbiol. Biotechnol.* **2003**, *61*, 10–20.

deVries, R. V. J. *Aspergillus* Enzymes Involved in Degradation of Plant Cell Wall Polysaccharides. *Microbiol. Mol. Biol. Rev.* **2001**, *65*, 497–522.

Dhananjaya, P.; Singh, R.; Prabha, S. R.; Sahu, P. K.; Singh, V. Agrowaste Bioconversion and Microbial Fortification Have Prospects for Soil Health, Crop Productivity, and Eco-Enterprising. *Int. J. Recycl. Org. Waste Agric.* **2019**, *8* (1), S457–S472.

Fenner, K.; Canonica, S.; Wackett, L. P.; Elsner, M. Evaluating Pesticide Degradation in the Environment: Blind Spots and Emerging Opportunities. *Science* **2013**, *341*, 752–758.

Ferraz, A.; Rodriguez, J.; Freer, J.; Baeza, J. Biodegradation of Pinusradiata Softwood by White and Brown-Rot Fungi. *World J. Microbiol. Biotechnol.* **2001**, *17*, 31–34.

Fissore, A.; Carrasco, L.; Reyes, P.; Rodriguez, J.; Freer, J. Evaluation of a Combined Brown Rot Decay-Chemical Delignification Process as a Pretreatment for Bioethanol Production from Pinus *radiata* Wood Chips. *J. Indust. Microbiol. Biotechnol.*. **2010**, *37*, 893–900.

Food and Agriculture Organization of the United Nations. en/#data/QC (accessed 19 Aug 2018).

Gao, Z.; Mori, T.; Kondo, R. The Pretreatment of Corn Stover with *Gloeophyllum trabeum* KU-41 for Enzymatic Hydrolysis. *Biotechnol. Biofuels* **2012**, *5*, 1–11.

Geng, X.; Li, K. Degradation of Non-Phenolic Lignin by the White-Rot Fungus *Pycnoporus cinnabarinus*. *Appl. Microbiol. Biotechnol.* **2002**, *60*, 342–346.

Ghosh, P.; Ghose, T. K. Bioethanol in India: Recent Past and Emerging Future. In *Advanced Biochemical Engineering/Biotechnology, Biotechnology in India*; Scheper, T., Ed.; II; Springer: New York, 2003; pp 20, 1–27.

Godheja, J.; Shekhar, S. K.; Siddiqui, S. A.; Modi, D. R. Xenobiotic Compounds Present in Soil and Water: A Review on Remediation Strategies. *J. Environ. Anal. Toxicol.* **2016**, *6*, 392.

Hai, H. T.; Tuyet, N. T. A. *Benefits of the 3R Approach for Agricultural Waste Management (AWM) in Vietnam. Under the Framework of Joint Project on Asia Resource Circulation Policy Research Working Paper Series.* Institute for Global Environmental Strategies Supported by the Ministry of Environment: Japan, 2010.

Hastrup, A. C. S.; Green, I. I. I. F.; Lebow, P. K.; Jensen, B. Enzymatic Oxalic Acid Regulation Correlated with Wood Degradation in Four Brown-Rot Fungi. *Int. J. Biodeterior. Biodegr.* **2012**, *75*, 109–114.

Hattaka, A. *Biodegradation of Lignin*; MHaASc., Ed.; Wiley-WCH, 2001.

Hendriks, A. T. W. M.; Zeeman, G. Pretreatments to Enhance the Digestibility of Lignocellulosic Biomass. *Bioresour. Technol.* **2009**, *100*, 10–18.

Irbe, I.; Andersone, I.; Andersons, B. Characterisation of the Initial Degradation Stage of Scots Pine (*Pinussylvestris* L.) Sapwood After Attack by Brown-Rot Fungus *Coniophora puteana*. *Biodegradation* **2011**, *22*, 719–728.

Isroi, I.; Millati, R.; Syamsiah, S.; Niklasson, C.; Cahyanto, M. N.; Lundquist, K.; Taherzadeh, M. J. Biological Pretreatment of Lignocelluloses with White-Rot Fungi and Its Applications: A Review. *Bioresources* **2011**, *6*, 5224–5259.

Itoh, H.; Wada, M.; Honda, Y.; Kuwahara, M.; Wantanabe, T. Bioorgano Solve Pretreatments for Simultaneous Saccharification and Fermentation of Beech Wood by Ethanolysis and White Rot Fungi. *J. Biotechnol.* **2003**, *103*, 273–280.

Jing, D. Improving the Simultaneous Production of Laccase and Lignin Peroxidise from *Streptomyces lavendulae* by Medium Optimization. *Bioresour. Technol.* **2010**, *101*, 7592–7597.

Kafilzadeh, F.; Ebrahimnezhad, M.; Tahery, Y. Isolation and Identification of Endosulfan-Degrading Bacteria and Evaluation of Their Bioremediation in Kor River, Iran. *Osong. Public Health Res. Perspect.* **2015**, *6*, 39–46.

Kato, T.; Haruki, M.; Imanaka, T. Isolation and Characterization of Long-Chain-Alkane Degrading *Bacillus thermoleovorans* from Deep Subterranean Petroleum Reservoirs. *J. Biosci. Bioeng.* **2001**, *91*, 64–70.

Kim, K. C.; Scung-soo, Y.; Young, O. A.; Scong-jun, K. Isolation and Characteristics of *Trichoderma harzianum* FJ1 Producing Cellulases and Xylanase. *J. Microbiol. Biotechnol.* **2003**, *13*, 1–8.

Knauf, M.; Moniruzzaman, M. Lignocellulosic Biomass Processing: A Perspective. *Int. Sugar J.* **2004**, *106*, 147–150.

Kumar, B. L.; Sai Gopal, D. V. R. Effective Role of Indigenous Microorganisms for Sustainable Environment. *Biotech* **2015**, *5*, 867–876.

Lalchandani, N. Times of India dt. 05/11/2012 Delhi Edition, 2012.

Lane, P. and Associates Ltd. *A Review of Odour Management Technologies for Use in Livestock Operations in Nova Scotia*; Agriculture Canada: unpublished, 1993.

Lee, J. W.; Gwak, K. S.; Park, J. Y.; Park, M. J.; Choi, D. H.; Kwon, M.; Choi, I. G. Biological Pretreatment of Softwood *Pinus densiflora* by Three White Rot Fungi. *J. Microbiol.* **2007**, *45*, 485–491.

Lee, S. M.; Koo, M. Y. Pilot-Scale Production of Cellulose Using *Trichoderma reesei* Rut C-30 in Fed-Batch Mode. *J. Microbiol. Biotechnol.* **2001**, *11*, 229–233.

Loow, Y. L.; Wu, T. Y.; Tan, K. A.; Lim, Y. S.; Siow, L. F.; MdJahim, J.; Mohammad, A. W.; Teoh, W. H. Recent Advances in the Application of Inorganic Salt Pretreatment for Transforming Lignocellulosic Biomass into Reducing Sugars. *J. Agric. Food Chem.* **2015**, *63*, 8349–8363.

Lu, C.; Wang, H.; Luo, Y.; Guo, L. An Efficient System for Pre-Delignification of Gramineous Biofuel Feedstock In Vitro: Application of a Laccase from *Pycnoporus sanguineus* H275. *Process Biochem.* **2010**, *45*, 1141–1147.

Lynd, L. R.; Weimer, P. J.; Vanzyl, W. H.; Pretorius, J. S. Microbial Cellulose Utilization: Fundamentals and Biotechnology. *Microbiol. Mol. Biol. Rev.* **2002**, *66*, 506–577.

Martens-Uzunova, E. S.; Schaap, P. J. Assessment of the Pectin Degrading Enzyme Network of *Aspergillus niger* by Functional Genomics. *Fungal Genet. Biol.* **2009**, *46* (1), S170–S179.

Mathieu, F.; Timmons, M. B. Techniques for Modern Aquaculture. In *American Society of Agricultural Engineers*; Wang, J. K., Ed.; St. Joseph: MI, 1995.

Mielenz, J. R. Ethanol Production from Biomass: Technology and Commercialization Status. *Curr. Opin. Microbiol.* **2001**, *4*, 324–325.

Miller, D.; Semmens, K. *Waste Management in Aquaculture*; Aquaculture information series, Extension Service: West Virginia University, 2002.

Mod, R. R.; Ory, R. L.; Morris, N. M.; Normand, F. L. Chemical Properties and Interactions of Rice Hemicellulose with Trace Minerals In Vitro. *J. Agric. Food Chem.* **1981**, *29*, 449–454.

Monrroy, M.; Ortega, I.; Ramirez, M.; Baeza, J.; Freer, J. Structural Change in Wood by Brown Rot Fungi and Effect on Enzymatic Hydrolysis. *Enzyme Microb. Technol.* **2011**, *49*, 472–477.

Narayanaswamy, N.; Dheeran, P.; Verma, S.; Kumar, S. Biological Pretreatment of Lignocellulosic Biomass for Enzymatic Saccharification. In: *Pretreatment Techniques for Biofuels and Biorefineries*; Fang, Z., Ed.; Green Energy and Technology; Springer: Verlag Berlin, Heidelberg, 2013; pp 3–34.

Nayak, S. K.; Dash, B.; Baliyarsingh, B. Microbial Remediation of Persistent Agrochemicals by Soil Bacteria: An Overview. *Microb. Biotechnol* **2018**, 275–301.

Nilsson, T.; Daniel, G.; Kirk, T. K.; Obst, J. R. Chemistry and Microscopy of Wood Decay by Some Higher Ascomycetes. *Holzfrschung* **1989**, *43*, 11–18.

Obi, F. O.; Ugwuishiwu, B. O.; Nwakaire, J. N. Agricultural Waste Concept, Generation, Utilization and Management. *Niger. J. Technol.* **2016**, *35* (4), 957–964.

Overcash, M. R. *Livestock Waste Management*. Humenik, F. J., Miner, J. R., Eds.; CRC Press: Boca Raton, 1973.

Pandey, A.; Biswas, S.; Sukumaran, R. K.; Kaushik, N. Study on Availability of Indian Biomass Resources for Exploitation: A Report Based on a Nationwide Survey TIFAC: New Delhi, 2009.

Patel, V. K.; Abhay, K. A.; Singh, A.; Anshuman, K. Integrated Agricultural Waste Management: A Solution of Many Problems. *Food Sci. Rep.* **2020**, *1*, 63–64.

Perclich, J. A.; Lockwood, J. L. Interaction of Atrazine with Soil Microorganisms: Population Changes and Accumulation. *Can. J. Microbiol.* **1978**, *24*, 1145–1152.

Perez, J.; Dorado, J, M.; Rubia, T. D. L.; Martinez, J. Biodegradation and Biological Treatments of Cellulose, Hemicellulose and Lignin: An overview. *Int. Microbiol.* **2002**, *5*, 53–63.

Petterson, R. C. The Chemical Composition of Wood. In *Chemistry of Solid Wood*; Rowell, R., Ed.; Adv. ChemSer, 207, Am Chem Soc: Washington, DC,1984; pp 57–126.

Pimentel, D.; McNair, S.; Janecka, J.; Wightman, J.; Simmonds, C.; O'connell, C.; Wong, E.; Russel, L.; Zern, J.; Aquino, T. et al. Economic and Environmental Threats of Alien Plant, Animal, and Microbe Invasions. *Agric. Ecosyst. Environ.* **2001**, *84*, 1–20.

Pollet, A.; Delcour, J. A.; Courtin, C. M. Structural Determinants of the Substrate Specificities of Xylanases from Different Glycoside Hydrolase Families. *Crit. Rev. Biotechnol.* **2010**, *30*, 176–191.

Pothiraj, C.; Kanmani, P.; Balaji, P. Bioconversion of Lignocellulose Materials. *Mycobiology* **2006**, *34*, 159–165.

Rasmussen, M. L.; Shrestha, P.; Khanal, S. K.; Pometto III, A. L.; van Leeuwen, J (Hans). Sequential Saccharification of Corn Fiber and Ethanol Production by the Brown-Rot Fungus *Gloeophyllum trabeum*. *Bioresour. Technol.* **2010**, *101*, 3526–3533.

Ridley, B. L.; O'Neil, M. A.; Mohnen, D. Pectins: Structure, Biosynthesis, and Oligogalacturonide-Related Signaling. *Phytochem.* **2001**, *57*, 929–967.

Robson, L. M.; Chambliss, G. H. Characterization of the Cellulolytic Activity of a Bacillus Isolate. *Appl. Environ. Microbiol.* **1984**, *47*, 1039.

Ruiz-Dueñas, F. J.; Martínez, A. T. Microbial Degradation of Lignin: How a Bulky Recalcitrant Polymer Is Efficiently Recycled in Nature and How We Can Take Advantage of This. *Microb. Biotechnol.* **2009**, *2*, 164–177.

Sadhu, S.; Saha, P.; Sen, S. K.; Mayilraj, S.; Maiti, T. K. Production, Purification and Characterization of a Novel Thermotolerant Endoglucanase (CMCase) from *Bacillus* Strain Isolated from Cow Dung. *Springer Plus* **2013**, *2*, 10.

Sahu, P. K.; Singh, D. P.; Prabha, R.; Meena, K. K.; Abhilash, P. C. Connecting Microbial Capabilities with the Soil and Plant Health: Options for Agricultural Sustainability. *Ecol. Indic.* 2018. https://doi.org/10.1016/j.ecolind.2018.05.084

Sasaki, C.; Takada, R.; Watanabe, T.; Honda, Y.; Karita, S.; Nakamura, Y.; Surface carbohydrate analysis and bioethanol production of sugarcane bagasse pretreated with the white rot fungus, *Ceriporiopsissubvermispora* and microwave hydrothermolysis. *Bioresource Technology.* **2011**, *102*, 9942–9946.

Schols, H. A.; Voragen, A. G. J. Complex Pectins: Structure Elucidation Using Enzymes. In *Progress Biotechnolol. Pectins Pectinases*; Visser, J., Voragen, A. G. J., Ed.; Elsevier Science: Amsterdam, The Netherlands, 1996; *14*, pp 3–19.

Shahzadi, T.; Mehmood, S.; Irshad, M.; Anwar, Z.; Afroz, A.; Zeeshan, N.; Sughra, K. Advances in Lignocellulosic Biotechnology: A Brief Review on Lignocellulosic Biomass and Cellulases. *Adv. Biosci. Biotechnol.* **2014**, *5*, 246.

Shi, J.; Chinn, M. S.; Sharma-Shivappa, R. R. Microbial Pretreatment of Cotton Stalks by Solid State Cultivation of *Phanerochaete chrysosporium. Bioresour. Technol.* **2008**, *99*, 6556–6564.

Shinoyama, H.; Ando, A.; Fujii, T.; Tahama, H. The Possibility of Enzymatic Synthesis of a Variety of -Xylosides Using the Transfer Reaction of *Aspergillus niger* Xylosidase. *Agric. Biol. Chem.* 1991, *55*, 849–850.

Sidhu, B.; Rupela, O.; Joshi, P. K.; Beri, V. Sustainability Implications of Burning Rice and Wheat Straw in Punjab. *Eco. Politic. Weekly* **1998**, *33* (39), 163–168.

Singh, B. K.; Walker, A. Microbial Degradation of Organophosphorus Compounds. *FEMS Microbiol. Rev.*2006, *30*, 428–471.

Singh, P.; Suman, A.; Tiwari, P. Biological Pretreatment of Sugarcane Trash for Its Conversion to Fermentable Sugars. *World J. Microbiol. Biotechnol.* **2008**, *24*, 667–673.

Singh, S.; Nain, L. Microorganisms in the Conversion of Agricultural Wastes to Compost. *Proc. Indian Natl. Sci. Acad.* **2014**, *80* (2), 473–481.

Sjostrom, E. *Wood Chemistry: Fundamentals and Applications*, II Ed.; Academic Press: San Diego, 1993.

Solà, M.; Riudavets, J.; Agustí, N. Detection and Identification of Five Common Internal Grain Insect Pests by Multiplex PCR. *Food Control.* 2018, *84*, 246–254.

Sorek, N.; Yeats, T. H.; Szemenyei, H.; Youngs, H.; Somerville, C. R. The Implications of Lignocellulosic Biomass Chemical Composition for the Production of Advanced Biofuels. *Bioscience* **2014**, *64*, 192–201.

Sun, F. H.; Li, J.; Yuan, Y. X.; Yan, Z. Y.; Liu, X. F. Effect of Biological Pretreatment with Tramates Hirsute yj9 on Enzymatic Hydrolysis of Corn Stover. *Int. Biodeterior. Biodegr.* **2011**, *65*, 931–938.

Sun, Y.; Cheng, J. Hydrolysis of Lignocellulosic Materials for Ethanol Production: A Review. *Bioresour. Technol.* 2002, *83*, 1–11.

Tamminga, S.; Jongbloed, A. W.; Van Eerdt, M. M.; Aarts, H. F. M.; Mandersloot, F.; Hoogervorst, N. J. P.; Westhoek, H. The forfaitaireexcretie van stikstof door landbouwhuisdieren [Standards for the Excretion of Nitrogen by Farm Animals], Rapport ID-Lelystad no. 00-2040R, InstituutvoorDierhouderijenDiergezondheid, Lelystad.*The Netherlands,* **2000**; p 71.

Taniguchi, M.; Suzuki, H.; Watanabe, D.; Tanaka, T.; Hoshino, K. Evaluation of Pretreatment with Pleurotusostreatus for Enzymatic Hydrolysis of Rice Straw. *J. Biosci. Bioenergy* **2005**, *100*, 637–643.

Thao, L. T. H. Nitrogen and Phosphorus in the Environment. *J. Survey Res.* **2003**, *15* (3), 56–62.

Van Dyk, J. S.; Gama, R.; Morrison, D.; Swart, S.; Pletschke, B. I. Food Processing Waste: Problems, Current Management and Prospects for Utilisation of the Lignocellulose Component through Enzyme Synergistic Degradation. *Renew. Sustain. Energy Rev.* **2013**, *26*, 521–531.

Vicuna, R. Ligninolysis. A Very Peculiar Microbial Process. *Mol. Biotechnol.* **2000**, *14*, 173–176.

Walia, U. S.; Brar, L. S.; Dhaliwal, B. K. Performance of Clodinafop and Fenoxaprop-p Ethyl for the Control of Phalaris Minor in Wheat. **1998**, *Indian J. Weed Sci.30*, 38–50.

Walker, A.; Perekh, N. R.; Roberts, S. J.; Welch, S. J. Evidence of the Enhanced Biodegradation of Napropamide in Soil. *Pesticide Sci.* **1993**, *39* (1), 55–60.

Walter, G. H.; Chandrasekaran, S.; Collins, P. J.; Jagadeesan, R.; Mohankumar, S.; Alagusundaram, K.; Ebert, P. R.; Daglish, G. J.; Nayak, M. K.; Mohan, S. et al. The Grand Challenge of Food Security: General Lessons from A Comprehensive Approach to Protecting Stored Grain from Insect Pests in Australia and India. *Indian J. Entomol.* **2016**, *78*, 7–16.

Wan, C.; Li, Y. Microbial Delignification of Corn Stover by *Ceriporiopsis subvermispora* for Improving Cellulose Digestibility. *Enzyme Microb. Technol.***2010**, *47*, 31–36.

Wen, Z.; Liao, W.; Chen, S. Production of Cellulase by *Trichoderma reesei* from Dairy Manure. *Bioresour. Technol.* **2005**, *96*, 491–499.

Ye, X.; Dong, F.; Lei, X. Microbial Resources and Ecology-Microbial Degradation of Pesticides. *Nat. Resour. Conserv. Res.* **2018**, 1.

Yu, H. B.; Du, W. Q.; Zhang, J.; Ma, F.; Zhang, X.; Zhong, W. Fungal Treatment of Cornstalks Enhances the Delignification and Xylan Loss During Mild Alkaline Pretreatment and Enzymatic Digestibility of Glucan. *Bioresour. Technol.***2010a**, *101*, 6728–6734.

Yu, H.; Zhang, X.; Song, L.; Ke, J.; Xu, C.; Du, W.; Zhang, J. Evaluation of White-Rot Fungi-Assisted Alkaline/Oxidative Pretreatment of Corn Straw Undergoing Enzymatic Hydrolysis by Cellulose. *J. Biosci. Bioeng.* **2010b**, *110*, 660–664.

Yu, J.; Zhang, J.; He, J.; Liu, Z.; Yu, Z. Combinations of Mild Physical or Chemical Pretreatment with Biological Pretreatment for Enzymatic Hydrolysis of Rice Hull. *Bioresour. Technol.* **2009**, *100*, 903–908.

Zaldivar, J.; Martinez, A.; Ingram, L. O. Effect of Selected Aldehydes on the Growth and Fermentation of Ethanologenic *Escherichia coli*. *Biotechnol. Bioeng.***1999**, *65*, 24–33.

Zhang, Y. H. P.; Lynd, L. R. Toward an Aggregated Understanding of Enzymatic Hydrolysis of Cellulose: Noncomplexed Cellulase Systems. *Biotechnol. Bioeng.* **2004**, *88*, 797–824.

CHAPTER 5

Microbial Consortium: A Biotechnological Tool for Enhanced Bioremediation in Pollution-Affected Environments

SABA MIR[1], MOONISA ASLAM DERVASH[2*], ASIF B. SHIKARI[3], and WASIA SHOWKAT[1]

[1]Division of Plant Biotechnology, SKUAST-K, Shalimar, Srinagar, J&K, India

[2]Division of Environmental Sciences, SKUAST-K, Shalimar, Srinagar, J&K, India

[3]MCRS, SKUAST-K, Sagam

*Corresponding author. E-mail: moonisadervash757@gmail.com

ABSTRACT

Unprecedented industrialization and population explosion have led to an extensive environmental contamination because of augmented acquittal of wastes laden with malevolent, perilous, and carcinogenic xenobiotics. These pollutants not only deteriorate the human health but also have a catastrophic effect on the environment. Physicochemical-based remediation means are expensive, produce derivative disposal problems, and possess limited valor for pollution mitigation because of the incessant surfacing of novel recalcitrant noxious wastes. As an outcome of eco-friendly approach, public approval, and trifling health hazards, microbial bioremediation has received substantial global consideration for pollution abatement. The current research trend has, therefore, oriented toward investigating innate ability of diverse microbes to breakdown these noxious wastes. Microbial-mediated bioremediation offers immense prospective to restore the polluted environs in an ecologically adequate strategy. A microbial consortium assists microbes to exploit a wide array of carbon sources. It equips microbes with sturdiness in retort to environmental stress factors. Microbes in a consortium can execute multifaceted tasks that are impracticable for a sole organism. With progression of technology, it is now feasible to comprehend microbial interface mechanism and construct consortia. In recent years, the significance of sophisticated tools, such as "genomics, proteomics, transcriptomics, metabolomics, and fluxomics" has amplified to devise the strategies to treat these noxious wastes in an eco-friendly comportment.

5.1 INTRODUCTION

Microorganisms are ubiquitous in nature and are usually found as complex communities known as "consortia." Microbes preferably remain in consortium as it makes them to accomplish the tasks that are otherwise impossible for individual ones. It has been documented that the microbes existing in community are more resilient to environmental fluctuations and resist the species invasion (Qian et al., 2019). Furthermore, the metabolic burden in a consortium is very low and the members can efficiently endure nutrient deficiency as a number of metabolic modes are available that help in sharing of metabolites within the community. These distinctive dispositions of a microbial syndicate rely upon two organizing doctrines, that is, "communication and division of labor between the members." Microbes communicate either through trading metabolites or by switching over to committed molecular signals within or between populations (McCarty and Ledesma-Amaro, 2019). Communication enables the division of labor among the members which in turn helps in the accomplishment of complex functions. Processing of complex reactions requiring multiple steps is another important feature of microbial consortia. Formerly, microbial exploration was subjugated by pure cultures, but nowadays the spotlight is poignant toward microbial consortia owing to their serviceable constancy and competence. This comes, in part, because of some limitations that come along with genetic engineering of a single cellular framework. For example, relocation of a multifarious alleyways in a single strain is restricted because multigene engineering in a single transformation event is lacking and requires an intricate pathway characterization, absence of precursors, and cofactors in the host cell, and the expression of heterologous enzymatic machinery causes huge metabolic expression. Moreover, monocultures are often more vulnerable to environmental changes and therefore require extremely controlled culture settings and sterilization protocols (McCarty and Ledesma-Amaro, 2019).

5.2 CLASSIFICATION OF MICROBIAL CONSORTIUM

Designing and constructing synthetic microbial consortia is an emerging field and, therefore, requires distinguished means of categorization that can aid us in understanding and thereby constructing a microbial consortium in a more efficient manner. Microbial consortia are classified into diverse kinds based on their mode of construction, function, and association.

5.2.1 MODE OF CONSTRUCTION

On the basis of construction, "Bernstein & Carlson" have classified microbial consortium into four types viz., "natural microbial consortium (NMC), artificial microbial consortium (AMC), semi-synthetic microbial consortium (SeMC), and synthetic microbial consortium (SyMC)."

5.2.2 NATURAL MICROBIAL CONSORTIUM

It refers to a group of microorganisms in nature living together in a symbiotic association. It is usually used for bioremediation, production of biogas, and wastewater treatment.

5.2.3 ARTIFICIAL MICROBIAL CONSORTIUM

In this type of consortium, diverse untamed microbes can be raised symbiotically together or artificially grown in a closed culture for economic production.

5.2.4   *SEMI-SYNTHETIC MICROBIAL CONSORTIUM*

Consortium in which untamed and artefact microbial entities are cultured together to perform a familiar objective is referred as *"semi-synthetic microbial consortium."*

5.2.5   *SYNTHETIC MICROBIAL CONSORTIUM*

In this type of consortium, metabolically engineered microbes are co-cultured to raise their efficiency.

5.3   MODE OF ASSOCIATION

Microbes that are found to interact with each other and maintain the growth can be co-cultured. These modes of association can be commensal, mutualist, and synergistic.

5.3.1   *COMMENSAL*

When one species derives benefit and the other is neither benefited nor harmed, commensalism exists. Although the word commensal means sharing the same table, other types of relations are equally common. The commensal may somehow obtain shelter, protection, or transportation by associating with another organism. This type of affiliation is exemplified by the interaction of *"Clostridium acetobutylicum"* and *"Bacillus cereus"* for butanol production (Wu et al., 2016). *"B. cereus"* acts as an oxygen scavenger and thereby generates a micro-anaerobic setting for the growth of *"C. acetobutylicum"* and production of butanol.

5.3.2   *MUTUALIST*

This type of interaction involves benefit to two associating species. It is a sort of obligatory association. In this type of affiliation, the organisms are dependent on each other for survival. If both species gain from an association, but are able to survive without it, the association is termed as proto-cooperation. Bernstein and Carlson (2012) have illustrated this interaction by exploiting two genetically engineered strains of "*E. coli*," where one of the strains is glucose +ve and the other is glucose –ve. The +ve one ferments the glucose and the –ve one devours fermented auxiliary products of the +ve strain, thereby makes the environment feasible for the growth of +ve strain.

5.3.3 SYNERGISTIC

When two species grow better than apart or when the by-products of one species enhance the survival of another without affecting the contributory specie, synergism exists. Carbon or energy sources are dispensed among community members in a noncompetitive comportment based on metabolic functionality and microbes follow distribution of work that enables them to processing of substrates in parallel. Eiteman et al. (2009) have illustrated this interaction by using two genetically relic strains of "*Escherichia coli*", where one strain utilizes glucose as a carbon source, whereas, the other strain can consume xylose only. However, co-culturing of these two enables them to consume a combined substrate more efficiently with increased productivity.

5.4 MODE OF FUNCTION

From the classification based on a mode of construction and association, it is complicated to envisage the actual role of a microbe in a consortium. Therefore, Bhatia et al. (2018) came up with another type of classification based on the functions performed by a microbe in a consortium.

5.4.1 ENVIRONMENT MAINTENANCE CONSORTIUM (EMC)

The category of consortium created with an aim to produce and maintain the most feasible environment for growth of partners. For example, "in an aerobic/anaerobic consortium, aerobes by scavenging the oxygen create an anaerobic setting for growth of its anaerobic collaborators" (Wu et al., 2016).

5.4.2   NUTRIENT EXCHANGE CONSORTIUM (NEC)

Some auxotrophic microbes fail to grow due to unavailability of some essential life-supporting nutrients. However, as suggested by Shou et al. (2007), the augmentation of these microbes is sustained by co-culturing the respective microbes along with auxiliary microbes that can manufacture such crucial nutrients.

5.4.3   SUBSTRATE FACILITATOR CONSORTIUM (SFC)

Since microbes fail to hydrolyze all kinds of polysaccharides or cannot utilize them as a carbon source. To overcome this limitation, Dwidar et al. (2013) illustrated the construction of consortia where some microbes produce various hydrolases and make the free sugars available for the growth of other microbes.

5.4.4   SIGNAL EXCHANGE CONSORTIUM (SEC)

In this type of consortium, microbes producing various signaling molecules are used to construct a consortium. These signaling molecules help to persuade the expression of cryptic genes in other microbes and to produce antibiotics and other metabolites (Perez et al., 2011).

5.5   CONSTRUCTION OF MICROBIAL CONSORTIUM

Grobkopf and Soyer (2014) have proposed two approaches for construction and designing of a SyMC. The first one is a top-down method, which involves the modification of a naturally occurring consortium. Here, various omics approaches are followed to elucidate the underlying molecular mechanisms and scrutinize the system principles. The second one is a bottom-up method, which involves the designing and construction of an artificial consortium. It focuses on the plan and edifice of indispensable natural entities such as genetic elements, modules, biocircuits, and biopathways or bionetworks underneath the regulation of bioengineering doctrine in a rational manner. Following the construction, the microbiological entities along with relevant assemblies are modeled scientifically, harmonized, optimized, and then tuned to attain explicit plan, rationale, and presentation criteria (Mukherji and Oudenaarden, 2009). Although there are no hard and fast rules for construction of a SyMC but some important points that should be taken into consideration for construction of an efficient microbial consortium are: "(i) There should be no competition for carbon source between the selected microbes which could otherwise affect their development and overall production"; "(ii) Standard inoculum ratio should be used to avoid the energy resource depletion over that dominance of one specie"; "(iii) Preferably microbes of the same species should be used for construction of a SyMC because they are often more compatible and display same behavior at cellular level as compared to the microbes from different species that compete and evolve the traits that are beneficial to a particular group (Hibbing et al., 2010)"; "(iv) growth parameters of microbes should be in a physiological range"; "(v) in order to engineer a medium, it is desirable to have prior knowledge regarding nutrient requirement of various microbes that support their growth without any biasing (Tan et al., 2015)"; and "(vi) various *in silico* approaches like constraint-based reconstruction and analysis (COBRA), flux-based analysis (FBA), and other systematic advances should be followed to explore the multifarious community interactions that supplementary aid in consortium design (Schellenberger et al., 2011)."

Furthermore, the accomplishment of a microbial syndicate not only relies on an individual capability of a microbe but also on their mutual interactions. Various types of morphological interactions shown by microbes are: "(i) **Distance inhibition**: in this type of interaction, the development of one microbe is ceased at a confident expanse from the competing one. In this, an antibiotic is secreted by one microbe that inhibits the development of other microbes"; "(ii) **Zone line**: this occurs when a microbial colony proliferates sufficiently and comes in contact with another one, a dark precipitate line also called as zone line is produced. This line also causes the production of a new substance"; "(iii) **Contact inhibition**: in this type of interaction, when two colonies grow and come in contact with each other, they cause the inhibition of each other's growth without the production of any new substance"; and "(iv) **Overgrowth**: in this type of interaction, there occurs breaching of one colony by the other colony (Bertrand et al., 2013)".

5.6 BIOREMEDIATION

From the past few decades, there has been an alarming escalation in environmental pollution because of augmented anthropogenic conduct on energy pools, insecure farming practices, and swift industrialization. A hotchpotch of disadvantageous noxious waste, such as "heavy metals, polychlorinated biphenyls, petroleum wastes, pesticides, and textile dyes" are released through various agencies into the environment (Liu et al., 2019). Complex degradation, elevated toxicity, and bioaccrual of these pollutants not only are responsible for toxic, teratogenic, mutagenic, and carcinogenic consequences on humans but also pose a severe hazard to the eco-sustainability (Bilal et al., 2019b). To overcome the horrible penalties of environmental pollution and to safeguard both environment and humans, novel advances need to be designed and bioremediation is one such strategy. Various progressions that have been widely used to address the environment-related pollutant issues include "filtration, oxidation and reduction, evaporation, solidifications, incineration, reverse osmosis, landfill deposition, electrochemical treatment, physiochemical treatment, lagooning treatment, and biological methods using microorganisms and their novel enzyme systems" (Liu et al., 2019). However, the use of these physio-chemical methods has largely been restricted due to the escalated processing expenditure, toxic and costly reagent requirements, and production of secondary pollutants. Considering the advantages of bioremediation such as squat investment, fine competence, absolute mineralization, and no resultant contamination, it has turned out to be of apex order among the chief advances for the dilapidation of noxious pollutants (Liu et al., 2019). In the recent precedent, it has got an increased interest as public venture to uncover sustainable strategies of remediating spoiled environs (Raghunandan et al., 2014, 2018; Kumar et al., 2016). Microbes restore and cleanup the polluted sites in an environmentally safer manner alongside the fabrication of safe end products (Pande et al., 2020). Bioremediation is characterized as a practice, in which the concentration of noxious waste is reduced (degraded, detoxified, mineralized, or transformed) to an innocuous state with the help of biological organisms. The process of remediation depends principally on the characteristic features of the noxious waste, which may consist of "plastics, agrochemicals, dyes, chlorinated compounds, heavy metals, greenhouse gases, nuclear waste, hydrocarbons, and sewage." In fact intriguing into consideration, the location of application, bioremediation techniques can be categorized as "*in situ* or *ex situ.*"

5.6.1 IN SITU BIOREMEDIATION

In this modus operandi, the polluted material is treated at the site of pollution. There is no need of any excavation; therefore, it hardly affects the soil structure. It is less costly as compared to "*ex situ* bioremediation" as no extra cost is requisite for excavation; nevertheless, expenditure of plan and on-site installation of various refined equipment to develop microbial commotions during bioremediation is of the foremost concern. It engages either stimulation of indigenous or naturally occurring microbial populations (by nourishing them with nutrients and oxygen to augment their metabolic commotion) or introduction of few definite engineered micro-organisms to the spot of contamination. "Polluted sites fouled with chlorinated solvents, dyes, heavy metals, and hydrocarbons have been successfully treated by using *in situ* bioremediation techniques (Azubuike et al., 2016)." The imperative ecological circumstances that require to be optimized for carrying out "*in situ* bioremediation" are "the eminence of electron acceptor, moisture content, nutrient availability, pH, and temperature" (Philp and Atlas 2005). Its relevance is also significantly predisposed by soil porosity, unlike "*ex situ* bioremediation."

5.6.2 EX SITU BIOREMEDIATION

In this modus operandi, the pollutants are excavated from polluted sites or the groundwater is pumped and subsequently transported to other sites for treatment. The selection criteria for "*ex situ* bioremediation techniques" are based on "treatment cost, pollutant type, depth of contamination, extent of pollution, physiographic, and geological aspects of the polluted site." Philp and Atlas 2005 have described performance as another criterion for selecting "*ex situ* bioremediation" techniques.

The advancement in "recombinant DNA technology" in microbial breeding has significantly accelerated the efficiency of pollutant degradation, with the production of a huge quantity of microbes with an amplified knack of contaminant degradation. "In addition to the screening of microbial strains, various strategies have been used for the construction of engineered strains are: (i) screening and cloning of highly effective degrading genes, (ii) escalating the expression of enzymes with degradation functions in microorganisms, (iii) expressing degradation genes for different pollutants in a recipient to construct super engineered bacteria, and (iv) protoplast fusion by blending the advantages of both parents for pollution degradation (Liu et al., 2019)."

5.7   BIOREMEDIATION OF TOXIC METALS USING MICROBIAL CONSORTIUM

Heavy metals constitute the most hazardous pollutants in the environment due to their ability of modifying the configuration of proteins and nucleic acids. Wang and Chen (2015) have reported that they can also impede with oxidative phosphorylation and osmotic processes. "Lead, chromium, nickel, zinc, and copper" are the different heavy metals, and they are mainly released into an environmental setting by industries. The various industries that release these metals into environment include "the electroplating, galvanizing, paint, mining, and metal-processing industries." Removal of heavy metals by traditional techniques, such as "coagulation, chemical precipitation, filtration, flocculation, ion exchange, reverse osmosis, biosorption, and membrane process" is considered extravagant and produce a gigantic degree of secondary wastes (Bilal et al., 2018c). On the other hand, the utilization of biological agents such as microorganisms presents an easy, efficient, and eco-friendly substitute tool for the purging of metal from various environs. "Heavy metals bioremediation traits of microorganisms" inescapably occur from their self-defense mechanisms, such as "cellular morphological alterations and secretion of enzymes." Such exceptional intrinsic traits of microbial isolates have been productively subjugated for diminution or degradation of detrimental contaminants. "*Aspergillus*," "*Bacillus*," "*Pseudomonas*," and "*Flavobacterium*" are some of the microbes that have the potential of metal reduction (Bhatia et al., 2018). Microbes carry out the adsorption of heavy metals by generating diverse extracellular materials (EPSs). These extracellular materials include diverse organic groups, for example, hydroxyl and carboxyl that possess the ability to form complexes with heavy metals. "An acclimatized mixed culture of *Enterobacter* sp., *Stenotrophomonas* sp., *Comamonas* sp., and *Ochrobactrum* sp" can confiscate heavy metals from wastewater stream have been reported by El-Bestawy et al. (2013). A mixed culture of algae "*Scenedesmus*" and "*Pseudokirchneriella*" can bioadsorb a mixture of metals, "Zn(II) and Ni(II)" (Kipigroch et al., 2012)." Ilamathi et al. (2014) have utilized an alginate-beads immobilized mixed culture ("yeast, *Pseudomonas aeruginosa, B. subtilis*, and *E. coli*") in a "liquid–solid fluidized bed for biosorption of copper, cadmium, chromium, and nickel." Employment of genetically artefact microbes for removal of heavy metals possesses the traits of strong compliance and high treatment competence, which is the major focus of current research. Scores of genetically "engineered bacteria and genetically engineered bacteria-based biosensors have been developed and accounted for the recognition and elimination of ecological pollutants."

5.8 BIOREMEDIATION OF DYESTUFF USING MICROBIAL CONSORTIUM

Various manufacturing categories, for example, "textile, leather, printing, dyeing, food, drugs, detergent, and cosmetics" utilize different kinds of artificial dyes. It is reported that "approximately 10%–15% of the dyes employed in the textile, tanning, paper, plastic, food, cosmetic, and pharmaceutical industries are discharged in the wastewater during the dyeing manoeuvres" (Bilal et al., 2018d). Advancement in research for the production of dyes with variant color shades and resistant fading are posing a severe menace to the environment because these dyes are more complex and imply resistant environmental degradation. "Huge amount of wastewater is produced by these industries, which if left untreated may be detrimental not only to environment but also to the human health. The toxicity of synthetic dyes is due to the aromatic compounds that are used for their production." The aromatic compounds, such as "benzidine and naphthalene," are potent carcinogens. Different physicochemical and biological interventions have been proposed for the handling of textile effluents. Physical and chemical treatments are not usually preferred due to the expensiveness and may lead to the generation of resultant noxious waste. Unlike these two, biological interventions are cheap, more effective, and have least consequences on the environment. "Exhaustive investigation efforts have been attempted to pioneer the novel, vastly competent, and environmental-friendly biological strategies coupled with a huge capability to eliminate dye pollutants in a more cost-effective way to protect ecosystems (Liu et al., 2019). Owing to different degradation effects following the overexpression of certain target dye-degrading genes, these microorganisms can notably amplify the decolorization efficiency of dyeing wastewater and thus has been principally employed in bioaugmentation systems (Rathod et al., 2017; Liang et al., 2018). Nascimento et al. (2011) have isolated a mixed culture from a semi-arid region of the Brazilian Northeast that can detoxify three textile azo dyes (Reactive Red 198, Reactive Red 141, and Reactive Blue 214) by degradation rather than adsorption. A mixed culture acquired from cow dung has also been exploited in a batch reactor to treat textile dye wastewater, resulting in 60.8% decolorization and 66.7% COD diminution (Babu et al., 2016)." Rathod et al. (2017) heterologously overexpressed "azo reductase encoding gene azoA" from "*Enterococcus* sp. L2" and "formate dehydrogenase encoding gene fdh" from "*Mycobacterium vaccae*" in diverse bacterial systems to augment the competence of azo dye decolorization progression of a whole cell.

5.9   BIOREMEDIATION OF PETROLEUM AND AROMATIC COMPOUNDS USING MICROBIAL CONSORTIUM

With the expansion of the marine shipping sector and the unremitting utilization of offshore oil, oil effluence has critically affected the marine environs and become one of the worldwide concerns (Carpenter, 2018). Aliphatic and polycyclic aromatic hydrocarbons (PAH) are the foremost key ingredients of wastes generated from petroleum industry. These PAH being carcinogenic in nature, therefore, affect the flora and fauna of spoiled sites by their mutagenic properties (Haritash and Kaushik, 2009). Biodegradation of petroleum-based hydrocarbons is a multifarious process and depends on the category and quantity of a hydrocarbon present and also on the conditions of an environment. Biodegradation of hydrocarbons in water is affected by "temperature, pH, nutrient availability, salinity, and pressure in the deep sea" (Das and Chandran, 2011). Microbial consortium is a promising approach for the biodegradation of these wastes as pure culture has a very limited range of degradation. As suggested by Kweon et al. (2011) that microbial consortium facilitates complete PAH degradation by the production of a wide range of enzymes simultaneously by mixed organisms. For the breakdown of crude oil, Komukai-Nakamura et al. (1996) have reported a microbial conglomerate of "*Acinetobacter* sp. T4" and *"Pseudomonasputida PB4."* In this *"Acinetobacter* sp. *T4"* degrades hydrocarbons and produce various metabolites, whereas *"P. putida PB4"* consumes these metabolites and is responsible for the degradation of aromatic compounds. "Wu et al. (2013) suggested that the use of bacteria and fungi is more effectual than pure cultures because a mixed culture has superior degradation and mineralization rates." A microbial syndicate possessing six bacterial species (*"B. cereus, Gordoniapolyisoprenivorans, Microbacterium* sp., *Mycobacteriumfortuitum, Microbacteriaceae* bacterium, and one fungus, *Fusariumoxysporum"*) has been designed by Jacques et al. (2008) that can degrade more than 95% of pyrene, phenanthrene, and anthracene.

5.10 BIOREMEDIATION OF AGRICULTURAL CHEMICALS USING MICROBIAL CONSORTIUM

Contemporary agriculture rehearsals have led to effluence of diverse media including "air, land, and water" due to inordinate use of pesticides and chemical fertilizers" (Craig, 2018; Li, 2018). It has also negatively affected human wellbeing and beleaguered animals by numerous conducts. According to the Environmental Protection Agency, "about, 45% of the whole pesticides consumed globally are organophosphorus insecticides that possess a momentous prospective to endanger the ecosystem" (Shabbir et al., 2018). Although various physicochemical techniques are available for pesticide degradation and remediation, but their use is often limited due to their high-cost processing and production of resultant pollutants. In comparison to this, biological treatments hold an exceptional status in pesticide remediation. An extensive diversity of microorganisms, that is, bacteria and fungi isolated and screened from varied sources has exposed a great potential to degrade several pesticide residues into undamaging products in the environment. Bioremediation brings about biostimulation that requires the incorporation of numerous nutrients in the media to promote the biosorption and degradation capability of microbial strains. Moreover, the indigenous microorganisms are relatively less efficient in biodegradation of diverse pesticide contaminants. The genetic modification of these microbes and construction of microbial consortium significantly improves their effectiveness of pollutant degradation and the capability of engineered bacteria to acclimatize to wider environmental setting. Organochlorines (OCs) degradation gene (linA) and organophosphates (OPs) degradation gene (mpd) were overexpressed in "*E. coli*" to construct a strain that could simultaneously degrade both OCs and OPs (Yang et al., 2012). A "pyrethroid hydrolase encoding *pytH* gene" responsible for fenpropathrin degradation was identified from "*Sphingobium sp. JZ-2*." Overexpression of this gene in "*Sphingobium sp. BA3*"significantly improved fenpropathrin degradation ability. In addition to the overexpression of specific genes, protoplast fusion technology, by merging the advantages of both parents, also acts as a potential approach for construction of engineered strains (Liu et al., 2019). For sustainable environment, production of biopesticides has emerged as an imperative inclination in protection of human health and sustainable progression of agriculture. Zhenhua et al. (2018) reported that "*Hrpn*, a nonspecific exciton encoding by gene hrpin," can induce resistance against phyto-pathology and also acts as a bioinsecticide.

5.11   STRATEGIES TO STUDY MICROBIAL MECHANISMS

5.11.1   CULTURE-BASED TECHNIQUES

Currently, most of the bioremediation discourses rely on the "treatability study," in which samples collected from polluted sites are classically incubated under controlled settings and the rates of their degradation and immobilization are documented (Head et al., 2003). Though these discourses offer an approximation of the impending metabolic perfor-mances of microbial syndicate but lack the information regarding microbes responsible for bioremediation. As suggested by Head et al. (2003), microbes responsible for remediation need to be isolated and characterized for precise exploration of the bioremediation process. One of the foremost limitations of culture-dependent techniques is that more than 99% of microbes that inhabit the varied natural environs are either unproductive or lack standard culture protocols (Malla et al., 2018). To trounce these precincts and shortcomings, various molecular approaches have been formulated to investigate the microorganisms accountable for bioremediation. It is invaluable to recover these microbes because their study explores an opportunity to inspect their biodegradation responses, along with other physiological aspects that are liable to manage growth and other activities in the contaminated environs.

5.11.2 OMICS-APPROACHES AND THEIR ROLE IN MICROBIAL BIOREMEDIATION

"Microbial bioremediation" employs the indigenous microbial consortia for environmental decontamination. The detoxification rate of contaminants relies on a range of aspects, for example, "the composition of the indigenous microbial communities, nature, and scope of the pollutant and environmental conditions" (Chakraborty et al., 2012). Therefore, it entails fusion of diverse multifaceted variables to optimize the process of bioremediation and to comprehend and foresee the providence of environmental contaminants. Molecular approaches, for example, "genomics, proteomics, transcriptomics, metabolomics, and fluxomics" are now incessantly recognizing their relevance in bioremediation process so as to comprehend the accurate mechanism involved. The exploitation of "omics tools" to study taxonomic and functional facets of the microbial communities from polluted sites has led to the breakthrough innovation of various novel bacteria that were not accessible otherwise by following the simple customary culturing techniques. Recently, colossal throughput "omics strategies" have been applied to elucidate the systems biology of the microbial syndicate in multitude of environs. Conversely, the victorious implementations of these complex bioremedial strategies necessitate a much comprehensive and inclusive perceptive of the aspects that manage the "growth, metabolism, structure, dynamics, and functions" of the native microbial syndicate of these sites. These biotechniques have made it virtually promising and cost-effectively viable to discover the "metagenomes" of spoiled environmental samples, harboring varied microbial syndicate. "This has not merely endowed with an insight regarding the variety but also recognized information about the meta-functionality of the microbial populations inhabiting the polluted environs." Even the amalgamation of data engendered via diverse "omic strategies" may be utilized to study the microbial metabolism during the bioremediation processes. Studies like these will afford a prospect to extend efficient strains of microbes so as to perk up metabolism of diverse xenobiotics (Desai et al., 2010). The competence of the bioremediation will certainly be augmented if the exact molecular strategies are appropriately used and systematically pursued. Collective interpretation from diverse "omics tools" has offered key insights regarding the survival, metabolism, and interaction of the microorganisms in their indigenous environs including gut microbiomes, deepsea sediments, groundwater and marine systems, and extreme milieus (Mella et al., 2018)." In order to accelerate the entire remediation of soiled environs, an inclusive comprehension of the "physiology, biochemistry, ecology, and phylogeny of the indigenous microbial consortia" of polluted sites is defensible. The relevance of genome-based practices in the exploration of both environmental samples and pure cultures makes it probable to construct the models that are obligatory to forecast the commotion of microbes under varied bioremediation strategies.

5.11.3 METAGENOMICS

It helps in shaping the bioremediation approaches in a number of ways, thereby holds great promise in the field of bioremediation (Tripathi et al., 2018). The focal point of metagenomics is "microbial diversity, evolution, and adaption," and it has revolutionized the field of microbiology along with other molecular techniques. Studies have prompted novel insights in the "metabolism, community structure, evolution, function, and genetic makeup of these communities" by investigating the microbial consortia from diverse environments, such as "acid mine drainage, soils, sediments and marine water, and human gut" (Mella et al., 2018). Bioremediation approaches based on metagenomics are considered most versatile and potent for carrying out the decontamination process (Mella et al., 2018). Its results are more reliable and offer better degradation ratios as compared to other bioremediation strategies. "First, it has significantly amplified our grasping trait regarding how microbes develop 'bucket-brigades' for the degradation of xenobiotics, thereby permitting the differentiation of spoiled environs into areas where the native microbiota are able to remediate the ecological status by using intensive '*ex situ* treatment' or by '*in situ* bioaugmentation.' Second, it helps in the identification of key microbial processes and also determines how the community composition could best be complemented to enable mineralization of a pollutant when metabolic crosstalk among different species is necessary, and is, therefore, carried out by bacterial consortia rather than by individual species (Mella et al., 2018). Third, it provides suitable databases that proffer a rich stock of genes for the construction of novel microbial strains for targeted utilization in bioremediation efforts."

5.11.4 MICROBIAL METABOLOMICS AND FLUXOMICS

"Metabolomics is the analysis of metabolite profiles of microbial communities under a specified set of circumstances. It is one of the most recent entries to the 'omics family.' In addition to genomics, transcriptomics, and proteomics, cutting-edge research is now intensifying toward the investigation of microbial cellular metabolites. Application of metabolome-based approaches to the environmental samples has made it possible to develop models that can foresee microbial activities under diverse bioremediation strategies. Metabolomics helps us in accepting the dynamic operations of the microbial communities and their functional contributions to the environs in an improved manner. Upon focus in a microbial cell to an environmental stress, a number of primary and secondary metabolites are released. The functions of these metabolites are explored by metabolomics approach." To probe the biological degradation of anthropogenic noxious waste, various metabolome studies have been carried out, for example, metabolome analysis of "*Sinorhizobium* sp." during phenanthrene degradation (Keum et al., 2008). "Carbon source metabolites in this study were put against the intercellular metabolomes, and the metabolites profiles (fatty acids, polyhydroxyalkanoates, and polar metabolites) analyzed with that an untargeted metabalome analysis. The potential of metabolomics in bioremediation research was elucidated by Villas-Boas and Bruheim (2007)." Many experimental and conceptual approaches have been developed by these scientists. "*Shewanella* is a marine bacterial species that is known to have the adjective of co-metabolic pathways that can biodegrade noxious metals, halogenated organic compounds, and radionuclides. Tang et al. (2007) appraised the fluxome profile of "*Shewanella*" using biochemical GC-MS, and statistical and genetic algorithms. Relatively supple metabolism flux when subjected to various carbon sources were presented in their results. Durand et al. (2010) used "liquid chromatography NMR" and "*ex situ* NMR" to define the metabolic pathways of this bacterial strain during the degradation of the herbicide mesotrione. Total of six metabolites of which the structures of four metabolites were recommended. Wharfe et al. (2010) applied FT-IR to examine the biochemical and phenotypic transformations in bacterial communities that could breakdown aromatic compounds (i.e., phenol) that had been released from industrial bioreactors.

5.12   CHALLENGES IN ENGINEERING MICROBIAL CONSORTIA

The various noteworthy confrontations linked with engineering microbial consortia are: "(i) First, natural microbial communities can sustain
homeostasis; members usually do not out-compete one another and do not
exhaust the resources in their environments. However, it is complicated
to devise either long-term homeostasis or long-term extermination into a
synthetic consortium because long-term behavior, and even the long-term
genetic composition of an engineered organism, is unpredictable. Thus,
engineered consortia should be designed for contexts in which members of
the consortium can be re-introduced or eliminated as required, and in which
their behavior can be tapped temporally"; "(ii) a second confrontation is
that, at least in nature, gene transfer between microbes is widespread. As a
result, engineered consortia should work despite horizontal gene transfer,
or even exploit it"; "(iii) a third confrontation will be to build up the
interventions for incorporating stable transformations into the genomes
of microbes that are not currently commonly engineered. Horizontal gene
transfer is limited when engineers make stable changes to the chromo-
some. In addition, organisms currently recalcitrant to genetic modification
methods often perform very useful functions that are difficult to engineer
into other organisms. For example, species of "*Clostridia*" (e.g., "*Clostridium thermocellum,*" for which there are no recognized genetic cloning
protocols, and "*Clostridium acetobutylicum,*" the protocols for which are
complex and proprietary) live in consortia with other microbes and naturally secrete powerful cellulases"; "(iv) a fourth major confrontation in
engineering consortia is fine-tuning the performance of numerous populations. Techniques such as directed evolution that can optimize the behavior
of a single population must be extended for relevance to multiple populations and varying environments. Elevated throughput screening methods
and economical gene-chip assay procedures will be tremendously helpful
for the proficient construction and appraisal of synthetic consortia."

5.13   CONCLUSION

Rapid industrialization and uncontrollable population growth have created
an immense threat to the environment by generating various pollutants,
therefore, calls for an urgent designing of novel degradation and cleanup
approaches. Since the interface between the microbial communities and

the environmental contaminants is far from being straightforward, and it is difficult to investigate the extreme natural habitat of these microbes. Environmental pollution can be viewed as an environmental upheaval and for such an upheaval, bioremediation can be approved as a "perfect remedy." However, advances in novel "omic approaches (genomics, transcriptomics, proteomics, metabolomics, and fluxomics)" offer a system-level understanding of microbial interactions and mechanisms involved in various bioremedial pathways. The relevance of these stratagems is still in their babbling stage; but the immensity of data that is constantly being generated by the present day "omic approaches" needs to be structured within the productive databases. "Omics approaches" show immense capability to predict organism's metabolism in polluted environments and to envisage the microbial mediated attenuation of the pollutants to fasten bioremediation. The study of molecular mechanisms behind the microbial transformations of the toxic pollutants using "omic approaches" to bioremediation would aid in tracking the caliber of organisms in efficient elimination of contaminants from contaminated environments.

KEYWORDS

- **agricultural chemicals**
- **environmental contamination**
- **cleanup intervention**
- **microbial consortium**
- **omics approach**

REFERENCES

Azubuike, C. C.; Chioma B. C.; Gideon, C. O. Bioremediation Techniques-Classification Based on Site of Application: Principles, Advantages, Limitations and Prospects. *World J. Microbiol. Biotechnolol.* **2016,** *32*, 180.

Babu, S. V.; Raghupathy, S.; Rajasimman, M. Anaerobic Treatment of Textile Dye Wastewater Using Mixed Culture in Batch Reactor. *J. Adv. Chem. Sci.* **2016,** *2*, 233–236.

Bernstein, H. C.; Carlson, R. P. Microbial Consortia Engineering for Cellular Factories: In Vitro to *In Silico* Systems. *J. Comput. Struct. Biotechnol.* **2012,** *3*, e201210017.

Bertrand, S.; Schumpp, O.; Bohni, N. et al. Detection of Metabolite Induction in Fungal Co-Cultures on Solid Media by High-Throughput Differential Ultra-High Pressure Liquid Chromatography–Time-of-Flight Mass Spectrometry Fingerprinting. *J. Chromatogr. A.* **2013,** *1292,* 219–228.

Bhatia, S. K.; Yong-Keun, C.; Eunsung, K.; YunGon, K.; YungHun, Y. Biotechnological Potential of Microbial Consortia and Future Perspectives. *Crit. Rev. Biotechnol.* **2018,** *38,* 1209–1229.

Bilal, M.; Rasheed, T.; Iqbal, H. M.; Yan, Y. Peroxidases-Assisted Removal of Environmentally-Related Hazardous Pollutants with Reference to the Reaction Mechanisms of Industrial Dyes. *Sci. Total Environ.* **2018d,** *644,* 1–13.

Bilal, M.; Rasheed, T.; Nabeel, F.; Iqbal, H. M.; Zhao, Y. Hazardous Contaminants in the Environment and Their Laccase-Assisted Degradation–a Review. *J. Environ. Manage.* **2019b,** *234,* 253–264.

Bilal, M.; Rasheed, T.; Sosa-Hernández, J.; Raza, A.; Nabeel, F.; Iqbal, H. Biosorption: An Interplay between Marine Algae and Potentially Toxic Elements—a Review. *Marine Drugs* **2018c,** *16* (2), 65.

Carpenter, A. Oil Pollution in the North Sea: The Impact of Governance Measures on Oil Pollution over Several Decades. *Hydrobiologia* **2018,** 1–19.

Chakraborty, R.; Wu, C. H.; Hazen, T. C. Systems Biology Approach to Bioremediation. *Curr. Opin. Biotechnol.* **2012,** *23,* 483–490.

Cosa, S.; Okoh, A. Bioflocculant Production by a Consortium of Two Bacterial Species and Its Potential Application in Industrial Wastewater and River Water Treatment. *Pol. J. Environ. Stud.* **2014,** *23,* 689–696.

Craig, K. A Review of the Chemistry, Pesticide Use, and Environmental Fate of Sulfur Dioxide, as Used in California. *Rev. Environ. Contam. Toxicol.* **2018,** *246,* 33–64.

Das, N.; Chandran, P. Microbial Degradation of Petroleum Hydrocarbon Contaminants: An Overview. *Biotechnol. Res. Int.* **2011,** 941810.

Desai, C.; Pathak, H.; Madamwar, D. Advances in Molecular and "-Omics" Technologies to Gauge Microbial Communities and Bioremediation at Xenobiotic/Anthropogen Contaminated Sites. *Bioresour. Technol.* **2010,** *101,* 1558–1569.

Doré, J.; Perraud, M.; Dieryckx, C.; Kohler, A.; Morin, E.; Henrissat, B. et al. Comparative Genomics, Proteomics and Transcriptomics Give New Insight into the Exoproteome of the Basidiomycete *Hebeloma cylindrosporum* and Its Involvement in Ectomycorrhizal Symbiosis. *New Phytol.* **2015,** *208,* 1169–1187.

Durand, S.; Sancelme, M.; Besse-Hoggan, P.; Combourieu, B. Biodegradation Pathway of Mesotrione: Complementarities of NMR, LC-NMR and LC-MS for Qualitative and Quantitative Metabolic Profiling. *Chemosphere* **2010,** *81,* 372–380.

Dwidar, M.; Kim, S.; Jeon, B. S., et al. Co-culturing a Novel Bacillus Strain with Clostridium tyrobutyricum ATCC 25755 to Produce Butyric Acid from Sucrose. *Biotechnol. Biofuels* **2013,** *6,* 1754–6834.

Ebn, S.; Kühner, S.; Wöhlbrand, L.; Fritz, I.; Wruck, W.; Hultschig, C. et al. Substrate-Dependent Regulation of Anaerobic Degradation Pathways for Toluene and Ethyl-benzenein a Denitrifying Substrate-Dependent Regulation of Anaerobic Degradation Pathways for Toluene and Ethylbenzene in A Denitrifying Bacterium. *J. Bacteriol.* **2005,** *187,* 1493–1503.

Eiteman, M. A.; Lee, S. A.; Altman, R.; et al. A Substrate Selective Co-Fermentation Strategy with *Escherichia coli* Produces Lactate by Simultaneously Consuming Xylose and Glucose. *Biotechnol Bioeng.* **2009,** *102*, 822–827.

El-Bestawy, E.; Sabir, J.; Mansy, A. H. et al. Isolation, Identification and Acclimatization of Atrazine-Resistant Soil Bacteria. *Ann. Agric. Sci.***2013,** *58*, 119–130.

Grobkopf, T.; Soyer, O. S. Synthetic Microbial Communities.*Curr. Opin. Microbiol.* **2014,** *18*, 72–77.

Haritash, A.; Kaushik, C. Biodegradation Aspects of Polycyclic Aromatic Hydrocarbons (PAHs): A Review. *J. Hazard. Mater.* **2009,** *169*, 1–15.

Head, I. M.; Singleton, I.; Milner, M. G. Bioremediation: A Critical Review. *Horizon Sci.* **2003.**

Hibbing, M. E.; Fuqua, C.; Parsek, M. R.; et al. Bacterial Competition: Surviving and Thriving in the Microbial Jungle. *Nat. Rev. Microbiol.* **2010,** *8*, 15–25.

Ilamathi, R.; Nirmala, G.; Muruganandam, L. Heavy Metals Biosorption in Liquid Solid Fluidized Bed by Immobilized Consortia in Alginate Beads. *J. Bioprocess Biotech.* **2014,** *4*, 1.

Jacob, J. M.; Karthik, C.; Saratale, R. G.; Kumar, S. S.; Prabakar, D.; Kadirvelu, K.; Pugazhendhi, A. Biological Approaches to Tackle Heavy Metal Pollution: A Survey of Literature. *J. Environ. Manag.* **2018,** *217*, 56–70.

Jacques, R. J.; Okeke, B. C.; Bento, F. M.; et al. Microbial Consortium Bioaugmentation of a Polycyclic Aromatic Hydrocarbons Contaminated Soil. *Bioresour. Technol.* **2008,** *99*, 2637–2643.

Keum, Y. S.; Seo, J. S.; Li, Q. X.; Kim, J. H. Comparative Metabolomic Analysis of Sinorhizobium sp. C4 During the Degradation of Phenanthrene. *Appl. Microbiol. Biotechnol.* **2008,** *80*, 863–872.

Komukai-Nakamura, S.; Sugiura, K.; Yamauchi-Inomata, Y., et al. Construction of Bacterial Consortia That Degrade Arabian Light Crude Oil. *J. Ferment. Bioeng.* **1996,** *82*, 570–574.

Konopka, A.; Wilkins, M. Application of Meta-Transcriptomics and Proteomics to Analysis of In Situ Physiological State. *Front. Microbiol.* **2012,** *3*, 184.

Kumar, A.; Chanderman, A.; Makolomakwa, M.; Perumal, K.; Singh, S. Microbial Production of Phytases for Combating Environmental Phosphate Pollution and Other Diverse Applications. *Crit. Rev. Environ. Sci. Technol.* **2016,** *46*, 556–591.

Kweon, O.; Kim, S. J.; Holland, R. D. et al. Polycyclic Aromatic Hydrocarbon Metabolic Network in *Mycobacterium vanbaalenii* PYR-1. *J. Bacteriol.* **2011,** *193*, 4326–4337.

Lacerda, C. M. R.; Choe, L. H.; Reardon, K. F. Metaproteomic Analysis of a Bacterial Community Response to Cadmium Exposure. *J. Proteome Res.* **2007,** *6*, 1145–1152.

Li, Z. Health Risk Characterization of Maximum Legal Exposures for Persistent Organic Pollutant (POP) Pesticides in Residential Soil: An Analysis. *J. Environ. Manag.* **2018,** *205*, 163–173.

Liang, Y.; Hou, J.; Liu, Y., Luo, Y.; Tang, J.; Cheng, J. J.; Daroch, M. Textile Dye Decolorizing Synechococcus PCC7942 Engineered with CotAlaccase. *Front. Bioeng. Biotechnol.* **2018,** *6*.

Liu, L.; Bilal, M.; Duan, X.; Hafiz, M. N. I. Mitigation of Environmental Pollution by Genetically Engineered Bacteria—Current Challenges and Future Perspectives. *Sci. Total Environ.* **2019,** *667*, 444–454.

Luvuyo, N.; Nwodo, U. U.; Mabinya, L. V. et al. Studies on Bioflocculant Production by a Mixed Culture of Methylobacterium sp. Obi and Actinobacterium sp. Mayor. *BMC Biotechnol.* **2013**, *13*, 62.

Malla, M. A.; Anamika, D.; Shweta, Y.; Ashwani, K.; Abeer, H.; Elsayed, F. A. A. Understanding and Designing the Strategies for the Microbe-Mediated Remediation of Environmental Contaminants Using Omics Approaches. *Front. Microbiol.* **2018**, *9*, 1132.

McCarty, N. S.; Ledesma-Amaro, R. Synthetic Biology Tools to Engineer Microbial Communities for Biotechnology. *Trends Biotechnol.* **2019**, *37*, 181–197.

Mukherji, S.; Oudenaarden, A. V. *Nat. Rev. Genet.* **2009**, *10*, 859–871.

Nascimento, C.; Magalhaes, D. P.; Brandao, M. et al. Degradation and Detoxification of Three Textile Azo Dyes by Mixed Fungal Cultures from Semi-Arid Region of Brazilian Northeast. *Braz. Arch. Biol. Technol.* **2011**, *54*, 621–628.

Nicolaisen, M. H.; Baelum, J.; Jacobsen, C. S.; Sorensen, J. Transcription Dynamics of the Functional tfdA Gene during MCPA Herbicide Degradation by Cupriavidusnecator AEO106 (pRO101) in Agricultural Soil. *Environ. Microbiol.* **2008**, *10*, 571–579.

Pande, V.; Satish, C. P.; Diksha, S.; Veena, P.; Mukesh, S. Bioremediation: An Emerging Effective Approach towards Environment Restoration. *Environ. Sustain.* 2020. https://doi.org/10.1007/s42398-020-00099-w

Perez, J.; MunozDorado, J.; Brana, A. F.; et al. Myxococcusxanthus Induces Actinorhodin Overproduction and Aerial Mycelium Formation by *Streptomyces coelicolor*. *Microb. Biotechnol.* **2011**, *4*, 175–183.

Philp, J. C.; Atlas, R. M. Bioremediation of Contaminated Soils and Aquifers. In *Bioremediation: Applied Microbial Solutions for Real-World Environmental Cleanup*; Atlas, R. M., Philp, J. C., Eds.; American Society for Microbiology (ASM) Press: Washington, 2005; pp 139–236.

Poretsky, R.; Ann Moran, M. Comparative Day/Night Metatranscriptomic Analysis of Microbial Communities in the North Pacific Subtropical Gyre. *Environ. Microbiol.* **2011**, *11*, 1358–1375.

Qian, X.; Lin, C.; Yuan, S.; Chong, C.; Wenming, Z.; Jie, Z.; Weiliang, D.; Min, J.; Fengxue, X.; Katrin, O. Biotechnological Potential and Applications of Microbial Consortia. *Biotechnol. Adv.* 2019. https://doi.org/10.1016/j.biotechadv.2019.107500.

Raghunandan, K.; Kumar, A.; Kumar, S.; Permaul, K.; Singh, S. Production of Gellan Gum, an Exopolysaccharide, from Biodiesel-Derived Waste Glycerol by Sphingomonas spp. *3 Biotech.* **2018**, *8*, 71.

Raghunandan, K.; McHunu, S.; Kumar, A.; Kumar, K. S.; Govender, A.; Permaul, K. et al. Biodegradation of Glycerol Using Bacterial Isolates from Soil Under Aerobic Conditions. *J. Environ. Sci. Heal. A Tox. Hazard. Subst. Environ. Eng.* **2014**, *49*, 85–92.

Rathod, J.; Dhebar, S.; Archana, G. Efficient Approach to Enhance Whole Cell Azo Dye Decolorization by Heterologous over Expression of Enterococcus sp. L2 azoreductase (azoA) and Mycobacterium Vaccae Formate Dehydrogenase (fdh) in Different Bacterial Systems. *Int. Biodeterior. Biodegr.* **2017**, *124*, 91–100.

Schellenberger, J.; Que, R.; Fleming, R. M.; et al. Quantitative Prediction of Cellular Metabolism with Constraint-Based Models: The COBRA Toolbox v2.0. *Nat. Protoc.* **2011**, *6*, 1290–1307.

Shabbir, M.; Singh, M.; Maiti, S.; Kumar, S.; Saha, S. K. Removal Enactment of Organo-phosphorous Pesticide Using Bacteria Isolated from Domestic Sewage. *Bioresour. Technol.* **2018**, *263*, 280–288.

Shou, W.; Ram, S.; Vilar, J. M. G. Synthetic Cooperation in Engineered Yeast Populations. *Proc. Natl. Acad. Sci. USA* **2007**, *104*, 1877–1882.

Tan, J.; Zuniga, C.; Zengler, K. Unraveling Interactions in Microbial Communities from Co-Cultures to Microbiomes. *J. Microbiol.* **2015**, *53*, 295–305.

Tang, Y. J.; Hwang, J. S.; Wemmer, D. E.; Keasling, J. D. Shewanellaoneidensis MR-1 Fluxome Under Various Oxygen Conditions. *Appl. Environ. Microbiol.* **2007**, *73*, 718–729.

Tripathi, M.; Singh, D.; Vikram, S.; Singh, V.; Kumar, S. Metagenomic Approach towards Bioprospection of Novel Biomolecule(s) and Environmental Bioremediation. *Annu. Res. Rev. Biol.* **2018**, *22*, 1–12.

Verberkmoes, N. C.; Russell, A. L.; Shah, M.; Godzik, A.; Rosenquist, M.; Halfvarson, J. et al. Shotgun Metaproteomics of the Human Distal Gut Microbiota. *ISME J.* **2009**, *3*, 179–189.

Villas-Boas, S. G.; Bruheim, P. The Potential of Metabolomics Tools in Bioremediation Studies. *Omi. A J. Integr. Biol.* **2007**, *11*, 305–313.

Wang, J.; Chen, C. The Current Status of Heavy Metal Pollution and Treatment Technology Development in China. *Environ. Technol. Rev.* **2015**, *4*, 39–53.

Wang, L.; Ma, F.; Qu, Y.; et al. Characterization of a Compound Bioflocculant Produced by Mixed Culture of Rhizobium radiobacter F2 and Bacillus sphaeicus F6. *World J. Microbiol. Biotechnol.* **2011**, *27*, 2559–2565.

Wharfe, E. S.; Jarvis, R. M.; Winder, C. L.; Whiteley, A. S.; Goodacre, R. Fourier Transform Infrared Spectroscopy as a Metabolite Fingerprinting Tool for Monitoring the Phenotypic Changes in Complex Bacterial Communities Capable of Degrading Phenol. *Environ. Microbiol.* **2010**, *12*, 3253–3263.

Wilmes, P.; Bond, P. L. The Application of Two-Dimensional Polyacrylamide Gel Electrophoresis and Downstream Analyses to a Mixed Community of Prokaryotic Microorganisms. *Environ. Microbiol.* **2004**, *6*, 911–920.

Wu, M.; Chen, L.; Tian, Y.; et al. Degradation of Polycyclic Aromatic Hydrocarbons by Microbial Consortia Enriched from Three Soils Using Two Different Culture Media. *Environ. Pollut.* **2013**, *178*, 152–158.

Wu, P.; Wang, G.; Wang, G. et al. Butanol Production under Microaerobic Conditions with a Symbiotic System of *Clostridium acetobutylicum* and *Bacillus Cereus*. *Microb. Cell Fact.* **2016**, *15*, 8.

CHAPTER 6

Bioaugmentation: A Way Out for Remediation of Polluted Environments

MOHAMMAD YASEEN MIR[1*], SAIMA HAMID[2], and
GULAB KHAN ROHELA[3]

[1]Centre of Research for Development, University of Kashmir, Hazratbal,
Srinagar, Jammu and Kashmir, India

[2]Department of Environmental Sciences, University of Kashmir,
Hazratbal, Srinagar, Jammu and Kashmir, India

[3]Biotechnology Section, Moriculture Division, Central Sericultural
Research & Training Institute, Central Silk Board, Ministry of Textiles,
Government of India, Pampore, Jammu and Kashmir, India

[*]Corresponding author. E-mail: yaseencord36@gmail.com

ABSTRACT

The unintended presence of agricultural waste in the aquatic environment
results to changes in the water characteristics, causing contamination and
producing severe damages to the freshwater bodies. Different agricultural
waste management technologies have been applied in the treatment of
agricultural waste before discharge, however, the use of biotechnology
will surely help in the management of these agricultural waste because
most agricultural waste management technologies are expensive, difficult
to access, or inefficient. Biotechnology is employed in the reduction of
risks associated with the pollution of soil, water, and air due to agricultural
processes, thereby serving as an effective tool for the achievement of
agricultural sustainability as it can be used for the monitoring of polluted
sites. Some tools employed in biotechnology include genetic engineering,
cell tissue culture, molecular techniques, bioinformatics, etc. Aquaculture,
livestock production and cultivation activities have been identified as

sources of agricultural waste. Different methods of agricultural waste management include compositing, recycling separation and sorting and incineration. The intervention of biotechnology in the treatment of agricultural waste has led to the conversion of the waste to bioethanol and biogas, which are eco-friendly, thus preventing the consequences of greenhouse effect. The utilization of agricultural waste as biomass for bioethanol and biogas production is advantageous because these biomasses are renewable substances which will reduce the rate of consumption of fossil fuel, for instance, the fuel needed by the entire transportation sector will not be satisfied by ethanol, but its combination with fossil fuels will reduce the pressure on fossil fuel and also reduce the environmental impact.

6.1 INTRODUCTION

The world is threatened due to the constant destruction of human activities, as excessive use of chemical fertilizers and pesticides in order to increase production of food grains but these chemicals along possess serious threat to marine and freshwater ecosystems after heavy washout from agricultural fields and gets accumulated in food chains of both terrestrial and aquatic ecosystems (Cherniwchan, 2012; Goudie, 2013). Oil refineries, gas stations, washout from farm fields sprayed with pesticides and fertilizers, and pharmaceutical industries are the most point sources of such contaminants in soil and water bodies as a huge list of pollutants has long residual period in various environs (Srirangan et al., 2012). As an example, polyethylene bags are the polymers that are highly recalcitrant in nature due to their severe toxic nature and its long persistency is due to three physical properties that are hydrophobic molecules, molecular weight, and complex structure. An estimate is about approximately 1trillion plastic bags are used globally with manufacturing rate of 12% per annum, and its single use leads them to directly get dump into landfills where they remain nonbiodegradable (Shimao, 2001; Leja and Lewandowicz, 2010; Harshavardhan and Jha, 2013). Due to offshore drilling rigs, leakage from tanker crude oils; refineries and common sources which are the causes for contamination of marine and freshwater sources. As till yet many accidents had happened in past in Miri, Malaysia, 2013 and in 2000, Clark County, Kentucky where due to oil spills many springs polluted that affected eighty thousand people in Coca, Ecuador (Godleads et al., 2015; Campbell and Clifford, 2010). While the accidents involving oil spills indicate comparatively a decreasing trend from 28.1 spill numbers in 1990 to 5.2 spill numbers in 2010, it has not been fully halted. In 2015, two major oil spillages have been recorded that occurred in Turkey and Singapore that are possessing threat to aquatic birds and animals as these chemicals are very persistent, carcinogenic, and mutagenic in nature, thus are the cause for bioaccumulation in food chains as till yet there are no such technologies that can be used to tackle oil spill pollution (Godleads et al., 2015; Srirangan et al., 2012).

6.2 BIOAUGMENTATION PRINCIPLE

There is a need to use some approaches that will degrade these chemicals in the environment as one of the important methods is to use microorganisms to deal with chemical or oil spills in marine and fresh environment that is known as the bioaugmentation. In this method of bioaugmentation, certain group of microbes can be used to degrade the complex pollutants by adding these microbes to contaminated site with the ample use of genetic diversity for faster way of degradation by using complex group of microbes along with certain physical parameters that need to check upon and it is the ultimate goal of this approach. By adding those potent microbial flora to the contaminate site along with native microbes will increase the rate of degradation with improved rate of removal efficiency along with to boost gene pool target by identifying certain strains with huge potential. Therefore, process of bioaugmentation enriches the genetic diversity at sites of contamination. Through increasing the microbial diversity, pollutants can be degraded more efficiently (Leahy and Colwell, 1990; Godleads et al., 2015).

6.3 FACTORS INFLUENCING BIOAUGMENTATION

The success of the bioaugmentation process is mostly depends on the type of microorganisms targeted to the polluted site for decontamination. It also depends on the introduced microbial species to cope with the existing natural flora of microorganisms, natural enemies, and several other factors including biotic and abiotic type (Godleads et al., 2015). Earlier researcher's have proved that by increasing the efficiency of the remediation method, it increases the biodegradation potential of the polluted soil (Mrozik and Piotrowska-Seget, 2010). Bioaugmentation process was carried mostly by targeting the polluted soils having a smaller number of natural flora and effectives are mostly depend on several factors like soil type, organic matter quality, nutrient content, pH, temperature, humidity, and aeration (Forsyth et al., 1995; Mrozik and Piotrowska-Seget, 2010). If any factor is not optimum then the bioaugmentation process will not be effective. The degrading abilities of microorganisms is due to the presence of specific genes and enzymes (Geize and Dijkhuizen, 2004; Jimenez et al., 2006; Rivelli et al., 2013). In addition, the adaptive abilities (changing membrane composition) of microorganisms to detoxify the contaminants via specific biological activities (Isken and Bont, 1998; Carvalho et al., 2009); the development biosurfactants, (Ron et al., 2002) and effective use of membrane pumps to reduce the toxic compounds in intracellular environment (Isken and Bont, 1998; Van Hamme et al., 2003) makes them as ideal organisms for bioremediation. Organic matter of soil is the most important soil parameter influencing the success of bioaugmentation process. Organic matter also influences the availability of pollutant, survivability of introduced microbial species, and also their degrading abilities.

Greer and Shelton in 1992 and Mrozik and Piotrowska in 2010 have conducted studies in which they assured that in high organic soil compared to soil containing lower organic matter where catabolic properties of microbes toward 2, 4-dichlorophenoxyacetic acid were found to be maximum (Greer and Shelton, 1992; Mrozik and Piotrowska-Seget, 2010). Moisture content of soil also plays an important role in bioaugmentation process (Ronen et al., 2000).

Soil moisture plays an important role as various studies has been conducted regarding survival rates of bacteria namely *Achromobacter piechaundii* TBPZ along with degradation rates of tribromophhenol and it has been reported that with increase in moisture content of 25% and 50% water content in the soil, the process of degradation increases and vice versa.

Availability of moisture content depends on the type of substrate in soil, and dehydration of bacteria cells is linked inversely with bioremediation process (Mashreghi and Prosser, 2006; Mrozik and Piotrowska-Seget, 2010).

Kim et al. (2008) concluded that *Pseudomonas* grows very rapidly in slurry systems modified with sterile soil as opposed to the soil that was previously combusted to extract organic matter. Hence content of insoluble organic compounds present in soil inversely affects the degradation of BTEX by microorganisms.

6.4 MICROBIAL CONSORTIA USED IN BIOAUGMENTATION

Current literature has shown the diversified roles of microorganisms in bioaugmentation process. *Burkholderia pickettii* was found to be suitable for biodegradation of quinoline (Wang et al., 2004; Xu et al., 2015). It was observed that the natural microflora were unable to degrade quinoline, where as in combination with *Burkholderia pickettii*, quinoline could be degraded within 8h. Tchelet et al. (1999) experiments demonstrated the effective degradation of chlorinated benzenes by P51 strain of *β-Proteobacterium Pseudomonas*. The results have shown the effective bioaugmentation process by adding pre-selected strains to the polluted site to degrade recalcitrant material like TCB. Contaminants in the soil not only influence the various properties of soil but also make microbial flora to perform several functions in the soil.

6.5   SELECTION OF MICROORGANISMS

In order to achieve success in the process of soil augmentation, it is very important to choose most suitable microbial consortium and for their growth certain conditions need to be kept in appropriate concentrations of microbes and other microbial growth features need to look upon and different methods were employed for selection of a suitable microbial strain. However in the process of bioremediation, initially microbial strains were first isolated from natural sites, multiplied in lab conditions and then at last inoculated at the initial site of contaminations. AKS2 strain of *Pseudomonas* sp. present in soil was characterized and found to degrade the pollutant polyethylene succinate (PES) under lab conditions (Tribedi et al. (2012). This strain degrades with PES without causing any damage to ecological balance of soil, proving an eco-friendly way for PES degradation (Tribedi et al., 2013). Microbes from various contaminated soil can be used to degrade pollutants of same category and after introducing into the polluted site (Goux et al., 2003; Ghazali et al., 2004). Table 6.1 shows the significance of bioaugmentation in polluted environments.

6.6   MEDIA

6.6.1   GROUNDWATER CONTAMINATION

By the process of diffusion, interchange of contaminants occurs through pores present in the aquifer and enters into the groundwater and thus makes it unfit for drinking purposes. Groundwater passes through the aquifer's pores. Pollutants are interchanged and dissolved through nonflowing bodies through diffusion. Such nonflowing sources are "sinks" for pollutants like NAPL (fuel which leaks) and organic matter.

TABLE 6.1 Utilization of Microorganisms in Bioaugmentation Studies.

Microorganism	Pollutants degraded	References
Pseudomonas putida KT2442	Naphthalene	Filonov et al. (2005)
Pseudomonas sp. *Pseudomonas putida* B13ST1(pPOB)	3-Phenoxybenzoic acid	Halden et al. (1999)
Pseudomonas putida PaW340/pDH5	4-Chlorobenzoic acid	Massa et al. (2009)
Pseudomonas fluorescens MP	2,4-Dinitrotoluene	Monti et al. (2005)
Rhodococcus sp. StrainRHA1	4-Chlorobenzoate	Rodrigues et al. (2001)
*Rhodococcus sp.*F92	Various petroleum products	Quek et al. (2006)
Cupriavidus necator RW112	Arochlor 1221 and 1232	Wittich and Wolff (2007)
Arthrobacter, Burkholderia, Pseudomonas, and Rhodococcus	Petroleum hydrocarbons	Adebusoye et al. (2007)
Phanerochaete chrysosporium	Biphenyl and triphenylmethane	Erika et al. (2013)
Aspergillus niger, A. fumigatus, F. solani, and P. funiculosum	Hydrocarbon	Al-Jawhari (2014)
Alcaligenes odorans, Bacillus subtilis, Corynebacterium propinquum, and Pseudomonas aeruginosa	Phenol	Singh et al. (2013)
Coprinellus radians	PAHs, methylnaphthalenes, and dibenzofurans	Aranda et al. (2010)
Tyromyces palustris, Gloeophyllum trabeum, and Trametes versicolor	Hydrocarbons	Karigar and Rao (2011)
Candida viswanathii	Phenanthrene and benzopyrene	Hesham et al. (2012)
Gleophyllum striatum	Striatum pyrene, anthracene, 9-methyl anthracene, Dibenzothiophene Lignin peroxidase	Yadav et al. (2011)
Penicillium chrysogenum	Monocyclic aromatic hydro carbons, benzene, toluene, ethyl benzene and xylene, and phenol compounds	Abdulsalam et al. (2013)

TABLE 6.1 *(Continued)*

Microorganism	Pollutants degraded	References
Bacillus cereus A	Diesel oil	Maliji et al. (2013)
B. brevis, P. aeruginosa KH6, B. licheniformis, and *B. sphaericus*	Crude oil	Aliaa et al. (2016)
Myrothecium roridum IM 6482	Industrial dyes	Anna et al. (2015)
Microbacterium profundi strain Shh49T	Fe	Wu et al. (2015)
Cunninghamella elegans	Heavy metals	Tigini et al. (2010)
Aerococcussp. and *Rhodopseudomonas palustris*	Pb, Cr, and Cd	Sinha and Paul (2014)
Saccharomyces cerevisiae	Heavy metals, lead, mercury, and nickel	Infante et al. 2014)
Lysinibacillus sphaericusCBAM5	Cobalt, copper, chromium, and lead	Peña-Montenegro et al. (2015)
Acenetobactor sp., Pseudomonas sp., Enterobacter sp., and Photobacterium sp.	Chlorpyrifos and methyl parathion	Ravi et al. (2015)

Such pollutants enter an aquifer and are regulated on the basis of the release of materials. The process for NAPL formation involves the flow of water into an aquifer and transforms into a medium of pollutant movement that spreads pollutants across the media. Therefore, groundwater pollution is problematic both inside and outside of the aquifer (Adeel et al., 2001). Aquifers may be a way to move pollutants within the aquifer. Dense non-aqueous phase liquids (DNAPLs) are the pollutants which find their way to water table via leaching from ground or soils, and these pollutants are heavier than water molecules and hence settle down to form a reservoir. DNAPLs get stick to soil and lead to plume formation in an aquifer (McCarty et al., 1998).

Two major processes exist for the treatment of polluted groundwater. First, adsorption of a thermodynamic product attaches pollutants, usually organic matter, to a particular binding location. The opposite, desorption and removal of pollutants away from the binding site in groundwater are combined with adsorption. They are very important to bioaugmentation in order to remove pollutants until they are trapped in. Absorption typically includes the interaction between the sorbed and aqueous stages. As the concentration of the sorbed-phase decreases the aqueous phase increases, until it reaches "equilibrium" (Suthersan and Payne, 2005).

In situ treatment is known as the treatment at the natural site of pollution and this method is commonly employed for groundwater treatment in which air sparging or extraction techniques are used. Air is being imparted into the aquifers in order to clean pollutant along with extraction methods that combine electron acceptors such as nitrate and oxygen and nutrients and then reinsert them back into the polluted site, and it needs to be done constantly in order to remove pollutants from groundwater via extraction process (EPA, 2004; Frazar, 2000). *In situ* treatment methods involve use of both aerobic and anaerobic microorganisms and correct amount of microbes has to be inoculated for degrading pollutants like methyl tert-butyl ether (Stocking et al., 2000; Prince, 2000).

Biologically enhanced reductive dechlorination (ERD) is one of the newer *in situ* treatment methods. Enhanced reductive dechlorination is the direct penetration of ethane and ethane into bacteria that are able to dechlorinate pollutants such as trichloroethylene and tetrachloroethylene. ERD can be utilized as substrate; hence, it favors microbial growth near the polluted site and hydrogen production under anaerobic conditions. However, during the degradation of ERD, harmful vinyl chlorides (VC) were produced as by-products. It is therefore necessary to apply strains such as *Dehalococcoides ethenenogenes* to the potential for the neutralizing the harmful compounds so that degradation process can continue (Zhu et al., 2001). Complete degradation of chloroform to harmless forms from polluted groundwater by *Dehalobacter* was achieved (Justice et al., 2014).

Arthrobacter sp. was used for degradation of nitroglycerin found in polluted groundwater by Husserl and Hughes (2013). How to correctly build something that can be incorporated directly into the groundwater is a vital factor to consider in the bioremediation treatment of ground water. Additional nature considerations include the hydraulic conductivity, the potential to degrade the waste and where the pollution occurs (US EPA, 2004). Overall, the use of *in situ* groundwater remediation has both benefits and drawbacks.

Ex situ method is among the treatment methods that involve use of heat to aquifer or microbial strains. Thermal treatment is effective against the pollutants like trichloroethylene (TCE) (Friis et al., 2005, 2006). Co-metabolism is another method that used specific strains in combination with specific substrates for effective degradation of pollutants (US EPA, 2004). A modified strain PR1301 of *Burkholderia cepacia* by utilizing phenol (carbon sources) and aquifer for degradation of TCE of polluted groundwater (McCarty et al., 1998).

6.6.2 SOIL CONTAMINATION

Organic soils consist of organic matter in the range of 5%–60% with carbohydrates, fats, amino acids, and lipids. Such material is unaltered due to environmental, physical, or chemical processes. The difference of humic and nonhumic materials is estimated by color (Aitken and Long, 2004). Contamination of organic soils occurs due to substrate consistency, pH, availability of nutrients, and mean time of cell life, and duration of pollutants remaining in the polluted soil.

First, properties of soil inhibit the analysis of pollutants' concentration. Contaminants of soils are diversified type such as soluble and insoluble forms and degradable and nondegradable forms (Aitken and Long, 2004). Contamination of organic soils occurs due to substrate stability, pH, availability of nutrients, and duration of pollutants remaining in the polluted soil. Organic soils are influenced by factors like it can join either in soil water particles or in soil particulate-generated areas. Properties of chemicals such as vapor pressure and molecular weight decide their solubility in water, based on the type of microbes available. Composition of chemical is also important for the pollutants to under degradation, in particular with primary microorganism degradation in the aqueous phase (Aitken and Long, 2004).

Additional organic material interactions have a relationship with the soils. Biodegradation process enhances the materials to be trapped in soil matrix. Soil interactions by sorption mechanism will separate microbes from the contaminants and also eliminate them from the aqueous phases. This causes a transition within microorganisms' ability to degrade. When sorption to humic substances occurs, many of the pollutants reduce the supply of microorganisms. This is an essential feature because pollutants are held within the soil during the sorption process. Cation exchange is another process where removal of pollutants was done more effectively (Aitken and Long, 2004).

Similar to groundwater treatment, *in situ* and *ex situ* treatments were employed in removal of pollutants from contaminated soil. During *in situ* treatment of contaminated soil, aeration bioventing or slurry-phase soil reactors were used. Bioventing uses air removed to reduce toxins in soils. Aeration of the slurry-phase soil reactors is done by mixing air and soil for removal of constituents (US EPA, 2001). There are many common uses of *in situ* soil bioremediation. In Ukiah, California, microorganisms mixing nutrients and chemicals were used for the treatment of formaldehyde.

Some effective treatments of *in situ* method includes removal of chemical pollutants like acetone, butyl alcohol, chromium, dimethylaniline, electroplating waste, and methylene chloride (EPA, 1990). Cutright (1995) was able to extract from the soil 96% of polycyclic aromatic hydrocarbons (PAHs). *Ex situ* methods are used as an alternative where a pollutant removal is ineffective by *in situ* treatment methods. In *ex situ* methods, contaminated soil is being transported to other areas (labs or open lands) where more viable growth of micro-organisms can be observed resulting in removal of pollutants. Land treatment includes a reactor which is built with liners and requires additional facilities like drainage and irrigation spaces. Using traveling gun sprinkling systems, the chemical fertilizers, agricultural wastes, and animal manures are used as carbon sources and nutrients to promote microbial growth within land treatment (Pope et al., 1993). Pollutants like PAHs are effectively treated by using sewage sludge (Paraiba et al., 2011).

Other approaches of the *ex situ* includes composting with manure, hay, and wood chips to increase organic content where the treated soils are placed in groups that have provided easy air passages (US EPA, 2001).

Finally, in slurry-phase soil bioreactors soils, sediments and sludge are combined to form slurry for maintaining the interaction between microbes and also for aeration. Shailaja et al. (2008) carried degradation polyvinyl chloride (PVC) plastics by slurry phase treatment. According to Valentin et al. (2007) for the treatment of PAH, slurry phase method is also used with a strain of fungus of *Bjerkandera adusta*, and same technique was employed with other consortium of microbes by Lewis (1993); Gamati et al. (1999), Cassidy and Hudak (2002). Anaerobic slurry reactor was used for treatment of a pesticide gamma-hexachlorocyclohexane (Quintero et al., 2006). Two-phase slurry bioreactors were used to degrade *n*-hexadecane (Partovinia et al., 2006), pendimethalin (Kao et al., 2001), 2,4,6-trinitrotoluene (Arienzo, 2000; Newcombe and Crawford, 2007; Shen et al., 1997), *n*-dodecane (Okuda et al., 2007), 2,4-dinitrotoluene (Daprato et al., 2005), diesel fuels and polychlorinated biphenyls (PCBs) (Cassidy and Hudak, 2001; Hudak and Cassidy, 2004), chlorpyrifos (Mohan et al., 2004), and other crude oils (Kuyukina et al., 2003).

Bioaugmented microbes have increased the treatment of diesel fuel pollutants in contaminated soil by 50% (Lebkowska et al., 2011). Similarly, bioleaching with *Acidithiobacillus thiooxidans* has increased removal of 92.7% lead (Pb) in polluted soils (Lee and Kim, 2010). Motor oil pollutant (65%) is removed by bioaugmentation process in polluted soils (Abdulsalam and Omale, 2009). Compared to biostimulation (81%) process, bioaugmentation (86%) process removed diesel oil pollutant more effectively at Marambio Station in Antarctica (Ruberto et al., 2009). Similarly among the microbes, *Pseudomonas putida* (99.997%) has removed more amount of diesel oil pollutant than *Acinetobacter lwoffi* (99.99%) (Kolwzan, 2008). Additionally, with *Geobacillus thermoleovorans* T80, contaminated soils were improved in pollutant removal by 70% (Perfumo et al., 2007). Andreoni et al. (1998) used the *Alcaligenes eutrophus* to degrade 2,4,6-trichlorophenol (TCP).

6.7 WASTE TREATMENT

6.7.1 HAZARDOUS WASTES

Since lethal wastes include hazardous substances, primarily hydrocarbon-based pollutants, thus in order to understand the relationship between environmental degradation of hydrocarbons and microbial metabolic pathways involved in it, an extensive research work was began way back in the 1950s. Prevalent study shows that many bacteria inside the hazardous waste system have a difficulty in timely prevention of harmful effects through fully destroying products. Microbes that are bioaugmented use the pollutants as sources of nutrient or for converting different substances that cannot be used to extract biomass, energy, or other nutrients and hence are useful for removal of toxic hydrocarbons (Atlas, 1993). Properly remediated wastewater can also be used for human activities (Rohela et al., 2019).

6.7.2   TREATMENT OF WASTEWATER FROM INDUSTRIES

Tanning leather industry, coke industries, metal grinding, wood industry, and other industrial processes (80) result in release of wastewater with materials that are toxic and highly harmful. These materials can include aromatic compounds, toxic dyes, chromic and other heavy metals, simple phenols and poyphenols, polycyclic compounds, and heterocyclic compounds (Kim et al., 2013; Bhattacharya and Gupta, 2013; Zhang et al., 2014). Bioaugmentation is useful in treatment of industrial-based wastewater because it can remove the pollutants without disturbing the natural sites of polluted industrial waterbodies. McLaughlin et al. (2006) established factors that allow microorganisms to different tasks. Within indigenous organisms, these activities help to increase physical activity and biological diversity. For instance, a subset of chlorophenol compounds such as chloroaromatic compounds is most commonly found in industrial based wastewater. *Pseudomonas putida*, has effectively removed chlorophene from wastewaters (McLaughlin et al., 2006). Mixing of inoculated bioaugmenting microbes has reduced the 2,4-dichlorophenol in chemical wastewater (Quan et al., 2004). Many researchers have considered the bioaugmentation useful in the treatment of industrial wastewater. *Phanerochaete chrysosporidium*–based bioaugmentation decreases hazardous lipid pollutants within olive oil-based industrial wastewater (Dhouib et al., 2006). Research also shows that strains from bioaugmented bacteria decreases the smell produced during anaerobic treatment. Bioaugmentation with microbial species *Nocardia, Micrococcus*, and *Pseudomonas* was ideal for cresol wastewater treatment (Pandya, 2007). *Acinetobacter* sp had recently been used to isolate chromium and nickel metals (Bhattacharya and Gupta, 2013). A consortium of *Firmicutes, Bacteroidetes*,

Proteobacteria, and *Deinococcus Thermus* was used for the degradation of TP, TN, and COD from wastewater (Kim et al., 2013).

6.8 BIOAUGMENTATION IN WASTEWATER TREATMENT PLANTS

Bioaugmentation includes culturing bacteria that are readily accessible to a wastewater treatment facility. Bioaugmentation is usually used in device startups. Main benefit of bioaugmentation in the wastewater treatment is microbial number decreases the development of harmful filamentous bacteria. This is due to diminishing concentrations of compound like liquid volatile suspended solid and means cell residence times (Gerardi, 2006). Next, coli is microorganisms present in fecal wastes and human gastrointestinal tract. Another group of microorganisms present in the substratum levels are saprophytic bacteria. One of the most important degradable constituents that use these microbes is demand for carbonaceous-based biochemical oxygen (cBOD). Cellulomonas a group of bacteria that releases exoenzymes that sever glucose bonds in exchange for cellulose is an example of bacteria within the network. Then, cellulose is a water-soluble compound used for breaking down soil and carbon dioxide. Bacteria produces enzymes that can utilize lipids to regulate the production of foam (Gerardi, 2006).

6.9 GENE BIOAUGMENTATION

In bioaugmentation, it was found that the inoculated bacteria may not survive for a longer time, and there is a chance of horizontal gene transfer into natural microbial flora of contaminated location. Recent advances in DNA sequencing revealed the role of horizontal gene transfer in microbial evolution and adjusting the microbes to highly toxic environment (Ochman et al., 2000). The horizontal gene transfer occurs through mediation by bacteriophage (transduction), by uptake of naked DNA (transformation), or by conjugation, that is, exchange of DNA material such as transposons or plasmids between different bacteria.

The impending benefits of gene bioaugmentation where remediation genes are in mobile form like self-transmissible plasmid over traditional approaches of cell bioaugmentation are:

1. In indigenous microorganisms that are able to proliferate and survive
    in the local environment, the remediation genes are introduced.
2. The long-term survival of the introduced host strain is not required.
    The most studied technology with respect to bioaugmentation is
    the transfer of plasmids, via conjugation (Sessitsch et al., 2002).

The comparison of bioaugmentation using two separate bacterial donors
for delivering plasmid pJP4 containing 2,4-D degrading genes was analyzed
by Newby et al. (155). The pJP4 plasmid was delivered either in *E. coli*
D11 or in original host *R. eutropha* JMP134 (Newby et al., 2000). The
R. eutropha JMP134 was able to mineralize 2,4-D but *E. Coli* D11 was
unable due to the lack of chromosomal genes that along with plasmid genes
cause complete mineralization of 2,4-D. Accordingly, in soils receiving
the *R. eutropha* JMP134, 2,4-D was degraded in 28d. However, it took
49d to degrade in soils receiving the *E. coli* D11 inoculant and nonbioaugmented soils. The numerous transconjugants were detected in the *E. coli*
D11 amended soil whereas most of 2,4-D degraders isolated from the soil
receiving *R. eutropha* JMP134 were identified as the inoculant organisms.
After initial 2,4-D amendment, additional 2,4-D was added to soil and it was
found that 2,4-D was degraded more rapidly in soils receiving *E. coli* D11
than in nonbioaugmented soil and soil receiving *R. eutropha*. Such findings
indicate the potential for degradation of specific pollutants by indigenous
microorganisms if furnished with the requisite genetic material via gene
bioaugmentation. Dejonghe et al. (2000) assessed the dissemination of
two different 2,4-D degradation plasmids in lower and upper horizons of
soil. The addition of an auxotrophic *Pseudomonas putida* strain containing
either of the two plasmids resulted in large transconjugant populations
($>105/g$) in both the lower and upper horizons. Upon adding to the soil,
donor populations decreased, while the growth of transconjugating populations correlated with the 2,4-D degradation. It is found that bioaugmentation
resulted in increased 2,4-D degradation in the B horizon that did not have an
indigenous degrading population; however, the A horizon that had an indigenous degrading population showed least effect. Gene bioaugmentation is
also found significant in metal-contaminated sites (Dong et al., 1998). An
important point to remember is that while considering gene bioaugmentation
technology is that United States Environmental Protection Agency (EPA)
had made a regulatory distinction between different hosts of plasmid even
if they are not genetically engineered (Thomas et al., 2003). EPA considers
microorganisms formed by combining genetic material from organisms of

different genera to be "new" organisms regulated under the Toxic Substances Control Act (TSCA). In case of organisms having mobile genetic elements like plasmids, EPA considers the recipient microorganism to be "new" and hence to be regulated by TSCA in case mobile genetic element was first reported from different generic microorganism. For instance, out of two hosts of pJP4, *E. coli* D11 would be regulated by TSCA as plasmid was transferred from *R. eutropha* JMP134 to *E. coli* D11 (Newby et al., 2000). All the countries do not consider these distinctions and some exclude natural processes particularly natural recombination or mating. It would be better to use the original host for bioaugmentation unless there is a compelling need for another host, in order to avoid these regulations.

6.10 CONCLUSION

Owing to rapid industrialization and urbanization, the world is at extreme risk, resulting in significant environmental problems that include global warming, acid rain, eutrophication, and much more. In addition, the disposal of hazardous waste into the soil and water sources such as rubber, polyethylene, PCB, etc., posed a serious danger to the microbial populations that are responsible for many significant environmental processes. A careful study of the bioaugmentation techniques as applied to different media allows the versatility of the application for various waste management purposes. The challenges currently facing proponents of this approach tend to be their inability to tolerate extreme environments outside sterile laboratories. Nevertheless, bioaugmentation has a sense of security; after all, it involves only the application of microorganisms that have been doing degradation work for millions of years.

KEYWORDS

- **agricultural waste**
- **biotechnology**
- **greenhouse gases**
- **algae blooms**

REFERENCES

Abdulsalam, S.; Adefila, S. S.; Bugaje, I. M.; Ibrahim, S. Bioremediation of Soil Contaminated with Used Motor Oil in a Closed System. *Bioremed. Biodegr.* **2013**, *3*,100–172.

Abdulsalam, S.; Omale, A. B. Comparison of Biostimulation and Bioaugmentation Techniques for the Remediation of Used Motor Oil Contaminated Soil. *Braz. Arch. Biol. Technol.* **2009**, *52* (3), 747–754.

Adebusoye, S. A.; Illori, M. A.; Amund, O. O. Microbial Degradation of Petroleum Hydrocarbons in a Polluted Tropical Strain. *J. Microbiol. Biotechnol.* **2007**, *23* (8), 1149–1159.

Adeel, Z.; Mercer, J. W.; Faust, C. R. Models for Describing Multiphase Flow and Transport of Contaminants (Chap. 1). In *Groundwater Contamination by Organic Pollutants: Analysis and Remediation*; Kaluarachchi, J. J., Ed.; American Society of Civil Engineers: Reston, VA, 2001.

Al-Jawhari, I. F. H. Ability of Some Soil Fungi in Biodegradation of Petroleum Hydrocarbon. *J. Appl. Environ. Microbiol.* **2014**, *2*, 46–52.

Aitken, M. D.; Long, T. C. Biotransformation, Biodegradation, and Bioremediation of Polycyclic Aromatic Hydrocarbons. In *Biodegradation and Bioremediation*; Singh, A., Ward, O. P., Eds.; Springer: Berlin, Germany, 2004.

Aliaa, M El-B.; Eltayeb, K. M.; Mostafa, A. R.; El-Assar, S. A. Biodegradation of Industrial Oil-Polluted Wastewater in Egypt by Bacterial Consortium Immobilized in Different Types of Carriers. *Pol. J. Environ. Stud.* **2016**, *25*, 1901–1909.

Amann, R. I.; Ludwig, W.; Schleifer, K. H. Phylogenetic Identification and In Situ Detection of Individual Cells without Cultivation. *Microbiol Rev.* **1995**, *59* (1), 143–169.

Andreoni, V.; Baggi, G.; Colombo, M.; Cavalca, L.; Zangrossi, M.; Bernasconi, S. Degradation of 2,4,6-trichlorophenol by a Specialized Organism and by Indigenous Soil Microflora: Bioaugmentation and Self-Remediability for Soil Restoration. *Lett. Appl. Microbiol.* **1998**, *27* (2), 86–92.

Anna, J. S.; Katarzyna, P.; Anna Sip, Jerzy, D. Malachite Green Decolorization by the Filamentous Fungus Myrothecium Roridum—Mechanistic Study and Process Optimization. *Bioresour. Technol.* **2015**, *194*, 43–48.

Aranda, E.; Ullrich, R.; Hofrichter, M. Conversion of Polycyclic Aromatic Hydrocarbons, Methyl Naphthalenes and Dibenzofuran by Two Fungal Peroxygenases. *Biodegradation* **2010**, *21*, 267–281.

Arienzo, M. Degradation of 2,4,6-Trinitrotoluene in Water and Soil Slurry Utilizing a Calcium Peroxide Compound. *Chemosphere*, **2000**, *40* (4), 331–337.

Atlas, R. M. Bioaugmentation to Enhance Bioremediation. In *Biotreatment of Industrial and Hazardous Waste*; Levin, M. A.; Gealt, M. A., Eds.; McGraw-Hill: New York, 1993.

Bhattacharya, A.; Gupta, A. Evaluation of Acinetobacter sp. B9 for Cr(VI) Resistance and Detoxification with Potential Application in Bioremediation of Heavymetals-Rich Industrial Wastewater. *Environ. Sci. Pollut. Res.* **2013**, *20* (9), 6628–6637.

Campbell, R.; Clifford, K. *Gulf Spill Is the Largest of Its Kind, Scientists Say.* The New York Times, **2010**.

Cassidy, D. P.; Hudak, A. J. Microorganism Selection and Biosurfactant Production in a Continuously and Periodically Operated Bioslurry Reactor. *J. Hazard. Mater.*, **2001**, *84* (2–3), 253–264.

Cassidy, D. R. Biological Surfactant Production in a Biological Slurry Reactor Treating Diesel Fuel Contaminated Soil. *Water Environ. Res.*, **2001**, *73* (1), 87–94.

Cassidy, D.; Hudak, A. Microorganism Selection and Performance in Bioslurry Reactors Treating PAH-Contaminated Soil. *Environ. Technol.*, **2002**, *23* (9), 1033–1042.

Cherniwchan, J. Economic Growth, Industrialization and the Environment. *Resour. Energy Eco.* **2012**, *34* (4), 442–467.

Cutright, T. A Feasible Approach to the Bioremediation of Contaminated Soil—From Lab-Scale to Field-Test. *Fresenius Environ. Bull.* **1995**, *4* (2), 67–73.

Daprato, R. C.; Zhang, C. L.; Spain, J. C.; Hughes, J. B. Modeling Aerobic Bioremediation of 2,4-Dinitrotoluene in a Bioslurry Reactor. *Environ. Eng. Sci.*, **2005**, *22* (5), 676–688.

de Carvalho, C. C. C. R.; Wick, L. Y.; Heipieper, H. J. Cell Wall Adaptations of Planktonic and Biofilm *Rhodococcus erythropolis* Cells to Growth on C5 to C16 n-Alkane Hydrocarbons. *Appl. Microbiol. Biotechnol.* **2009**, *82* (2), 311–320.

Dejonghe, W.; Goris, J.; El Fantroussi, S.; Hofte, M.; De Vos, P.; Verstraete, W.; Top, E. M. Effect of Dissemination of 2,4-dichlorophenoxyacetic acid (2,4- D) Degradation Plasmids on 2,4-D Degradation and on Bacterial Community Structure in Two Different Soil Horizons. *Appl. Environ. Microbiol.* **2000**, *66*, 3297.

der Geize, R.; Dijkhuizen, L. Harnessing the Catabolic Diversity of *Rhodococci* for Environmental and Biotechnological Applications. *Curr. Opin. Microbiol.* **2004**, *7* (3), 255–261.

Dhouib, A.; Ellouz, M.; Aloui, F.; Sayadi, S. Effect of Bioaugmentation of Activated Sludge with White Rot Fungi on Olive Mill Wastewater Detoxification. *Lett. Appl. Microbiol.*, **2006**, *42* (4), 405–411.

Dong, Q.; Springeal, D.; Schoeters, J.; Nuyts, G.; Mergeay, M.; Diels, L. Horizontal Transfer of Bacterial Heavy Metal Resistance Genes and Its Applications to Activated Sludge Systems. *Water Sci. Tech.* **1998**, *37*, 465.

Environmental Protection Agency. Handbook on In Situ Treatment of Hazardous Waste-Contaminated Soils (EPA/540/2-90/002). Cincinnati, 2004.

Erika, A. W.; Vivian, B.; Claudia, C.; Jorge, F. G. Biodegradation of Phenol in Static Cultures by *Penicillium chrysogenum*erk1: Catalytic Abilities and Residual Phytotoxicity. *Rev. Argent. Mcrobiol.* **2013**, *44*, 113–121.

Filonov, A. E.; Akhmetov, L.; Puntus, I.; Esikova, T.; Gafarov, A. B.; Izmalkova, T.; Sokolov, S.; Kosheleva, I. A.; Boronin, A. The Construction and Monitoring of Genetically Tagged, Plasmid-Containing, Naphthalene Degrading Strains in Soil. *Microbiology* **2005**, *74* (4), 453–458.

Forsyth, J. V.; Tsao, Y. M.; Bleam, R. D. Bioremediation: When Is Augmentation Needed? In *Bioaugmentation for Site Remediation*; Hinchee, R. E., Fredrickso, J., Alleman, B. C., Eds.; Battelle Press: Columbus, OH, 1995; p 14.

Frazar, C. *The Bioremediation and Phytoremediation of Pesticide-Contaminated Sites*; Environmental Protection Agency: Washington, DC, 2000.

Friis, A. K.; Albrechtsen, H. J.; Cox, E.; Bjerg, P. L. The Need for Bioaugmentation After Thermal Treatment of a TCE-Contaminated Aquifer: Laboratory Experiments. *J. Contam. Hydrol.* **2006**, *88* (3–4), 235–248.

Friis, A.; Albrechtsen, H.; Heron, G.; Bjerg, P. Redox Processes and Release of Organic Matter After Thermal Treatment of a TCE-Contaminated Aquifer. *Environ. Sci. Technol.* **2005**, *39* (15), 5787–5795.

Gamati, S.; Gosselin, C.; Bergeron, E.; Chenier, M.; Truong, T.; Bisaillon, J. New Plug Flow Slurry Bioreactor for Polycyclic Aromatic Hydrocarbon Degradation. In *Third Americana Pan-American Environmental Technology Trade Show and Conference*, Montreal, Canada, 24–26 March 1999.

Gerardi, M. H. *Wastewater Bacteria*; Wiley: Hoboken, NJ, 2006.

Ghazali, F. M.; Rahman, R. N. Z. A.; Salleh, A. B. et al. Biodegradation of Hydrocarbons in Soil by Microbial Consortium. *Int. Biodeterior. Biodegrad.* **2004,** *54* (1), 61–67.

Godleads, A.; Prekeyi, T. F.; Samson, O.; Igelenyah, E. Bioremediation, Biostimulation and Bioaugmention: A Review. *Int. J. Environ. Bioremed. Biodegr.* **2015,** *3* (1), 28–39.

Goudie, A. S. *The Human Impact on the Natural Environment: Past, Present and Future*; John Wiley and Sons, 2013; p 424.

Goux, S.; Shapir, N.; Fantroussi, S. Lelong, S.; Agathos, S.; Pussemier, L. Long Term Maintenance of Rapid Atrazine Degradation in Soils Inoculated with Atrazine Degraders. *Water Air Soil Pollut. Focus.* **2003,** *3*, 131–142.

Greer, L. E.; Shelton, D. R. Effect of Inoculants Strain and Organic Matter Content on Kinetics of 2,4-Dichloro-Phenoxyacetic Acid Degradation in Soil. *Appl. Environ. Microbiol.* **1992,** *58* (5), 1459–1465.

Halden, R.; Tepp, S.; Halden, B.; Dwyer, D. Degradation of 3-Phenoxybenzoic Acid in Soil by Pseudomonas Pseudoalcaligenes POB310 (pPOB) and Two Modified Pseudomonas Strains. *Appl. Environ. Microbiol.* **1999,** *65* (8), 3354–3359.

Harshavardhan, K.; Jha, B. Biodegradation of Low-Density Polyethylene by Marine Bacteria from Pelagic Waters, Arabian Sea, India. *Mar. Pollut. Bull.* **2013,** *77* (1–2),100–106.

Hesham, A.; Khan, S.; Tao, Y.; Li, D.; Zhang, Y.; Yang, M. Biodegradation of High Molecular Weight PAHs Using Isolated Yeast Mixtures: Application of Metagenomic Methods for Community Structure Analyses. *Environ. Sci. Pollut. Res. Int.* **2012,** *19*, 3568–3578.

Hudak, A. J.; Cassidy, D. P. Stimulating In-Soil Rhamnolipid Production in a Bioslurry Reactor by Limiting Nitrogen. *Biotechnol. Bioeng.* **2004,** *88* (7), 861–868.

Husserl, J.; Hughes, J. B. Biodegradation of Nitroglycerin in Porous Media and Potential for Bioaugmentation with Arthrobacter sp. Strain JBH1. *Chemosphere,* **2013,** *92* (6), 721–724.

Infante, J. C.; De Arco, R. D.; Angulo, M. E. Removal of Lead, Mercury and Nickel Using the Yeast *Saccharomyces cerevisiae*. *Revista MVZ Córdoba* **2014,** *19*, 4141–4149.

Isken, S.; de Bont, J. A. Bacteria Tolerant to Organic Solvents. *Extremophiles* **1998,** *2* (3), 229–238.

Jimenez, N.; Vinas, M.; Sabate, J. Díez, S.; Bayona, J.; Solanas, A.; Albaiges, J. The Prestige Oil Spill. 2. Enhanced Biodegradation of a Heavy Fuel Oil under Field Conditions by the Use of an Oleophilic Fertilizer. *Environ. Sci. Technol.* **2006,** *40* (8), 2578–2585.

Justice, S.; Higgins, S.; Mack, E.; Griffiths, D.; Tang, S.; Edwards, E.; Löffler, F. Bioaugmentation with Distinct Dehalobacter Strains Achieves Chloroform Detoxification in Microorganisms. *Environ. Sci. Technol.* **2014,** *48* (3), 1851–1858.

Kao, C. M.; Chen, S. C.; Liu, J. K.; Wu, M. J. Evaluation of TCDD Biodegradability under Different Redox Conditions. *Chemosphere* **2001,** *44* (6), 1447–1454.

Karigar, C. S.; Rao, S. S. Role of Microbial Enzymes in the Bioremediation of Pollutants: A Review. *Enzyme Res.* **2011,** 1–11.

Kim, I. N.; Ekpeghere, K.; Ha, S. Y.; Kim, S. H.; Kim, B. S.; Song, B.; et al. (2013). An Eco-Friendly Treatment of Tannery Wastewater Using Bioaugmentation with a Novel

Microbial Consortium. *J. Environ. Sci. Health. Part A Tox./Hazard. Subst. Environ. Eng.* **2013**, *48* (13), 1732–1739.

Kim, J.; Le, N.; Chung, B.; Park, J. H.; Bae, J. W.; Madsen, E.; Jeon, C. Influence of Soil Components on the Biodegradation of Benzene, Toluene, Ethylbenzene and o-, m-, and p-xylenes by the Newly Isolated Bacterium *Pseudoxanthomonas spadix* BD-a59. *Appl. Environ. Microbiol.* **2008**, *74* (23), 7313–7320.

Kolwzan, B. Assessment and Choice of Inoculants for the Bioremediation of Soil Contaminated with Petroleum Products. *Ochrona S'rodowiska* **2008**, *30* (4), 3–14.

Kuyukina, M. S.; Ivshina, I. B.; Ritchkova, M. I.; Philp, J. C.; Cunningham, C. J.;Christofi, N. Bioremediation of Crude Oil-Contaminated Soil Using Slurry-Phase Biological Treatment and Land Farming Techniques. *Soil Sediment Contam.* **2003**, *12* (1), 85–99.

Leadbetter, J. R. Cultivation of Recalcitrant Microbes: Cells Are Alive, Well and Revealing Their Secrets in the 21st Century Laboratory. *Curr. Opin. Microbiol.* **2003**, *6* (3), 274–281.

Leahy, J. G.; Colwell, R. R. Microbial Degradation of Hydrocarbons in the environment. *Microbial Rev.* **1990**, *54* (3), 305–315.

Lebkowska, M.; Zborowska, E.; Karwowska, E.; Miaskiewicz-Peska, E.; Muszynski, A.; Tabernacka, A. et al. Bioremediation of Soil Polluted with Fuels by Sequential Multiple Injection of Native Microorganisms: Field-Scale Processes in Poland. *Ecol. Eng.* **2011**, *37* (11), 1895–1900.

Lee, K.; Kim, K. Heavy Metal Removal from Shooting Range Soil by Hybrid Electrokinetics with Bacteria and Enhancing Agents. *Environ. Sci. Technol.* **2010**, *44* (24), 9482–9487.

Leja, K.; Lewandowicz, G. Polymer Biodegradation and Biodegradable Polymers-a Review. *Polish J. Environ. Stud.* **2010**, *19* (2), 255–266.

Lewis, R. F. Site Demonstration of Slurry-Phase Biodegradation of PAH Contaminated Soil. *J. Air Waste Manage. Assoc.* **1993**, *43* (4), 503–508.

Maliji, D.; Olama, Z.; Holail, H. Environmental Studies on the Microbial Degradation of Oil Hydrocarbons and Its Application in Lebanese Oil Polluted Coastal and Marine Ecosystem. *Int. J. Curr. Microbiol. App. Sci.* **2013**, *2*, 1–18.

Mashreghi, M.; Prosser, J. L. Survival and Activity of Lux-Marked Phenanthrene-Degrading *Pseudomonas stutzeri* P16 under Different Conditions. *Iran. J. Sci. Technol.* **2006**, *30* (1), 71–80.

Massa, V.; Infantino, A.; Radice, F. Orlandi, V.; Tavecchio, F.; Giudici, R.; Conti, Fabio.; Urbini, G.; Di Guardo, A.; Barbieri, P. Efficiency of Natural and Engineered Bacterial Strains in the Degradation of 4-Chlorobenzoic Acid in Soil Slurry. *Int. Biodeterior. Biodegr.* **2009**, *63* (1), 112–115.

McCarty, P. L.; Hopkins, G. D.; Munakata-Marr, J.; Mathwaon, V. C.; Dolan, M. E.; Dion, L. B.; Shields, M. S.; Forney, L. J.; Tiedje, J. M. *Bioaugmentation with Burkholderia cepacia PR1301 for in Situ Bioremediation of Trichloroethylene Contaminated Groundwater* (EPA/600/S-98/001), Gulf Breeze, FL, 1998.

McLaughlin, H.; Farrell, A.; Quilty, B. Bioaugmentation of Activated Sludge with Two *Pseudomonas putidastrains* for the Degradation of 4-Chlorophenol. *J. Environ. Sci. Health. Part A Tox./Hazard. Subst. Environ. Eng.* **2006**, *41* (5), 763–777.

Mohan, S. V.; Sirisha, K.; Rao, N. C.; Sarma, P. N.; Reddy, S. J. Degradation of Chlorpyrifos Contaminated Soil by Bioslurry Reactor Operated in Sequencing Batch Mode: Bioprocess Monitoring. *J. Hazard. Mater.* **2004**, *116* (1–2), 39–48.

Monti, M.; Smania, A.; Fabro, G. Alvarez, M.; Argaraña, C. Engineering Pseudomonas Fluorescens for Biodegradation of 2, 4-dinitrotoluene. *Appl. Environ. Microbiol.* **2005,** *71* (12), 8864–8872.

Mrozik, A.; Piotrowska-Seget, Z. Bioaugmentation as a Strategy for Cleaning Up of Soils Contaminated with Aromatic Compounds. *Microbiol. Res.* **2010,** *165* (5), 363–375.

Nesbo, C. L.; Boucher, Y.; Doolittle, W. F. Defining the Core of Non Transferable Prokaryotic Genes: The Euryarchaeal Core. *J. Mol. Evol.* **2001,** *53* (4–5), 340–350.

Newby, D. T.; Josephson, K. L.; Pepper, I. L. Detection and Characterization of Plasmid pJP4 transfer to Indigenous Soil Bacteria. *Appl. Environ. Microbiol.* **2000,** *66,* 290.

Newcombe, D. A.; Crawford, R. L. Transformation and Fate of 2,4,6-Trinitrotoluene (TNT) in Anaerobic Bioslurry Reactors under Various Aeration Schemes: Implications for the Decontamination of Soils. *Biodegradation* **2007,** *18* (6), 741–754.

Ochman, H.; Lawrence, J. G.; Groisman, E. A. Lateral Gene Transfer and the Nature of Bacterial Innovation. *Nature* **2000,** *405,* 299.

Okuda, T.; Alcantara-Garduno, M. E.; Suzuki, M.; Matsui, C.; Kose, T.; Nishijima, W.; Okada, M. Enhancement of Biodegradation of Oil Adsorbed on Fine Soils in a Bioslurryreactor. *Chemosphere* **2007,** *68* (2), 281–286.

Pandya, M. T. Treatment of Industrial Wastewater Using Photooxidation and Bioaugmentation Technology. *Water Sci. Technol.* **2007,** *56* (7), 117–124.

Paraiba, L. C.; Queiroz, S. C. N.; de Souza, D. R. C.; Saito, M. L. Risk Simulation of Soil Contamination by Polycyclic Aromatic Hydrocarbons from Sewage Sludge Used as Fertilizers. *J. Braz. Chem. Soc.* **2011,** *22* (6).

Partovinia, A.; Naeimpoor, F.; Hejazi, P. Carbon Content Reduction in a Model Reluctant Clayey Soil: Slurry Phase n-Hexadecane Bioremediation. *J. Hazard. Mater.* **2010,** *181* (1–3), 133–139.

Peña-Montenegro, T. D.; Lozano, L.; Dussán, J. Genome Sequence and Description of the Mosquitocidal and Heavy Metal Tolerant Strain *Lysinibacillus sphaericus* CBAM5. *Stand Genomic Sic* **2015,** *10,* 1–10.

Perfumo, A.; Banat, I. M.; Marchant, R.; Vezzulli, L. Thermally Enhanced Approaches for Bioremediation of Hydrocarbon-Contaminated Soils. *Chemosphere* **2007,** *66* (1), 179–184.

Pope, D. F.; Matthews, J. E. Soil Treatment: Land Treatment. In *Bioremediation of Hazardous Waste Sites: Practical Applications to Implementation* (EPA/600/K-93/002). Washington, DC, 1993, April.

Prince, R. Biodegradation of Methyl Tertiary-Butyl Ether (MTBE) and Other Fuel Oxygenates. *Crit. Rev. Microbiol.* **2000,** *26* (3), 163–178.

Quaiser, A; Ochsenreiter, T.; Lanz, C.; Schuster, S.; Treusch, A.; Eck, J.; Schleper, C. Acidobacteria form a Coherent But Highly Diverse Group within the Bacterial Domain: Evidence from Environmental Genomics. *Mol. Microbiol.* **2003,** *50* (2), 563–575.

Quan, X. C.; Hanchang, S. L.; Hong, W.; Jianlong, Q. Y. Removal of 2,4-Dichlorophenol in a Conventional Activated Sludge System through Bioaugmentation. *Process Biochem.,* **2004,** *39* (11), 1701–1707.

Quek, E.; Ting, Y. P.; Tan, H. M. Rhodococcus sp. F92 Immobilized on Polyurethane foam Shows Ability to Degrade Various Petroleum Products. *Bioresour. Technol.* **2006,** *97* (1), 32–38.

Quintero, J.; Moreira, M.; Lema, J.; Feijoo, G. An Anaerobic Bioreactor Allows the Efficient Degradation of HCH Isomers in Soil Slurry. *Chemosphere* **2006,** *63* (6), 1005–1013.

Ramakrishna, M.; Mohan, S.; Venkata, S.; Shailaja, S.; Narashima, R.; Sharma, P. N. Identification of Metabolites during Biodegradation of Pendimethalin in Bioslurry Reactor. *J. Hazard. Mater.*, **2008**, *151* (2–3), 658–661.

Ravi, R. K.; Pathak, B.; Fulekar, M. H. Bioremediation of Persistent Pesticides in Rice Field Soil Environment Using Surface Soil Treatment Reactor. *Int. J. Curr. Microbiol. App. Sci.* **2015**, *4*, 359–369.

Rivelli, V.; Franzetti, A.; Gandolfi, I. Cordoni, S.; Bestetti, G. Persistence and Degrading Activity of Free and Immobilised Allochthonous Bacteria during Bioremediation of Hydrocarbon-Contaminated Soils. *Biodegradation* **2013**, *24* (1), 1–11.

Rodrigues, J. L.; Maltseva, O.; Tsoi, T.; Helton, R.; Quensen, J.; Fukuda, M.; Tiedje, J. Development of a Rhodococcus Recombinant Strain for Degradation of Products from Anaerobic Dechlorination of PCBs. *Environ. Sci. Technol.* **2001**, *35* (4), 663–668.

Ronen, Z.; Vasiluk, L.; Abeliovich, A.; Nejidat, A. Activity and Survival of Tribromophenol-Degrading Bacteria in a Contaminated Desert Soil. *Soil Biol. Biochem.* **2000**, *32* (11), 1643–1650.

Roy, P. K.; Surekha, P.; Tulsi, E.; Deshmukh, C.; Rajagopal, C. Degradation of Abiotically Aged LDPE Films Containing Pro-Oxidant by Bacterial Consortium. *Polym. Degrad. Stab.* **2008**, *93* (10), 1917–1922.

Ruberto, L.; Dias, R.; Lo Balbo, A.; Vazquez, S. C.; Herrnandez, E. A.; Mac Cormack, W. P. Influence of Nutrients Addition and Bioaugmentation on the Hydrocarbon Biodegradation of a Chronically Contaminated Antarctic Soil. *J. Appl. Microbiol.* **2009**, *106* (4), 1101–1110.

Schloss, P. D.; Handelsman, J. Biotechnological Prospects from Metagenomics. *Curr. Opin. Biotechnol.* **2003**, *14* (3), 303–310.

Sessitsch, A.; Gyamfi, S.; Stralis-Pavese, N.; Weilharter, A.; Pfeifer, U. RNA Isolation from Soil for Bacterial Community and Functional Analysis: Evaluation of Different Extraction and Soil Conservation Protocols. *J. Microbiol. Methods.* **2002**, *51*, 171.

Shailaja, S.; Mohan, S. V.; Krishna, M. R.; Sarma, P. N. Degradation of Di-ethylhexyl Phthalate (DEHP) in Bioslurry Phase Reactor and Identification of Metabolites by HPLC and MS. *Int. Biodeterior. Biodegr.* **2008**, *62* (2), 143–152.

Shen, C. F.; Guiot, S. R.; Thiboutot, S.; Ampleman, G.; Hawari, J. Fate of Explosives and Their Metabolites in Bioslurry Treatment Processes. *Biodegradation* **1997**, *8* (5), 339–347.

Shimao, M. Biodegradation of Plastics. *Curr. Opin. Biotechnol.* **2001**, *12* (3), 242–247.

Singh, A.; Kumar, V.; Srivastava, J. N. Assessment of Bioremediation of Oil and Phenol Contents in Refinery Waste Water via Bacterial Consortium. *J. Pet. Environ. Biotechnol.* **2013**, *4*, 1–4.

Sinha, S. N.; Paul, D. Heavy Metal Tolerance and Accumulation by Bacterial Strains Isolated from Waste Water. *J. Chem. Biol. Phys. Sci.* **2014**, *4*, 812–817.

Srirangan, K.; Akawi, L.; Young, M. M.; Chou, C. P. Towards Sustainable Production of Clean Energy Carriers from Biomass Resources. *Appl. Energy* **2012**, *100*, 172–186.

Stocking, A.; Deeb, R.; Flores, A.; Stringfellow, W.; Talley, J.; Brownell, R.; Kavanaugh, M. Bioremediation of MTBE: A Review from a Practical Perspective. *Biodegradation* **2000**, *11* (2–3), 187–201.

Suthersan, S. S.; Payne, F. C. *In Situ Remediation Engineering*; CRC Press: Boca Raton, FL, **2005**.

Tchelet, R.; Meckenstock, R.; Steinle, P.; Meer, J. Population Dynamics of an Introduced Bacterium Degrading Chlorinated Benzenes in a Soil Column and in Sewage Sludge. *Biodegrdation* **1999,** *10* (2), 113–125.

Thomas, J. C.; Davies, E. C.; Malick, F. K.; Endreszl, C.; Williams, C. R.; Abbas, M.; Petrella, S.; Swisher, K.; Perron, M.; Edwards, R.; Ostenkowski, P.; Urbanczyk, N.; Wiesend, W. N.; Murray, K. S. Yeast Metallothionein in Transgenic Tobacco Promotes Copper Uptake from Contaminated Soils. *Biotechnol. Prog.* **2003,** *19,* 273.

Tigini, V.; Prigione, V.; Giansanti, P.; Mangiavillano, A.; Pannocchia, A. Varese, G. Fungal Biosorption, an Innovative Treatment for the Decolourisation and Detoxification of Textile Effluents. *Water* **2010,** *2,* 550–565.

Tribedi, P.; Sarkar, S.; Mukherjee, K.; Sil, A. Isolation of a Novel *Pseudomonas* sp from Soil That Can Efficiently Degrade Polyethylene Succinate. *Environ. Sci. Pollut. Res.* **2012,** *19* (6), 2115–2124.

US Environmental Protection Agency. In-situ groundwater bioremediation (chap. 10). In How to Evaluate Alternative Cleanup Technologies for Underground Storage Tank Sites: A Guide for Corrective Action Plan Reviewers (EPA 510-R-04-002), Washington, DC, 2004.

US Environmental Protection Agency. Use of Bioremediation At Superfund Sites (EPA 542-R-01-019). Washington, DC, 2001.

Valentin, L.; Lu-Chau, T. A.; Lopez, C.; Feijoo, G.; Moreira, M. T.; Lema, J. M. Biodegradation of Dibenzothiophene, Fluoranthene, Pyrene and Chrysene in a Soil Slurry Reactor by the White-Rot Fungus Bjerkandera sp. BOS55. *Process Biochem.* **2007,** *42* (4), 641–648.

Van Hamme, J. D.; Singh, A.; Ward, O. P. Recent Advances in Petroleum Microbiology. *Microbiol. Mol. Biol. Rev.* **2003,** *67* (4), 503–549.

Wang, J. L.; Mao, Z. Y.; Han, L. P. Qian, Y. Bioremediation of Quinoline- Contaminated Soil Using Bioaugmentation in Slurry-Phase Reactor. *Biomed. Environ. Sci.* **2004,** *17* (2), 187–195.

Whitman, W. B.; Coleman, D. C.; Wiebe, W. J. Prokaryotes: The Unseen Majority. *Proc. Natl. Acad. Sci. USA.* **1998,** *95* (12), 6578–6583.

Wittich, R. M.; Wolff, P. Growth of the Genetically Engineered Strain Cupriavidus Necator RW112 with Chlorobenzoates and Technical Chlorobiphenyls. *Microbiology* **2007,** *153* (Pt 1), 186–195.

Wu, Y. H.; Zhou, P.; Cheng, H.; Wang, C. S.; Wu. M. Draft Genome Sequence of *Microbacterium profundi*Shh49T, an Actinobacterium Isolated from Deep-Sea Sediment of a Polymetallic Nodule Environment. *Genome Announc.* **2015,** *3,* 1–2.

Xu, P.; Ma W.; Han H. Hou, B.; Jia, S. Biodegradation and Interaction of Quinoline and Glucose in Dual Substrates System. *Bull. Environ. Contam. Toxicol.* **2015,** *94* (3), 365–369.

Yadav, M.; Singh, S.; Sharma, J.; Deo, S. K. Oxidation of Polyaromatic Hydrocarbons in Systems Containing Water Miscible Organic Solvents by the Lignin Peroxidase of Gleophyllum Striatum MTCC-1117. *Environ. Technol.***2011,** *32,* 1287–1294.

Zhang, J.; Wen, D. Zhao, C.; Tang, X. Bioaugmentation Accelerates the Shift of Bacterial Community Structure Against Shock Load: A Case Study of Coking Wastewater Treatment by Zeolite-Sequencing Batch Reactor. *Appl. Microbiol. Biotechnol.* **2014,** *98* (2), 863–873.

Zhu, X.; Venosa, A. D.; Suidan, M. T.; Lee, K. Guidelines for the Bioremediation of Marine Shorelines and Freshwater Wetlands, 2001.

CHAPTER 7

Biodegradable Waste: Renewable Energy Source

MONICA BUTNARIU[1*] and ELENA BONCIU[2]

[1]Banat's University of Agricultural Sciences and Veterinary Medicine "King Michael I of Romania" from Timisoara, Calea Aradului 119, Timis, Romania

[2]University of Craiova, A.I. Cuza 13, Craiova, Romania

*Corresponding author. E-mail: monicabutnariu@yahoo.com

ABSTRACT

Reducing environmental pollution is a theme of environmental policy, of economic–ecological policy, and is becoming a main theme. This problem is in fact an economic, social, developmental theme, and a human rights theme, which touches all aspects of human and nature. An important aspect in achieving the priorities of the relationship between ecology and economy is the knowledge and awareness of the waste management mechanism, including the biodegradable ones, of the factors that contribute to their generation and of the methods of combating the pollution of the environmental factors. The biomass as a secondary raw material being considered as agricultural waste presents a green solution, which could produce electricity, technological steam, and thermal agent based on the exploitation of high-yield woody species and agricultural biomass. This type of solution is combined with the requirements of green energy.

7.1 INTRODUCTION

Agriculture makes a major contribution to the supply of raw materials for the production of solid fuel, carbon retention, and the reduction of the greenhouse effect. The valorization of biomass is an opportunity for agriculture both in opening up new opportunities and in reducing carbon from agricultural activities. For biofuels, the target of 10% represents a reasonable, ambitious, and future design according to the rural development policy and the legislation regarding the Community agricultural policy (How et al., 2020).

This not only covers the topic of renewable energy but also extends the problem of adaptation to the consequences of climate change, which is the key to the challenges, especially for the agro-food sector. In the context of concerns for improving the quality of the environment, but also for increasing the costs in the energy sector, while reducing fossil fuel resources, identifying and implementing new technologies for energy treatment and recovery of waste has become a priority especially among the of developed countries.

Research and applications regarding the use of unconventional, renewable energy are growing from year to year in the developed countries. The energy crisis, the limited conventional resources, the forecast of their exhaustion in the near future require the finding of new alternatives in time. The biogas, although it may not seem to have a special weight, will still have its share of contribution. In parallel with the production of biogas, the manure from fermentation constitutes a very good organic fertilizer, comparable qualitatively with humus. Simply put, solid fuel is made from valuable vegetable biomass (Purohit et al., 2016).

Due to this increased accessibility to bioregenerable energy sources, solid biodegradable fuels are a solution to reduce the country's dependence on the import of energy resources, and from a strategic point of view, harnessing the potential of renewable raw materials must become a starting point for energy and energy policies. An energy species can be the energetic willow, which has an alternative energy source similar to fossil fuels such as coal, fuel, oil, etc. The big difference between willow and coal are the pollutant emissions released into the atmosphere. Burning willow in raw or pelleted form has near-zero emissions.

The biological waste is defined as biodegradable waste from gardens and parks, food or kitchen waste from households, restaurants, catering companies, or retail stores, as well as similar waste from food processing plants. These do not include forest or agricultural residues, manure, sewage sludge or other biodegradable waste, such as natural textile materials, paper or processed wood (How et al., 2020).

The definition does not include those by-products from the food industry that never become waste. According to estimates, the total annual amount of biological waste in the EU is between 76.5 and 102 Mt for food and garden waste that is part of mixed municipal solid waste, reaching up to 37 Mt for waste from the food and beverage industry.

Biological waste is rotten and generally humid. There are two major streams of waste—vegetable waste from parks, gardens, etc. and kitchen waste.

Vegetable waste usually contains between 50% and 60% water and more wood (lignocellulose), and kitchen waste does not contain wood, but contains 80% water.

Biological waste management options include, in addition to prevention at source, collection (separately or in combination with mixed waste), anaerobic digestion and composting, incineration, and storage of waste (Purohit et al., 2016).

The ecological and economic benefits of the different treatment methods depend significantly on local conditions, such as demographic density, infrastructure and climate, as well as existing markets for associated products (energy and compost).

Currently, Member States apply very different national waste management policies, some taking very little action and others adopting ambitious policies. This can aggravate the negative impact on the environment and prevent or delay the maximum use of biological waste management techniques.

The Green Paper aims to explore options for improving biological waste management. It contains a summary of important general information regarding current biological waste management policies, as well as new research findings in the field, outlines key issues to be discussed, and invites stakeholders to contribute by sharing their knowledge and views with on the path to follow.

The Green Paper aims to prepare a debate on the possible need for future policy measures, trying to gather opinions on improving biological waste management, taking into account the hierarchy of waste management options, the possible economic, social and environmental benefits, as well as the most effective policy tools needed to achieve this goal.

As indicated in the Green Paper, there are great difficulties and uncertainties regarding the data regarding the options for biological waste management (Sharma et al., 2020).

Therefore, the Commission wishes to invite all interested parties to provide the available data to facilitate the elaboration of the subsequent impact assessment on the various options for managing biological waste.

7.2   BIODEGRADABLE WASTE—RENEWABLE SOURCE

Biomass is the biodegradable part of agricultural products, waste and residues, including plant and animal, forestry and related industries, as well as the biodegradable part of industrial, municipal and household waste from peasant households.

This is one of the most important renewable energy resources of the present, as well as of the future, due to the great potential and different benefits offered economically, socially, and ecologically.

In order to prevent an energy crisis, biomass becomes a solution of national energy utility. It entails the property of regeneration itself, representing all biodegradable products and waste. Biomass contains stored chemical energy, which derives from solar energy and is formed by the process of photosynthesis, when plants store solar energy in the form of chemical compounds. Today, the biomass is a generic term that refers to any organic matter of plant and/or animal origin, available and regenerates through natural processes or as a product/by-product of a human action (Fatma et al., 2018).

7.2.1   BIOMASS—AN IMPORTANT ENERGY SOURCE

Biomass is the biodegradable part of agricultural products, waste and residues, including plant and animal substances, forestry and related industries, as well as the biodegradable part of industrial and urban waste.

Organic matter of biological origin. A cogeneration thermal power plant can operate with solid biomass, more precisely with any vegetable organic matter, metabolic residues of animal origin (garbage), as well as microorganisms.

For example, agricultural biomass includes by-products of plants, such as straws, chocolates, stems (sunflower, soy), leaves (beet), pods (soy, beans), shells (nuts, peanuts) and seeds (plum, peach, and garbage from animal farms.

Besides the sources of agricultural biomass, biomass wastes are also produced by forestry, the primary and secondary material waste produced from the exploitation of the forests and due to the plantations of resinous and deciduous species (Shikinaka et al., 2018).

It is important to mention that biomass represents organic matter of biological origin. Therefore, it represents the organic component of nature. This energy resource is also an important source of renewable energy, playing an important role in the context of strategies for achieving energy independence through the use of renewable energy. The promotion of biomass as an energy source is a priority of economic, energy, and environmental policies. Biomass serves as a raw material also in the formation of biogas.

The technologies used in the EU for the production of biogas from biomass and organic waste are also welcome for our republic. These technologies would contribute positively to solving many problems in rural areas, such as disposal of biodegradable waste, including animal waste, electricity and thermal energy supply, etc.

According to some studies, the available technologies allow the application of the cost-efficient methods of biogas production from the individual peasant farms to the large livestock farms, including the poultry factories.

The first such technologies are about to be implemented.

Waste obtained after anaerobic fermentation will have excellent organic fertilizers for soil fertilization, which will be produced in large quantities, such as solids (50 tons/24 h) and liquids (373 tons/24 h). Great attention will also be paid to bioethanol production. As a raw material for the production of bioethanol, sugar sorghum can be used.

This crop can ensure a crop harvest of 70–80 tons/ha in a period of 70–80 days, with a sugar content of 13–18% of the stem mass. From the amount of sorghum harvested per hectare, approx. 4–5 tons of bioethanol or bioethanol is produced 2–3 times more compared with cereal crops. To produce 1 ton of ethanol, 2–4 tons of dry wood material or approx. 3 tons of cereals is required.

Approximately, 450 L of bioethanol is obtained from 1 ton of dry corn. Different technologies will be used in the production of bioethanol from cereals and other materials, which differ according to the energy sources used in the conversion processes.

This mixed fuel production complex is the only one of its kind in the republic and its success will set the beginning of a development phase in the field of using renewable waste. Previously, other attempts were made to produce biogas from different wastes.

Thus, with the financial assistance granted by the Netherlands in 2002 (Netherlands Program for Cooperation with Central and Eastern Europe), the first industrial plant for the production of biogas was built on the poultry farm, the capacity of the fermenter being 700 m³. The captured biogas was used for a limited time as a cogeneration fuel—87 kW electric powers and 116 kW thermal powers (European Commission, 2012).

Unfortunately, the system went out of operation, was dismantled and eventually considered as used metal.

7.2.1.1   THE SOURCES, CLASSES AND FORMS OF BIOMASS

Obviously, the biomass generating sources are the sectors of economic activity, but also individual, household, from which the biodegradable waste results.

Cereal crops, sugar plants, oil plants are important sources of biomass. The biomass classification is performed depending on the sectors of origin, the agricultural crops practiced, the forms and the fields of use. Below are presented the main branches of the economy, whose activity results in biomass generation.

Forestry and wood processing industry: Trees, trunks, wood processing wastes, stumps; bark, forest biomass resulting from tree dressing, wood residues, fibrous residues from the pulp and paper industry, chemically untreated fibrous residues.

Agriculture and the agro-food industry: Agricultural products (grains, seeds, pods, roots, etc.), agricultural wastes (straw, maize stalks and corn, sunflower stalks, shells, etc.), agro-food waste, waste from industrial plants.

Animal husbandry: Animal waste (Ozbayram et al., 2020).

Communal household: Solid and liquid household waste, organic waste from industrial processes.

7.2.1.2   TYPES OF BIOMASS ACCORDING TO THE SOURCES AND SECTORS OF PRODUCTION

Forest biomass: Forests and plantations: Trees, trunks, wood processing wastes, bark, forest biomass resulting from tree dressing, etc.

Grass biomass: Agricultural and horticultural plants: Cereal, oilseed, root crops, vegetables, herbs, flowers, grass biomass from land management.

Fruits biomass: Fruit from orchards, horticulture, fruit trees, fruits with pips or kernels, industry of processing of plants and fruits, products, and waste, grass residues, fruit residues.

As biomass resources represent diversity, the classification of biomass can also be done from the point of view of primary, secondary, and tertiary waste and biomass grown exclusively for energy purposes.

- Primary wastes (residues) are produced from plants or forest products. Such biomass is available "in the field" and must be collected for its later use.
- Secondary waste is a product of biomass processing in the agro-food and timber sectors and is available at paper or food processing factories, etc.
- Tertiary waste becomes available after a biomass product has been used. These represent wastes that vary from the point of view of the organic fraction, including household waste, wood waste, wastes from wastewater treatment, etc.
- Forestry waste includes unusable waste, commercially imperfect trees, dry trees, and other nonmarketable trees and must be cut to clean the forest. Cutting down trees in the forest leads not only to the healing of the forest, but also to the production of residues that can be used to produce energy.

Due to the fact that these residues are spread over large areas and in difficult to reach places, they are difficult to recover and the costs are high.

Some species of energy plants are in the category of woody biomass, for example, trees that grow very fast. The harvesting period of such plants varies between 3 and 10 years depending on the species of the tree, and the period between two plantations can be even more than 20 years. The willow is a good example of a plant for a short rotation of the plantation that can be harvested every 2–5 years over a period of 20–25 years (long-lasting plant) (Karmee et al., 2015).

The energy crops. Trees with high growth speed: poplar, willow, eucalyptus. Agricultural crops: sugar cane, rapeseed, sugar beet. Perennial crops: *Miscanthus*. High-growth herbaceous plants: *Switchgrass* or *Panicum virgatum* (a perennial growing in North America), *Miscanthus* or elephant grass (Ugandan grass).

Residues. Wood from the dressing of trees and from constructions. Straws and stalks of cereals. Other residues from the processing of some food products (sugar cane, tea, coffee, nuts, olives).

Waste and by-products. Wastes from wood processing: Shavings, sawdust, paper waste, organic fraction of municipal waste, used vegetable oils and animal fats, methane captured from landfills, wastewater treatment plants and from manure.

As the development of the bioethanol industry in cereals has led to their price increase, research is promoted to obtain biofuels from lignocelluloses biomass (straw, cocoons, nonfeed and nonfood plants, etc.), or from manure and waste (manure, wastewater, waste city, industrial waste, etc.).

The following biofuels were called the second generation of biofuels.

Solid biofuels are obtained from the simplest, from valuable plant biomass. There are fixed, or even mobile, briquettes (pellets) production equipment that convert the cellulose waste (sawdust, straw, other plant products, not otherwise used, or is simply burned in the field without using released energy) into a saleable merchandise.

Liquid biofuels are biodiesel and bioethanol. Biodiesel is made very simply from oil plants. In contrast, second-generation bioethanol (the one obtained from cellulose, not from cereals), requires a more complex manufacturing process (Hašková, 2017).

7.2.1.3   WET BIOMASS AND DRY BIOMASS

The wet biomass is a biomass with relatively high water content and low lignin content. Wet biomass is suitable for the production of biogas by anaerobic conversion (fermentation) due to the composition properties.

The dry biomass is a biomass with high lignin content and low water content. This type of biomass is not suitable for anaerobic treatment for the purpose of biogas production, because the lignin content cannot be converted anaerobically (Pelletier et al., 2017).

Due to the low water content, these wastes are ideal for thermal recovery, the most representative examples being forest and agricultural waste (straw, plant stems, etc.).

7.2.1.4   THE CONVERSION OF BIOMASS INTO ENERGY

The potential of energy crops depends on the available land surfaces, environmental protection policies and sustainable soil and water management.

Biomass can be used to produce car fuels, electricity, and heat.

The biomass resources are divided into three categories: primary, secondary, and tertiary. Primary resources are produced directly by photosynthesis and include perennial plants, woody crops, herbaceous plants, oilseed plant seeds, residues from the exploitation of agricultural and forestry crops (straw, corncobs, and tree bark).

Secondary resources are the result of the processing of primary biomass resources, either physically (sawdust) or chemically (the liquid obtained from cellulose processing) or by biological method (manure). Tertiary biomass resources are continuous flows resulting from consumption: animal fats, waste vegetable oils, and packaging waste.

Figure 7.1 presents the most common categories of biomass from agriculture, forestry, and organic waste and conversion pathways.

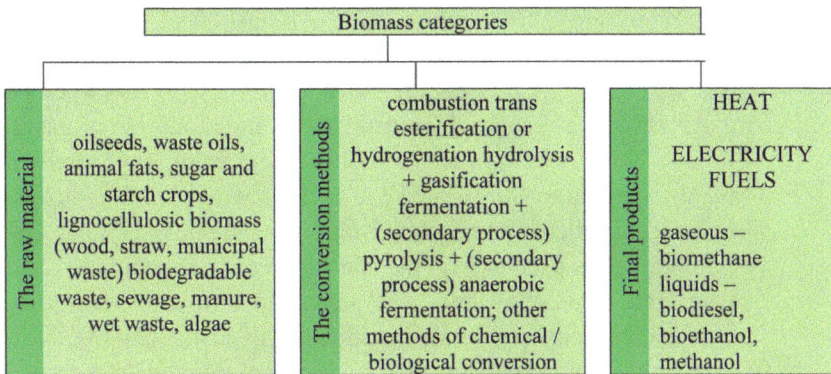

FIGURE 7.1 The biomass conversion into energy.

Biomass can be transformed into biofuels by different methods, depending on its physicochemical properties (Vendruscolo et al., 2016).

The most common conversion methods consist of refining, hydrolysis, pyrolysis, extraction, fermentation, gasification, liquefaction processes.

7.2.2 METHODS OF BIOMASS HARNESSING FOR ENERGY PURPOSES

The biomass recovery is done by conversion and combustion, the latter being the most widespread use of solid biomass. This process includes the production of solid, liquid, and gaseous biofuels or intermediate fuel products, comprising a number of technologies, or direct combustion of biomass for the purpose of heat and electricity generation.

The methods of using biomass for energy purposes, in terms of conversion processes, are the following:

- Physical—grinding, separation, drying, pressing, briquetting.
- Thermochemical—combustion, pyrolysis, gasification, hydrogenation.
- Chemical conversion—esterification.
- Biochemical—anaerobic, aerobic, alcoholic fermentation (Gerber Van Doren et al., 2017).

Thermochemical conversion includes a number of complex biomass degeneration reactions under certain conditions:

a) In the ordinary sense, it is the most widespread way of producing biomass energy. The combustion or combustion of biomass is a thermochemical process with the release of heat and light. The main stages of this process are—drying, formation of carbonate by pyrolysis, gasification by burning coal and oxidation of gases. Biomass combustion is used to produce heat in small or medium capacity installations (<3–5 MW), such as wood burning stoves, log boilers, pellet burners, wood chip ovens, straw boilers. The heat obtained is used to heat the spaces, to produce hot water and steam, and to prepare food. The steam, in turn, can be used to produce electricity within the thermoelectric power stations.

b) Gasification. Subject to the thermochemical conversion process, the solid biomass is converted into combustible gases, resulting in the synthesis or syngas (the generator gas). This syngas contains carbon oxide (CO or tile gas, popular), hydrogen, methane, and inert gases such as nitrogen. It serves as a fuel for the production of electricity, heat, and can also be used to produce other types of fuel.

Gasification is an old technology, in the years 1840–1860, the first successful commercial gas station was built, and in 1900–1920, a large number of gas installations were produced and sold to produce electricity. Gasification technology represents an energy alternative for rural areas because it involves the production of electricity in small and medium installations based on solid biomass. The solid biomass is burned in cogeneration or trigeneration plants to produce electricity, the only energy obtained from gasification (Chan et al., 2019).

c) Pyrolysis is a thermochemical process of decomposition of organic matter at high temperatures and pressures, in an inert environment, in the total or partial absence of the oxidizing agent (air, oxygen). In practice, this process is also called degassing. Under the effect of the high temperature, splitting and different structuring of the organic molecules occur, after pyrolysis, the biomass is transformed into gaseous fuels (syngas), liquids (pyrolysis oil), and solids (manganese).

d) The physicochemical conversion includes the production of vegetable oil from seeds and the esterification of this oil is carried out until a methyl ester fatty acid, a substitute for diesel fuel is obtained. This technology is used in the production of biodiesel, being a basic one.

e) Biochemical conversion includes biomass energy conversion technologies based on the application of biological processes. These include:

- Aerobic fermentation (ethanol production).
- Anaerobic fermentation (biogas production).

Anaerobic fermentation is a biochemical process that occurs naturally on Earth and consists of the decomposition of organic materials under the action of microorganisms in the absence of oxygen.

The gas mixture resulting from this process is known as biogas and mainly contains 50–75% CH_4 and 20%–45% CO_2, together with low percentage of H_2S, N_2, H_2, and O_2.

In industrial practice, the anaerobic fermentation process takes place in special reactors (anaerobic digesters) under controlled working conditions.

At the global level, projects for the anaerobic treatment of solid municipal waste initially appeared to solve the problem of these wastes that present a high level of humidity and salt, which does not make them suitable for disposal by methods, such as incineration or composting. Due to the high degree of rotting, the organic waste generated in the home can provide a good feed substrate for anaerobic fermentation reactors (Chen et al., 2015).

In order to be recycled, it is necessary to first separate the organic waste at the source. The experience of the states that have developed this method of treatment/recovery of municipal waste has indicated that by separating at source the fermentable fraction, the highest quality organic substrate for the populations of bacteria involved in the process is ensured. In addition, the resulting material, since it does not contain indigestible components for bacteria, such as plastic, glass, fine metal debris, is a good fertilizer and conditioner for the soil, especially after a subsequent aerobic treatment.

Anaerobic fermentation is a very complex biochemical process, which involves several different types of bacteria working together to break down the complex organic substances existing in waste to the final gaseous products (CH_4, CO_2, and H_2O).

Anaerobic fermentation consists mainly of the following four biochemical stages:

1. Hydrolysis, carried out by bacteria that convert insoluble carbohydrates, proteins, and fats into simple sugars, fatty acids, amino acids, and peptides.
2. Acidogenesis, where acidogenic bacteria transform hydrolysis products into simple organic acids, alcohols, CO_2, and hydrogen.
3. Acetogenesis, in which acetogenic bacteria convert fatty acids with more than two carbon atoms into acetate and hydrogen.
4. Methanogenesis, the final step in the anaerobic fermentation process, in which methanogenic bacteria produce biogas from acetic acid or from hydrogen and CO_2.

Acidogenic bacteria reproduce very quickly and are not very sensitive to physical changes in the environment (temperature, acidity), compared with methanogenic bacteria that reproduce much slower and are very sensitive to such variations.

Methanogens cease their enzymatic activity or even die if the conditions in the fermentation reactor are not adequate (Soo et al., 2016).

When methanogens begin to disappear from the organic mass, the acidogenic bacteria continue to multiply, increasing the volatile acid concentration and implicitly decreasing the pH of the reactor mass, stopping the production of biogas.

Although anaerobic fermentation is a process that occurs naturally in the marshy areas since the formation of the planet, the technological process of obtaining the biogas requires special attention and monitoring in order to maintain optimal conditions of temperature, pH, humidity, agitation in the reactor, and composition of the organic mass, with a view to complete the decomposition of organic substances until the final stage of biogas production.

The treatment of the biodegradable municipal waste by anaerobic fermentation has net advantages compared with the other ways of treating and eliminating the waste.

In addition to the benefits to the environment and the health of the population by preventing contamination of environmental factors, such as water–air–soil, reducing the outbreaks of infection and eliminating unpleasant odors, this method of waste treatment generates products with high energy and economic value, such as biogas and eco-friendly fertilizing materials.

The most important advantage offered by this method of anaerobic treatment of municipal organic waste is the degradation of the organic fraction, that is, of degradation of biomass with the help of special bacteria in the absence of air or oxygen (Steinberg et al., 2017).

The biological degradation of the organic fraction includes the following phases:

- Hydrolysis: Organic macromolecules, such as proteins, fats, or cellulose are cleaved into smaller molecules, such as saccharides, amino acids, fatty acids, and water. The process is possible due to the activity of hydrolytic microorganisms.
- Acidogenesis: Small organic molecules (sugars, amino acids, fatty acids, and water) are cleaved into organic acids, carbon dioxide, hydrogen sulfite and ammonia.
- Methanogenesis: Methane, carbon dioxide, and water are produced from acetates.

Anaerobic fermentation of biomass produces biogas with a methane content of 60–70%, which is used as fuel for cogeneration and trigeneration energy units (electricity production, heat, and cooling agent).

Following the anaerobic treatment, the commercial compost and the water are reused for irrigation or waste in rivers and lakes without harming the environment. This process represents not only a technology for producing energy, but also for waste treatment.

f) Cogeneration represents the combined production of thermal and mechanical energy based on a single fuel. Mechanical energy is often used to drive an alternator and produce electricity. Thermal energy is used for the production of hot water or steam. For certain applications, it allows the simultaneous production of heat and cold.

g) Trigeneration can be applied mainly in hospitals, clinics, airports, offices, governments, and heating networks, and in industries using hot water, steam or hot air, and in greenhouses (Rosero–Henao et al., 2019).

7.2.3   *PRODUCTION OF ENERGY FROM BIOMASS*

Biomass can be converted into energy or biofuels through a variety of processes and technologies. Many of these are applied in the clean energy industry, but others are in the nascent phase, that is, they are trying to prove the usefulness of these technologies.

On the market, there are combustion installations for solid biomass (pellets, briquettes), biogas, bioethanol, biodiesel, for syngas, and synthetic fuels production (synthetic natural gas, synthetic ethanol, etc.).

7.3   THE SOLID, LIQUID, AND GASEOUS BIOFUELS FROM BIODEGRADABLE WASTE

The term biofuel refers to the solid, liquid, and gaseous fuels that are produced from biomass. Biofuels are nonpolluting, locally accessible, sustainable, and bioregenerable.

7.3.1   *PRODUCTION OF BIOFUELS*

The recent global energy crisis has set in motion the international scientific community. The price of crude oil is increasingly difficult to control. Therefore, new ways of obtaining natural fuels are being sought. The most sustainable solution is the replacement of conventional fuels with biomass obtained from biomass. Although it is part of the largest biofuel producer group, the EU is far ahead of countries like Brazil or the US. In 2003, 30 years after Brazil launched the Pro Alcohol program, the EU established the legal and fiscal framework to encourage the production and use of biofuels in the member countries.

The legislative package consists of two directives.

The Promotional Directive has set indicators and targets to encourage Member States to use 2% biofuels from total consumption by 2006 and 5.75% by 2011.

The second directive refers to the taxation of energy products. Member States shall exempt, in whole or in part, products containing bioregenerable substances. In 2002, fuel alcohol accounted for about 15% of the alcohol produced in the EU (396 million liters) and was produced only in three countries: Spain (56%), France (30%), and Sweden (14%).

If in the EU it was necessary to add at least 5.75% biofuel by 2011, in Brazil, for example, the law required the introduction of at least 26% ethanol in gasoline.

However, in Brazil, ethanol is produced from bagasse (exhausted sugar cane, resulting from the extraction of sugar), and only a small part, from corn).

Regarding the promotion of biofuels in the EU, the economic factors and the long- and medium-term economic policies, by granting subsidies and funds for the use of renewable energy sources, create an economic context favorable to increase their degree of energy use (European Environment Agency, 2006).

7.3.2   THE IMPORTANCE OF BIOFUELS

The data below show the importance of using biofuels, specifying the sectors in which biofuels have a certain weight and effects for each sector.
Energy security:

a) Increasing energy security by diversifying used energy resources and limiting dependence on imported energy resources.
b) Reduction of imports of petroleum products.
c) Enlargement of the energy resource base by using the national potential of bioenergetics resources.
d) Increasing the efficiency level of the technologies.

Durability:

a) Improving the energy efficiency of technologies for the production and use of biofuels.
b) Rational use of the country's soil resources without a negative impact on biodiversity and drying of carbon-rich soils.
c) Reduction of greenhouse gas emissions throughout the production and use cycle of biofuels (The first step was the signing of the Kyoto Protocol, which implies a 70% reduction in greenhouse gas emissions).
d) Use of agro-food, municipal waste, and forestry waste.
e) Supporting the research and development, dissemination, and technological transfer activities of the results of the applicable researches regarding the production and use of biofuels.

Competitiveness:

a) Developing competitive markets for fuels in accordance with their environmental impact throughout the life cycle.
b) Widening the range of energy plants grown in accordance with pedoclimatic conditions.
c) Development of technologies for the cultivation of energy plants by maximizing the level of production and increasing the energy efficiency.
d) Optimization of the technologies of generation of biofuels of generation I (biodiesel and biogasoline).
e) Development of second-generation fuel production technologies.

Socioeconomic development of rural areas:

a) Use at full capacity of the agricultural potential existing in rural areas.
b) Development of small and medium-sized enterprises in rural areas.
c) Market expansion of agricultural products (food and nonfood).
d) Increasing the degree of employment available in rural areas.
e) The promotion of a renewable energy market will contribute to the labor employment growth and the efficiency of the efforts in the field of research and innovation (Kumar et al., 2018).

Types of biofuels:

According to Directive EC/2003/30 of the Council and the European Parliament of 8 May 2003, regarding the promotion of the use of biofuels or other renewable fuels for transport, biofuels are of the following types:

Bioethanol: Ethanol extracted from biomass and/or from the biodegradable part of the waste, which can be used as biofuel.

Biodiesel: A methyl ester extracted from vegetable or animal oil, of diesel quality, which can be used as biofuel.

Biogas: A gaseous fuel resulting from biomass and/or from the biodegradable part of the waste that can be purified to the quality of the pure gas, which can be used as biofuel or wood gas.

Biomethanol: Dimethylester extracted from biomass for use as a biofuel.

Biodimethylester: Dimethylester extracted from biomass for use as a biofuel.

Bio-ETBE (ethyl-tert-butyl-ester): ETBE is a bioethanol product. The percentage by volume of bio-ETBE considered as biofuel is 47%.

Bio-MTBE (methyl-tert-butyl-ester): A fuel based on biomethanol. The percentage by volume of bio-MTBE considered as biofuel is 36%.

Synthetic biofuels: Synthetic hydrocarbons or mixtures of synthetic hydrocarbons that have been extracted from biomass.

Biohydrogen: Hydrogen extracted from biomass and/or from the biodegradable part of the waste, for use as a biofuel.

Pure vegetable oil: Oil produced from oily plants by pressing, extraction or comparable procedures, raw or refined, but not chemically modified, when it is compatible with the engines to which it is used and when it is in accordance with the requirements of the rules regarding the NOx emissions (Oriez et al., 2019).

Biofuels can also be classified as follows:

A. **Biofuels of the generation I.** These biofuels are made from carbohydrates, starch, vegetable oil, and animal fats by conventional technologies.

     The major disadvantage of these biofuels is that the biomass used is common to that used for food.

B. **Biofuels of the generation II.** Biomass is represented by lignocellulosic materials obtained by harvesting nonfood plants and residual biomass.

     Thus, cellulose bioethanol, synthetic biofuels, biogas from lignocellulosic material, and biohydrogen from lignocellulosic material can be produced.

C. **Biofuels of the generation III.** Biomass is represented by genetically modified raw materials: oil plants with high oil productivity, woody biomass with a lower lignin content to improve the processing process. Specialists have added new poplar species with lower lignin content to improve the processing process.

Researchers have already made the genetic map of sorghum and maize, which allows agronomists to modify genetic information in order to regulate oil production.

Some American company that has been developing such plant varieties for years.

The American company is developing varieties of trees that are intended for the production of biofuel and timber (Albanez et al., 2016).

The biomass growth using microorganisms (such as phytoplankton, micro–algae, bacteria) to produce lipids for conversion to biodiesel is done in open basins, photobioreactors or hybrid systems. The CO_2 produced in power station and industrial installations can be used to fuel the process (CO_2 recycling and bio–fixing).

In Tables 7.1 and 7.2 are presented the biofuels of generation I and II, as well as the raw materials and production processes used.

TABLE 7.1   Biofuels of the I–Generation.

Biofuel	Specific name	Raw materials	Production process
Bioethanol	Conventional Bioethanol	Sugar cane, cereals	Hydrolysis and fermentation
Vegetable oil	Pure plants oil	Oilseed plants	Cold pressing
Biodiesel	Rapeseed methyl ester (RME), methyl/ethyl ester of fatty acids (FAME/FAEE)	Oilseed plants	Cold pressing + transesterification
Biodiesel	Biodiesel from waste	Burnt cooking oil, animal fats	Transesterification
Biogas	Cleaned biogas	Biomass (wet)	Fermentation
Bio-ETBE		Bioethanol	Chemical synthesis

The main product is biofuel obtained from algae. Other products are bioethanol from forest plant crops rotated by cellulose hydrolysis, biooil or biodiesel from algae cultivation using CO_2 from thermoelectric power stations, biodiesel from biomass waste gasification, bio-n-butanol from biomass fermentation (as co–solvent for ethanol / methanol–gasoline mixtures or as a chemical). The third-generation biofuel is based on technologies that are not yet commercialized. They will require new infrastructure including distribution networks, power stations and machines, as well as political and technical support once they are placed on the market (Wang et al., 2017).

TABLE 7.2 Biofuels of the II–Generation.

Biofuel	Specific name	Raw materials	Production process
Bioethanol	Cellulose bioethanol	Lignocellulose	Hydrolysis and synthetic fermentation
Biofuels	Biomass to liquid (BTL):	Lignocellulose	Gasification and synthesis
Biodiesel	Hydro-treated biodiesel	Vegetable oils	Hydro-treatment
Biohydrogen		Lignocellulose	Gasification and synthesis or biological processes

D. **Biofuels of the generation IV.** The fourth-generation technology combines genetically optimized feedstock, which is developed to capture large amounts of carbon with the help of genetically modified microorganisms, which are designed to increase fuel efficiency. Biomass is based on cross-breeding or genetically modified crops that specifically absorb very high amounts of CO_2. Biohydrogen is obtained from the fermentation of the selected biomass and biohydrogen from the photolysis of water using microorganisms as catalyst. These biofuels can be obtained by rapid pyrolysis—technology that uses burnt biomass at 400–600°C in the absence of air (Grover et al., 2014).

7.3.2.1 TECHNOLOGICAL CHARACTERISTICS OF OBTAINING BIOFUELS

Biofuels are divided into three broad categories: solids, liquids, and gases.

Solid biofuels. Many solid plant materials can provide thermal energy through burning. From the point of view of provenance, solid biofuels can be classified into combustible wood materials, cereal straws, corncobs, and cereal grains.

The most representative categories of combustible wood materials are firewood, tree bark, branches from forest exploitation, branches cut from the maintenance of orchards, vines, sawdust, logs, small pieces of lumber, and other residues from wood processing. Usually, trees from logging are a high-quality homogeneous biofuel. From the energy point of view, the combustible wood materials have average energy content between 14 and 19 MJ/kg.

A very important category of solid biofuels used for the production of thermal energy by combustion is straw.

Their energy content is quite high and varies depending on the humidity: 14.5 MJ/kg at 15% humidity, 12.6 MJ/kg at 25% humidity, 10.8 MJ/kg at 35% humidity. The corncobs used as biofuel to obtain thermal energy have a good energy content of about 18.5 MJ/kg.

The energy power of the corncobs varies between 15.3 and 21.7 MJ/kg, depending on the humidity. In recent years, cereal grains have also started to be used as biofuels for the production of thermal energy. The caloric value of cereals is between 3.95 and 4.28 kWh/kg, depending on the type of cereal, which means that 2.5 kg of grains can replace approximately 1 liter of liquid fuel when heated (Valdivia et al., 2016).

The following negative effects are possible as a result of burning fossil fuels: air pollution, water pollution, soil pollution due to solid residues, and sonic pollution of the surroundings. The size of the pollution depends on the fuel used to generate the thermal energy on the one hand, as well as on the ways of burning the fuels on the other.

Solid fuel in the form of pellets and briquettes that are produced from agricultural residues (straw, corncobs, corn stalks, soybean residues, rape and tobacco, vine shoots, and technological remnants of orchard mainte-nance, etc.) is an environmentally friendly fuel and represents an efficient alternative to conventional fuels for thermal boilers (natural gas, liquid fuel, coal, firewood, etc.).

The major difference, compared with the classic ones, is the small size and the regular shape of the pellets, which allows their use as a fuel for the automated thermal power stations.

The advantages of this fuel are:

- The problem of environmental pollution is solved with sawdust and wood waste or by burning stubble and vegetable debris.
- The agricultural dry biomass and energy plants represent an inex-haustible resource of raw materials.

The production of pellets and briquettes (environmentally friendly, nonpolluting products) is undertaken with the application of a technology with a high degree of mechanization, low manufacturing costs; it allows to obtain the thermal energy at advantageous costs.

The main potential beneficiaries of the results of the application of the mentioned technologies will be the small and medium farmers, the associations of owners of agricultural lands, the economic agents that carry out activities in the agricultural field and who want to provide partially or totally the thermal energy by using their own sources of renewable energy, the marketing of pellets and briquettes in order to ensure an energy independence.

Pelletizing represents the operation of conversion into fuel of agricultural biomass and energy plants specially prepared by the component equipment of the technological flow of manufacture, being performed by extrusion that is by the forced and continuous passage of a very large quantity of material through a very small orifice (Wang et al., 2018).

This technology for the manufacture of pellets and briquettes requires different special equipment in the technological flow of manufacture for chopping, sorting, drying, transporting, and pelleting/briquetting of solid agricultural biomass or energy plants. Of the energy plants, cultivated especially for obtaining the thermal energy by combustion, the energy willow (*Salix viminalis*) is considered the most efficient, having a calorific power of 20.5 MJ/kg. Other energetic plants, such as the "Szarvasi-1" energy herb are of interest.

The researches carried out with the energetic grass of the last decade, as well as the practical results obtained justify us to give the name of industrial grass. At the same time, the possibilities of using this plant are found in different industrial branches.

The energy grass can provide the basic material for both ordinary heating systems (stoves) and for automated installations (Lavoie et al., 2010).

Among the varieties of the registered energy, willow of UPOV Swedish origin can be listed:

- **Tora (EU 627).** Tora is a Siberian willow from the cross of SW and *Snake* species. The species has long branches and a lower trunk than other species. The stem is dark brown and glossy. *Tora* is often curved, inheriting this particular feature of the *Snake* species. Bending may vary from year to year, depending on the weather and the treatment with chemicals. It has a high production performance, which makes it a favorite species. *Tora* resists to rusting of the leaves and to pests. This species is not liked by wild animals, being avoided by rabbits, deer, and stags.

- **Tordis (EU 9288).** Tordis comes from the crossing of the *Tora* and *Ulv* species. It develops exceptionally from the first year (reaching heights of over 4 m) in the areas of southern Sweden and Poland. *Tordis* is resistant to leaf rust.
- **Sven (EU 11635).** Sven comes from the crossing of the species *Jorun* and *Bjorn*. Sven has lancelets leaves (straight up), straight trunk with rarer branches, just like *Tora*. The species has a high production performance, like Tora, it withstands rust of the leaf, but the pests attack the shoots of this plant.
- **Inger (EU 11635).** Inger comes from the crossing of a Russian species (from Novosibirsk area) with the species *Jorr*. It grows better on dry soil than other species. At harvest, it is drier than the Tora and grows more often due to the higher number of secondary shoots. These secondary shoots (sylleptic shoots) are not resistant and fall into the preparation of the propagation matter. Inger resists to leaf rust and to pests.
- **Jorr (EU 0626).** *Jorr* is a *Dutch* species, with good resistance to leaf rust. The species is characterized by rapid growth during the planting period. The plant with a dark green and bushy stem is a "gray" but safe species. It is characterized by an average yield and an average resistance to leaf rust. *Jorr* is successfully used for wastewater treatment, developing optimally in such environment too (Brännvall et al., 2015).

Liquid biofuels. These fuels are obtained by processing plants specially grown for energy purposes. The chemical–biological production of liquid fuels is based on a series of chemical reactions and biological processes.

The raw material consists of biomass with a high content of starch and carbohydrate elements. Rapeseed oil is a very good fuel for Diesel engines and is also known as biodiesel.

Rapeseed oil can completely replace diesel, without the need for special engines, and existing engines can be used without modification or with very few modifications. Biodiesel can be mixed with conventional diesel fuel. The parameters of rapeseed oil as biofuel are close to those of diesel. The viscosity in biodiesel is slightly higher than that in diesel, but problems can occur only in very cold weather. Biodiesel is a mixture of alkyl esters of fatty, methyl or ethyl acids. Raw materials are subjected to transesterification with methanol or ethanol in the presence of catalysts. According to Committee D–2 ASTM (American Society of Testing and

Materials), biodiesel is defined as "the compound consisting of monoalkyl esters of the fatty acids with hydrocarbonate long-chain derived from lipids (fats) that are renewable as vegetable oils and animal fats, used as fuel for Diesel engines."

- The raw materials used to obtain biodiesel are as follows:
- Vegetable raw materials: soybeans, canola (rape-like plant), sunflower, palm, cottonseed, linseed and rapeseed, seaweed, mustard and saffron.
- Used vegetable oil: used for roasting by restaurants and industrial producers.
- Raw material of animal origin: animal fat from slaughterhouses.

Biodiesel brands with a different content of esters are sold on the European market, from 5% (B-5) to 100% (B-100). The use of B-20 biodiesel allows the emission of greenhouse gases (carbon dioxide, carbon monoxide, and methane) to be reduced by an average of 15% compared with diesel, and the B-100 emission reduction of 32% solid particles, with 35% carbon monoxide and 8% sulfur oxides. In this way, the elimination of pollutants with a strong impact on climate and health is considerably reduced (Li et al., 2018).

Advantages of biodiesel:

- Reduces pollutant emissions, because it does not contain sulfur and aromatic substances.
- It is nontoxic and is four times more biodegradable than the classic diesel.
- Greater safety in terms of storage, handling and use, as it has a higher flash point (130°C compared with 60°C for diesel).
- Reduces energy dependence on the global oil market.
- Makes it possible to comply with the requirements of EURO III and EURO IV levels, regarding exhaust gas emissions.
- It has a higher combustion in the diesel engine, resulting in a reduction of carbon monoxide emissions by 50–65%, smoke by 42–57%, and nitrogen monoxide by 20%.
- Carbon dioxide resulting from the combustion of biodiesel does not contribute to the greenhouse effect, because it comes from vegetable oils (renewable sources), which closes the cycle of carbon dioxide, because the plants that produce oils, consume it through the process of photosynthesis.

- It burns 75% cleaner than diesel, so the nonburnt hydrocarbons, CO, and particulates in the exhaust gas are substantially reduced.
- The ozone formation potential when the engine runs on biodiesel is 50% lower than on conventional diesel fuel.
- Exhaust gases from biodiesel are not harmful and do not irritate the eyes (smell similar to fried potatoes).
- It can be used in any diesel engine and is a much better lubricant than diesel and increases engine life—a German truck entered in the Record Book covering more than 1.25 million km with biodiesel alone, with the original engine (Mueller et al., 2015).

Disadvantages of biodiesel:

- Requires some minor engine modifications and adjustments.
- It has a higher viscosity, so pumping is more difficult and deposits appear at the injectors.
- The use of large quantities of biodiesel may require additional measures to be taken to protect parts that come into contact with pure biodiesel.
- The nominal power of the engine is reduced by about 5–7%, due to its lower calorific value compared to diesel.
- Higher values for the disturbance temperature and the liquefaction point, so problems with starting the engine at lower temperatures (cold may solidify) and higher consumption.
- Higher percentage of nitrogen oxides (NOx) emissions.
- It has lower stability to oxidizing agents and can cause problems with long-term storage.
- For synthesis, we work with methanol and NaOH which are toxic.
- After synthesis, glycerol is obtained as a by-product which must be used.
- At present, the factory price is higher than that of diesel, but in some countries of the European Community, it is subsidized by the state.
- Requires large areas of land for the cultivation of oil plants (Ge et al., 2014).

The biodiesel obtaining: Methanol is used for transesterification of oils and obtaining biodiesel in this way, but due to its toxicity and high price, lately experiments have been made to replace it with ethanol, which in turn is produced in large quantities from renewable natural sources. Glycerine, an important product, can be obtained after the transesterification process, which after further purification can be used in the pharmaceutical or cosmetic industry.

Cold pressed oil is the oil obtained from oil plants by pressing, extraction or similar procedures, raw or refined but not chemically modified. It can be used as biofuel in special cases where its use is compatible with the type of engine used and the requirements regarding the protection of the environment. The use of 100% raw vegetable oil must meet certain specifications due to its lower calorific value and higher viscosity than diesel. It is recommended to use diesel fuel at the start up to 75°C, and before switching off the engine, switch to diesel again to degrease the injection equipment. A mixture of vegetable oil can also be used as a fuel (Suriapparao et al., 2020).

Theoretically, all edible oils and fats—vegetable and animal—can be converted into biodiesel. In order to obtain the fuel-specific properties, in Europe, oils pretreated with rapeseed or sunflower seeds are preferred, but other oils, such as soybean, palm and palm kernel, coconut and cotton seeds can be used, as well as animal fats.

Transesterification is based on the chemical reaction between triglycerides and methanol to form methyl esters (biodiesel) and glycerine in the presence of an alkaline catalyst. The reaction is produced in a double-stage mixing–decanter (mixer–decanter).

Transesterification takes place in the mixing section (floor), while the next section, the decanting one, allows the separation of methyl esters as a dispersed phase from the glycerine water as the concentrated phase. The next counter current washing step for methyl ester removes the tiny (derivative) by-products, and after the final drying step, results in a ready-to-use biodiesel fuel. The surplus of methanol contained in glycerinated water is removed in a rectifying column, which collects methanol at a purity level and in a state in which it is ready for use for the recycling stream (Tanger et al., 2013).

To further purify glycerinated water, additional steps of chemical treatment, evaporation, distillation, and discoloration can be optionally followed to obtain qualitative pharmaceutical or technical glycerine.

Bioethanol is another liquid biofuel used to fuel internal combustion engines. Bioethanol is obtained by fermentation of renewable energy sources, followed by distillation and dehydration processes. Typical raw materials include starches and cereals (wheat, corn, rye, cassava, rice, and potatoes), sugars (sugar cane molasses, sugar beet molasses, sugar syrup, fructose, and whey). The raw material from which bioethanol is obtained is sugar derived from starch through an enzymatic reaction. Fermentation generates ethyl alcohol with the help of yeast. This ethyl alcohol (bioethanol) is then separated by distillation. In order to obtain a bioethanol of gasoline quality, the alcohol must be dehydrated, until the bioethanol concentration is increased to more than 99%. The by-product of the process by which ethanol is obtained is distillation residue. It can be used to feed biogas stations for power generation or can be converted into dried distillers grains with solubles (DDGS) (wheat and dry soluble components obtained from distillation), a rich protein food for animals (Amorim et al., 2011).

Thus, it can be obtained from many types of agricultural products, for example, from sugar beet, potato, and cereals, but for practice are of interest those energetic plants that can be cultivated on soils with more modest properties, plants with high yields per hectare, whose product is not important for food. Of these, sugar sorghum occupies a special place.

Bioethanol cannot completely replace gasoline, but only partially, in spark ignition engines. In unmodified engines, the share of bioethanol may be only 5–6%.

Cellulose bioethanol is obtained from lignocellulosic biomass which requires preliminary treatment before enzymatic hydrolysis and fermentation. The purpose of pretreatment of lignocellulosic biomass is to make cellulose more accessible by enzymatic hydrolysis and to solubilize sugars from the hemicellulose constitution.

Biomass pretreatment is performed by thermal, enzymatic, or acid methods. Enzymatic hydrolysis is performed with cellulose which initially hydrolyses cellulose to cellobiose (glucose disaccharide) and subsequently cellulose is hydrolyzed to glucose (Khan et al., 2018).

Algae fuel, the so-called Oilgae, is a biofuel made from algae. Algae are the raw materials with low substrate consumption, but with high biofuel productivity.

Algae biofuel is biodegradable, and given the high price of fossil fuels, there is a growing interest in algal farming. The selection of algae must take into account the climate, the solar energy, the water quality, the growth rate of the algae, the oil content, the composition of the algae oil, the requirements for the growing environment, and the possibility of growing in bioreactors. Important parameters for algae growth include solar energy level, light exposure time (day/night cycle), process water temperature and flow, CO_2 content, the macroelements content in the growing environment (C, N, P, Mg, Ca, K, Na, and Cl), the content of microelements in the growth medium (Fe, B, Zn, Mn, Mo, Cu, Co, Cd, V, Al, Ni, Cr, Br, I, etc.), the content of vitamins (Tu et al., 2017).

Algae oil extraction is performed by commercially available classic processes (cold pressing extraction), new technologies (microwave, ultrasonic, and supercritical fluid extraction).

The effective production of bioethanol through the use of agricultural energy resources implies three basic principles according to Figure 7.2.

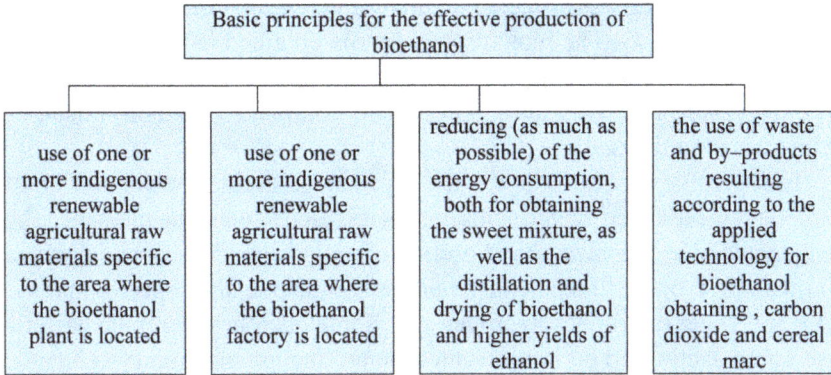

FIGURE 7.2   Basic principles for the effective production of bioethanol through the use of agricultural energetic resources.

The production of "energy crops" is conditioned by the environmental conditions in a given region and by the agricultural management, which can ensure large and constant productions per hectare, with bioethanol availability (Margaritopoulou et al., 2016).

Gaseous biofuels. Biogas is a product obtained by anaerobic fermentation of animal manure, biomass, and wastewater that have a high content of organic substances.

The energy resource of biogas is variable and it depends on the methane content of the biogas. It was agreed for the unification of the expression mode that the standard biogas should be considered as having 60% methane content.

As biogas crop plants, corn, soft cereals, sunflower, fodder sorghum, Sudan grass are suitable. Biohydrogen is the biofuel with the highest amount of energy per mass and causes zero emission in vehicles. Hydrogen can be obtained by steam reforming methane from natural gas, partial oxidation/reforming other carbon-based fuels, gasification of coal or biomass, pyrolysis, dissociation of methanol or ammonia, electrolysis of water (if the source of electricity is renewable energy, then the net carbon dioxide emissions are zero), thermochemical decomposition of water, biochemical photosynthesis or fermentation and other electrochemical or photochemical processes (Show et al., 2018).

The biological production of hydrogen is a viable alternative for obtaining hydrogen. In agreement with the support of the development and the minimization of the residues, the production of the biohydrogen using these sources has attracted, in the last years, a considerable attention from the specialists. The biohydrogen can be obtained with anaerobic and photosynthetic microorganisms, using either solid residues from agriculture or wastewater from agrozootechnical complexes or food industries with high organic load.

In anaerobic processes, hydrogen is produced during the conversion of organic residues into organic acids, which are used to generate methane. The acidogenic phase of anaerobic digestion of organic residues can be directed to obtain hydrogen. Photosynthetic processes include algal species that use CO_2 and water to produce hydrogen gas. Some photoheterotrophic bacteria use acetic, lactic, and butyric acids for the production of H_2 and CO_2.

The production of biohydrogen by anaerobic fermentation of organic residues represents an alternative energy source for the future that can lead to partial or total replacement of fossil fuels (Santos et al., 2018).

The specific biological processes used to produce hydrogen are shown in Table 7.3.

7.3.2.2   THE HYDROGEN PRODUCTION USING ALGAE

Algae cleave water molecules to hydrogen and oxygen ions through photosynthesis. The hydrogen ions generated are converted to hydrogen gas by the hydrogenating enzymes. Hydrogen can be produced using green and blue-green algae, depending on the reaction:

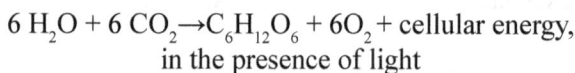

$$6\ H_2O + 6\ CO_2 \rightarrow C_6H_{12}O_6 + 6O_2 + \text{cellular energy},$$
in the presence of light

TABLE 7.3   Specific Biological Processes for Hydrogen Production.

Specific processes	General reactions	Usual microorganisms
Direct biophotolysis	$2 H_2O + light \rightarrow 2 H_2 + O2$	Microalgae
Indirect biophotolysis	$6 CO_2 + 6 H_2O + light \rightarrow C_6H_{12}O_6 + 6 O_2$	Microalgae, cyanobacteria
	$C_6H_{12}O_6 + 2H_2O \rightarrow 4H_2 + 2CH_3COOH + 2CO_2$	
	$CH_3COOH + 4 H_2O + light \rightarrow 8 H_2 + 4 CO_2$	
Photofermentation	$CH_3COOH + H_2O + light \rightarrow 4 H_2 + CO_2$	Microalgae, red bacteria
Fermentation phases	$C_6H_{12}O_6 + 2H_2O \rightarrow 4H_2 + 2CH_3COOH + 2CO_2$	Fermentation bacteria + methanogenic bacteria
	$2 CH_3COOH \rightarrow 2 CH_4 + 2 CO_2$	
Fermentation in the absence of light	$C_6H_{12}O_6 + 6 H_2O \rightarrow 12 H_2 + 6 CO_2$	Fermentation bacteria
The reaction of the transfer of pond gas	$CO + H_2O \rightarrow CO_2 + H2$	Fermentation bacteria and photosynthetic bacteria

Bacterial photosynthesis involves the reduction of organic compounds in cellular material and cellular energy. *Clamydomonas reinhardtii* is one of the algae responsible for obtaining hydrogen. Hydrogen-forming activity was detected in green algae, *Scenedesmus obliquus*, in green seaweed, *Chlorococcum littoral*, *Playtmonas subcordiformis*, *Chlorella fusca*. Hydrogen production using algae can be considered an economical method, within the limits of the use of water as a renewable resource and of CO_2 consumption as one of the air pollutants. The major limitation of the process is the inhibitory effect of the oxygen generated for the enzymes involved in the hydrogen generator (Liu et al., 2015).

7.3.2.3   THE PRODUCTION OF GASEOUS HYDROGEN BY ANAEROBIC FERMENTATION IN THE ABSENCE OF LIGHT

Organisms belonging to the genus *Clostridium*, such as *C. butyricum*, *C. therolacticum*, *C. pasteurianum*, *C. paraputrificum* M–21, and *C. bifermentants* are required anaerobically. *Clostridium* species produce gaseous hydrogen following an exponential law. Dominant culture of *Clostridia* can be easily obtained by heat treatment of biological sludge. Spores formed at high temperatures can be activated when the required environmental conditions provide hydrogen gas. Other microorganisms that can supply hydrogen can be *Enterobacter aerogenes*, *Enterobacter cloacae* ITT–BY, *E. coli*, *Hafnia alvei*, etc.

The amount of hydrogen varies from 1 to 1.2 mmol/mol glucose, when the cultures were grown under anaerobic conditions (Arumugam et al., 2014).

Also, the hydrogen produced by the species *Thermotogales* and *Bacillus* sp. has been detected in acidogenic mesophilic cultures. Laboratory research has shown the ability of microorganisms to produce gaseous hydrogen using anaerobic thermophilic microorganisms belonging to the genus *Thermoanaerobacterium*, *Clostridium thermolacticum* that can produce lactose hydrogen at a temperature of 58°C. Environmental conditions are the essential parameters in the processes of hydrogen production. The pH affects hydrogen production, biogas content, the types of organic acids contained, and the specific rate of hydrogen production.

The pH limit for maximum of hydrogen or the specific rate of hydrogen production is between five and six and for thermophilic cultures of 4.5. The gradual decrease of the pH inhibits the production of hydrogen, because it affects the activity of the iron contained in the hydrogenating enzymes. Therefore, pH control is important. Substrate composition, temperature, and type of microbial culture are the most important parameters that affect the process of obtaining hydrogen. The major products in the production of hydrogen by anaerobic fermentation in the dark of carbohydrates are acetic, butyric, and propionic acids.

The formation of acids is determined by the size of the pH. Large amounts of butyric acid can be obtained at a pH = 4.0–6.0. The concentrations of acetate and butyrate can be almost equal at a pH = 6.5–7.0. Gaseous hydrogen is produced by strictly anaerobic organisms. Therefore, reducing agents, such as argon, nitrogen, hydrogen gas, and L–cysteine, HCl are used to remove oxygen traces present in the environment.

The use of reducing agents for the production of industrial hydrogen gas leads to expensive and therefore uneconomic solutions (Nobre et al., 2013).

Enterobacter aerogenes is optionally anaerobic and the amount of hydrogen produced with this culture is comparable to *Clostridium* sp. The food industry is the main supplier of wastewater with medium and large organic loads. The high content of carbohydrates in the form of glucose, starch, and cellulose makes solid food residues a potential substrate for the production of biological hydrogen.

Following the experimental researches, it was possible to ascertain the great potential of hydrogen production from food residues, in thermophilic conditions compared with mesophilic ones. The sludge from wastewater treatment plants contains large quantities of polysaccharides and proteins. The hydrogen amounts of 1.2 mg H_2/g COD and 0.6 mol/kg COD were found when the sludge was used as a substrate.

However, large amounts of hydrogen (15 mgH_2/g COD) were obtained from the filtered material. If the sludge is pretreated, the conversion intensity of hydrogen soluble COD (0.9 mmol/g sludge) increases (Jitrwung and Yargeau, 2015).

7.3.2.4 THE PRODUCTION OF HYDROGEN BY PHOTOFERMENTATION

Certain photoheterotrophic bacteria are capable of converting organic acids (acetic, lactic, and butyric) to hydrogen (H_2) and CO_2, under anaerobic conditions, in the presence of light. Therefore, the organic acids produced during the acidogenic phase, the anaerobic digestion of organic residues, can be converted to H_2 and CO_2 with these types of anaerobic photosynthetic bacteria. Preferred organic acids, as a carbon source include acetic, butyric, propionic, lactic, and malic acid.

Red photosynthetic bacteria, such as *Rhodobacter sphaeroides, Rhodobacter capsulatus, Rhodovulum sulfidophilum* W-1S and *Rhodopseudomonas palustris* have been investigated to varying degrees in hydrogen production. Following experimental research, it was found that the optimum temperature and pH for the bacteria involved in the process are known to be: for temperature, within 30°C–35°C, and for pH, pHopt = 7.0. The rate of hydrogen production varies with light intensity, carbon source, and microbial crop types.

One of the parameters that affect the performance of the photo fermentation is the light intensity. Increased light intensity has a stimulating effect on hydrogen production, but has a reverse effect on light conversion efficiency. Butyrate has been found to require high light intensity (4000 lux) compared with acetate and propionate, and hydrogen production in dark conditions is lower than that produced under lighting conditions. The alternation of cycles of 14 h light/10 h dark has been shown to be much more efficient, as the rates of hydrogen production are high and the cellular concentration is comparable to the continuous illumination.

It is reported that the rate of obtaining hydrogen during a cycle of 30 min darkness / light is 22 L/m²/day, being twice as high as that obtained when lighting a cycle of 12 h, under the same conditions. The use of industrial effluents for the production of hydrogen with photosynthetic bacteria is possible, although these cultures prefer organic acids as a carbon source. One of the major problems in hydrogen production is the color of wastewater, which can reduce the penetration of light. High concentration of ammonia inhibits the enzyme nitrogenase, reducing the productivity of hydrogen. The high content of organic matter (COD) and the presence of toxic compounds (heavy metals, phenols, and PAH) in industrial effluents require a preliminary treatment, before obtaining hydrogen (Uyar et al., 2015).

The hydrogen, in combination with fuel cells, is considered to be an energy source for both transportation and stationary uses. Hydrogen has the significant advantage that it does not produce pollutants when it is converted to energy, and hydrogen fuel cells offer increased efficiency in generating energy. Currently, the hydrogen is produced almost entirely from fossil fuels, such as natural gas, oil, and coal, based on well-established conversion processes.

In these cases, the carbon dioxide released into the atmosphere during the hydrogen production process is smaller than that resulting from the direct combustion of these fuels, for the production of equal amounts of energy. The use of hydrogen produced from generable sources, such as biomass, substantially reduces the amount of CO_2 released into the atmosphere (Cosnier et al., 2014).

The fuel cells. The shortcomings of storing electricity in batteries (in the case of electric powered vehicles) can be overcome by generating electricity from fuel cells. The first hydrogen fuel cells were used to generate electricity in Apollo space missions, after which numerous other applications were found. Subsequent research has led to the idea that hydrogen fuel cell vehicles can be a viable alternative. Fuel cells work by combining hydrogen and oxygen to create electricity without the need for conventional engines that are noisy and polluting. Generally speaking, a fuel cell functions as a battery. Fuel cells do not have movable components that require lubricating or lubricating oils; they are characterized by zero emissions in the sense of greenhouse gas emissions and limited the oxide emissions.

The major criteria for the selection of residues used in the production of biohydrogen are the cost price, the carbohydrate content, and the biodegradability. Carbohydrates, such as glucose and lactose are more readily biodegradable, preferred for hydrogen production.

The materials used for the production of hydrogen can be starch and cellulose from agricultural and food residues, industrial wastewater, materials high in carbohydrates, sludges from wastewater treatment. Agricultural and food residues contain starch and/or cellulose. Starch can be hydrolyzed to glucose and maltose by acid or enzymatic hydrolysis, followed by the conversion of carbohydrates to organic acids and then to hydrogen gas. Cellulose contained in agricultural residues requires additional pretreatment. Cellulose and hemicellulose from organic residues can be hydrolyzed into carbohydrates which are then processed into organic acids and gaseous hydrogen (Wang et al., 2009).

7.3.3 THE POLLUTION PREVENTION—ECONOMIC AND ECOLOGICAL OBJECTIVE

Planting one hectare of land with energetic willow costs about 1700–2000 euros. This investment is made only once, the operating time is 25–30 years. The average production per hectare is 30–40 tons, reaching up to 60 tons under intensive irrigation conditions.

First of all, the initiators and investors of this implementation process must be the energy companies together with the producers, who must finance this investment. Such an action is an upward curve for both business and population health and the environment. Replacing oil with biofuel is a solution.

Another fuel alternative for car transport can also be ethyl alcohol, still being considered in 1925 by Henry Ford "the fuel of the future."

This fuel can be extracted from fruits, sawdust, almost anything. There is fuel in every form of vegetable substance that can ferment.

Another solution to prevent pollution is to consider nonproductive land afforestation. This issue requires primary attention. First of all, it is necessary to convince the local public authorities, the owners who have nonproductive land to lease these areas to the competent bodies for the planting of new massive forests and forest strips or under the plantation of the energy willow. The forest is considered the "lung of the planet" because it continuously recycles carbon dioxide, the evaporation of the trees turns into rain. And on the contrary, because of the lack of trees, the localities will become deserts, less oxygen will be produced and all the carbon dioxide existing in the soil will be released into the atmosphere.

First of all, there is a need for a real energy policy, which includes the creation of an autonomous energy system (solar, wind, safe hydrographic centers), using the practice of EU countries. From the point of view of today's technologies, providing localities with solar and wind energy represents the safest alternative to traditional energy. At the mentioned ones, we find that the nuclear aspect is permanently influenced by the nuclear emissions. Natural radiation sources account for 82% of the average amount of exposure to humans annually, while nuclear power represents less than 1%.

To use the composting process effectively, separate collection of biodegradable waste is required. The composting of the municipal waste collected in the mix should be avoided, as they have a high content of heavy metals, such as Cd, Pb, Cu, Zn, and Hg. Separate collection of biodegradable material can be done in all urban areas of the country by carrying out individual composting (reuse of biodegradable materials in their own households). Under the existing situation, in the less-dense urban environment, it is recommended to introduce separate collection of biodegradable waste for composting in pilot stations. Separate collection of biodegradable waste in order to obtain compost is a first, useful, and efficient step for the recovery and reduction of the amount of organic waste deposited (Chen et al., 2015).

Depending on the type of material and the time required for the composting process, it is necessary to apply various schemes, methods, and technologies for composting.

Separate collection of biodegradable waste and composting solves only a small part of the problem of biodegradable waste management. Only certain streams of biodegradable waste can be collected separately and treated by composting, most of which are found in domestic and similar waste.

The experience of other states has shown that the implementation of separate collection systems is not cost-effective and efficient in crowded urban areas. The composition of nonsorted household waste, as has been practiced in other countries (e.g., Germany), has proved to be defective, because of the high content of impurities and harmful substances in the waste, lower quality composts are produced, and these lead to problems with their use.

In one way or another, waste, even those stored on specialized ramps, penetrate (directly or through their decomposition products) into surface and underground waters, soil, and air.

Air pollution with unpleasant odors and with wind-driven suspensions is particularly evident in the storage area of current dumps where cell exploitation and inert materials coverage are not practiced. The current practices of collecting, transporting/storing urban and rural waste facilitate the multiplication and dissemination of pathogens and their vectors. Leaks from the slopes of deposits near surface waters contribute to their pollution with organic substances and suspensions (Mahidhara et al., 2019).

Nonwaterproofed landfills have the source of infestation of groundwater with nitrates and nitrites, but also with other polluting elements. Both the seepage from the deposits and the runoff waters influence the quality of the surrounding soils, which has an impact on their use. In turn, the soil can be directly and indirectly affected by waste and pollutants. Direct pollution is caused by the unauthorized location or dispersion of different wastes in the soil or on the ground, including toxic, industrial, agricultural, and hospital wastes. Part of the waste and polluted water is found in the soil—causing some damage through the pollutants they contain.

Indirectly, the soil is polluted by polluted atmospheric precipitation. Atmospheric pollutants, for the most part, are finally deposited in the soil, either with "acid rain" or by direct precipitation. Thus, in this way, the soil is indirectly subjected to the harmful impact of the activity of the pollution sources. The polluted soil is the source of pollution of foodstuffs, which then reach the area of use of humans and animals, as well as the pollution of the waters of the groundwater, which are used as drinking water (Szczesny et al., 2018).

There are many potential impacts associated with the storage of biodegradable waste, including the generation of leachate and gas (with a strong greenhouse effect), odors, insects, pests, and land use in landfills. First of all, the leakages of nutrients into the surface waters that can cause the water body overloaded. Then leaks from and improper storage of agricultural waste can greatly threaten the environment if the waste reaches surface waters.

Also, activities on agricultural farms can cause ammonia and methane emissions to cause acidification and contribute to greenhouse gas emissions. Incineration of biodegradable waste and landfilling leads to the emergence of gases directly responsible for the emergence of the greenhouse effect. Their combustion produces carbon dioxide, and the storage of waste leads to the formation of methane, an even more harmful gas for the environment and human health. This happens due to the imperfection of these processes (the burning at low temperatures due to the low caloric coefficient, humidity, and rot). The uncontrolled emission of the gases resulting from the waste landfills brings serious damage to the environment and to the society. Emissions spread rapidly through air, water, and soil in different forms, being through toxic solid parts or in the form of gas (Pasternak et al., 2016).

At present, most of the old garbage dumps in the country emit large quantities of methane gas into the air. According to laboratory investigations, the methane produced at the landfills constitutes about 12–15% of the total methane emitted into the atmosphere, having a greenhouse effect 11 times higher than the capacity of carbon dioxide. The methane represents an important energetically resource, which needs to be used by different methods. It has been estimated that for a period of 20 years, from 1 ton of household waste stored at a ramp, about 15–200 m^3 of biogas is emitted, which could be used for energy production. On average, 1 m^3 of biogas is equivalent to about 0.5 L of oil (Khetkorn et al., 2017).

Depending on the age, the composition of the biogas differs. The methane appears after a period of anaerobic fermentation, while the carbon dioxide immediately appears after the storage of waste. A negative aspect is also that many biodegradable materials are stored together with nonrecyclable ones, being mixed and contaminated from a chemical and biological point of view, and their recovery is difficult.

7.4   CONCLUSIONS AND REMARKS

Due to the continuous increase in oil prices and due to the evolution of biofuel technologies, their use is becoming more attractive. The future of the use of biofuels depends on the policy of each country and on the development of production technologies. The first generation of biofuels, which already exists on the market, is obtained by processing the seeds of plants. This generation of biofuels requires the cultivation of large areas of land. As a result, habitat and biodiversity of the environment may be affected. On the other hand, the production of first-generation biofuels has an adverse impact on the production of plants needed for the food industry. The second generation of biofuels is still in the technology development phase. For the production of second-generation synthetic fuels, complex research activities are carried out. On the other hand, plant crops for biofuels must also become profitable. Therefore, the need to identify and implement new clean technologies for waste treatment and disposal emerged, with energy recovery and transformation into useful products, so that waste becomes a valuable resource.

KEYWORDS

- **biodegradable waste**
- **agricultural waste**
- **biomass**

REFERENCES

Albanez, R.; Chiaranda, B. C.; Ferreira, R. G.; França, A. L.; Honório, C. D.; Rodrigues, J. A.; Ratusznei, S. M.; Zaiat, M. Anaerobic Biological Treatment of Vinasse for

Environmental Compliance and Methane Production. *Appl. Biochem. Biotechnol.* **2016**, *178* (1), 21–43.

Amorim, H. V.; Lopes, M. L.; de Castro Oliveira, J. V.; Buckeridge, M. S.; Goldman, G. H. Scientific Challenges of Bioethanol Production in Brazil. *Appl. Microbiol. Biotechnol.* **2011**, *91* (5), 1267–1275.

Arumugam, A.; Sandhya, M.; Ponnusami, V. Biohydrogen and Polyhydroxyalkanoate Co–Production by *Enterobacter aerogenes* and *Rhodobacter sphaeroides* from *Calophyllum inophyllum* Oil Cake. *Bioresour. Technol.* **2014**, *164*, 170–176.

Brännvall, E.; Wolters, M.; Sjöblom, R.; Kumpiene, J. Elements Availability in Soil Fertilized with Pelletized Fly Ash and Biosolids. *J. Environ. Manage.* **2015**, *159*, 27–36.

Chan, Y. H.; Cheah, K. W.; How, B. S.; Loy, A. C. M.; Shahbaz, M.; Singh, H. K. G.; Yusuf, N. R.; Shuhaili, A. F. A.; Yusup, S.; Ghani, W. A. W. A. K.; Rambli, J.; Kansha, Y.; Lam, H. L.; Hong, B. H.; Ngan, S. L. An Overview of Biomass Thermochemical Conversion Technologies in Malaysia. *Sci. Total Environ.* **2019**, *680*, 105–123.

Chen, W. H.; Lin, B. J.; Huang, M. Y.; Chang, J. S. Thermochemical Conversion of Microalgal Biomass Into Biofuels: A Review. *Bioresour. Technol.* **2015a**, *184*, 314–327.

Chen, Y.; Bellini, M.; Bevilacqua, M.; Fornasiero, P.; Lavacchi, A.; Miller, H. A.; Wang, L.; Vizza, F. Direct Alcohol Fuel Cells: Toward the Power Densities of Hydrogen–Fed Proton Exchange Membrane Fuel Cells. *Chem. Sus. Chem.* **2015b**, *8* (3), 524–533.

Cosnier, S.; Holzinger, M.; Le Goff, A. Recent Advances in Carbon Nanotube–Based Enzymatic Fuel Cells. *Front. Bioeng. Biotechnol.* **2014**, *2*, 45. doi: 10.3389/fbioe.2014.00045.

European Commission. Proposal for a Directive of the European Parliament and of the Council Amending Directive 98/70/EC Relating to the Quality of Petrol and Diesel Fuels and Amending Directive 2009/28/EC on the Promotion of the Use of Energy from Renewable Sources, COM (**2012**) 595 Final. European Commission, Brussels, p 23.

European Environment Agency (EEA). How Much Bioenergy Can Europe Produce without Harming the Environment? EEA Report No. 7, 2006.

Fatma, S.; Hameed, A.; Noman, M.; Ahmed, T.; Shahid, M.; Tariq, M.; Sohail, I.; Tabassum, R. Lignocellulosic Biomass: A Sustainable Bioenergy Source for the Future. *Protein Pept Lett.* **2018**, *25* (2), 148–163.

Ge, X.; Yang, L.; Sheets, J. P.; Yu, Z.; Li, Y. Biological Conversion of Methane to Liquid Fuels: Status and Opportunities. *Biotechnol. Adv.* **2014**, *32* (8), 1460–1475.

Gerber, Van Doren, L.; Posmanik, R.; Bicalho, F. A.; Tester, J. W.; Sills, D. L. Prospects for Energy Recovery during Hydrothermal and Biological Processing of Waste Biomass. *Bioresour. Technol.* **2017**, *225*, 67–74.

Grover, A.; Singh, S.; Pandey, P.; Patade, V. Y.; Gupta, S. M.; Nasim, M. Overexpression of NAC Gene from *Lepidium latifolium* L. Enhances Biomass, Shortens Life Cycle and Induces Cold Stress Tolerance in Tobacco: Potential for Engineering Fourth Generation Biofuel Crops. *Mol. Biol. Rep.* **2014**, *41* (11), 7479–7489.

Hašková S. Holistic Assessment and Ethical Disputation on a New Trend in Solid Biofuels. *Sci. Eng. Ethics.* **2017**, *23* (2), 509–519.

How, S. W.; Sin, J. H.; Wong, S. Y. Y.; Lim, P. B.; Mohd Aris. A.; Ngoh, G. C.; Shoji, T.; Curtis, T. P.; Chua, A. S. M. Characterization of Slowly–Biodegradable Organic Compounds and Hydrolysis Kinetics in Tropical Wastewater for Biological Nitrogen Removal. *Water Sci. Technol.* **2020**, *81* (1), 71–80.

Jitrwung, R.; Yargeau, V. Biohydrogen and Bioethanol Production from Biodiesel–Based Glycerol by *Enterobacter aerogenes* in a Continuous Stir Tank Reactor. *Int. J. Mol. Sci.* **2015**, *16* (5), 10650–10664.

Karmee, S. K.; Patria, R. D.; Lin, C. S. K. Techno–Economic Evaluation of Biodiesel Production from Waste Cooking Oil—A Case Study of Hong Kong. *Int. J. Mol. Sci.* **2015**, *16*, 4362–4371.

Khan, M. I.; Shin, J. H.; Kim, J. D. The Promising Future of Microalgae: Current Status, Challenges, and Optimization of a Sustainable and Renewable Industry for Biofuels, Feed, and Other Products. *Microb. Cell Fact.* **2018**, *17* (1), 36.

Khetkorn, W.; Rastogi, R. P.; Incharoensakdi, A.; Lindblad, P.; Madamwar, D.; Pandey, A.; Larroche, C. Microalgal Hydrogen Production—A Review. *Bioresour. Technol.* **2017**, *243*, 1194–1206.

Kumar, A.; Kaushal, S.; Saraf, S. A.; Singh, J. S. Microbial Bio–Fuels: A Solution to Carbon Emissions and Energy Crisis. *Front. Biosci.* (Landmark Ed). **2018**, *23*, 1789–1802.

Lavoie, J. M.; Capek–Menard, E.; Gauvin, H.; Chornet, E. Production of Pulp from Salix Viminalis Energy Crops Using the FIRSST Process. *Bioresour. Technol.* **2010**, *101* (13), 4940–4946.

Li, H.; Wang, M.; Wang, X.; Zhang, Y.; Lu, H.; Duan, N.; Li, B.; Zhang, D.; Dong, T.; Liu, Z. Biogas Liquid Digestate Grown Chlorella sp. for Biocrude Oil Production via Hydrothermal Liquefaction. *Sci. Total Environ.* **2018**, *635*, 70–77.

Liu, Y.; Ghosh, D.; Hallenbeck, P. C. Biological Reformation of Ethanol to Hydrogen by *Rhodopseudomonas palustris* CGA009. *Bioresour. Technol.* **2015**, *176*, 189–195.

Mahidhara, G.; Burrow, H.; Sasikala, C.; Ramana, C. V. Biological Hydrogen Production: Molecular and Electrolytic Perspectives. *World J. Microbiol. Biotechnol.* **2019**, *35* (8), 116.

Margaritopoulou, T.; Roka, L.; Alexopoulou, E.; Christou, M.; Rigas, S.; Haralampidis, K.; Milioni, D. Biotechnology towards Energy Crops. *Mol. Biotechnol.* **2016**, *58* (3), 149–158.

Mueller, T. J.; Grisewood, M. J.; Nazem–Bokaee, H.; Gopalakrishnan, S.; Ferry, J. G.; Wood, T. K.; Maranas, C. D. Methane Oxidation by Anaerobic Archaea for Conversion to Liquid Fuels. *J. Ind. Microbiol. Biotechnol.* **2015**, *42* (3), 391–401.

Nobre, B. P.; Villalobos, F.; Barragán, B. E.; Oliveira, A. C.; Batista, A. P.; Marques, P. A.; Mendes, R. L.; Sovová, H.; Palavra, A. F.; Gouveia, L. A Biorefinery from Nannochloropsis sp. Microalga—Extraction of Oils and Pigments. Production of Biohydrogen from the Leftover Biomass. *Bioresour. Technol.* **2013**, *135*, 128–136.

Oriez, V.; Peydecastaing, J.; Pontalier, P. Y. Lignocellulosic Biomass Fractionation by Mineral Acids and Resulting Extract Purification Processes: Conditions, Yields, and Purities. *Molecules* **2019**, *24* (23).

Ozbayram, E. G.; Kleinsteuber, S.; Nikolausz, M. Biotechnological Utilization of Animal Gut Microbiota for Valorization of Lignocellulosic Biomass. *Appl. Microbiol. Biotechnol.* **2020**, *104* (2), 489–508.

Pasternak, G.; Greenman, J.; Ieropoulos, I. Comprehensive Study on Ceramic Membranes for Low–Cost Microbial Fuel Cells. *Chem. Sus. Chem.* **2016**, *9* (1), 88–96.

Pelletier, J.; Siampale, A.; Legendre, P.; Jantz, P.; Laporte, N. T.; Goetz, S. J. Human and Natural Controls of the Variation in Aboveground Tree Biomass in African Dry Tropical Forests. *Ecol. Appl.* **2017**, *27* (5), 1578–1593.

Purohit, H. J.; Kapley, A.; Khardenavis, A.; Qureshi, A.; Dafale, N. A. Insights in Waste Management Bioprocesses Using Genomic Tools. *Adv. Appl. Microbiol.* **2016**, *97*, 121–170.

Rosero–Henao, J. C.; Bueno, B. E.; de Souza, R.; Ribeiro, R.; Lopes de Oliveira, A.; Gomide, C. A.; Gomes, T. M.; Tommaso, G. Potential Benefits of Near Critical and Supercritical Pre–Treatment of Lignocellulosic Biomass towards Anaerobic Digestion. *Waste Manage. Res.* **2019**, *37* (1), 74–82.

Santos, L. C. D.; Adarme, O. F. H.; Baêta, B. E. L.; Gurgel, L. V. A.; Aquino, S. F. Production of Biogas (Methane and Hydrogen) from Anaerobic Digestion of Hemicellulosic Hydrolysate Generated in the Oxidative Pretreatment of Coffee Husks. *Bioresour. Technol.* **2018**, *263*, 601–612.

Sharma, M.; Joshi, S.; Kumar, A. Assessing Enablers of E–Waste Management in Circular Economy Using DEMATEL Method: An Indian Perspective. *Environ. Sci. Pollut. Res. Int.* **2020**, *27* (12), 13325–13338.

Shikinaka, K.; Otsuka, Y.; Nakamura, M.; Masai, E.; Katayama, Y. Utilization of Lignocellulosic Biomass via Novel Sustainable Process. *J. Oleo Sci.* **2018**, *67* (9), 1059–1070.

Show, K. Y.; Yan, Y.; Ling, M.; Ye, G.; Li, T.; Lee, D. J. Hydrogen Production from Algal Biomass—Advances, Challenges and Prospects. *Bioresour. Technol.* **2018**, *257*, 290–300.

Soo, V. W.; McAnulty, M. J.; Tripathi, A.; Zhu, F.; Zhang, L.; Hatzakis, E.; Smith, P. B.; Agrawal, S.; Nazem–Bokaee, H.; Gopalakrishnan, S.; Salis, H. M.; Ferry, J. G.; Maranas, C. D.; Patterson, A. D.; Wood, T. K. Reversing Methanogenesis to Capture Methane for Liquid Biofuel Precursors. *Microb. Cell Fact.* **2016**, *15*, 11.

Steinberg, L. M.; Kronyak, R. E.; House, C. H. Coupling of Anaerobic Waste Treatment to Produce Protein– and Lipid–Rich Bacterial Biomass. *Life Sci. Space Res.* (Amst). **2017**, *15*, 32–42.

Suriapparao, D. V.; Vinu, R.; Shukla, A.; Haldar, S. Effective Deoxygenation for the Production of Liquid Biofuels via Microwave Assisted Co–Pyrolysis of Agro Residues and Waste Plastics Combined with Catalytic Upgradation. *Bioresour. Technol.* **2020**, *302*, 122775.

Szczesny, J.; Marković, N.; Conzuelo, F.; Zacarias, S.; Pereira, I. A. C.; Lubitz, W.; Plumeré, N.; Schuhmann, W.; Ruff, A. A Gas Breathing Hydrogen/Air Biofuel Cell Comprising a Redox Polymer/Hydrogenase–Based Bioanode. *Nat. Commun.* **2018**, *9* (1), 4715.

Tanger, P.; Field, J. L.; Jahn, C. E.; Defoort, M. W.; Leach, J. E. Biomass for Thermochemical Conversion: Targets and Challenges. *Front. Plant Sci.* **2013**, *4*, 218.

Tu, Q.; Eckelman, M.; Zimmerman, J. Meta–analysis and Harmonization of Life Cycle Assessment Studies for Algae Biofuels. *Environ. Sci. Technol.* **2017**, *5*, *51* (17), 9419–9432.

Uyar, B.; Gürgan, M.; Özgür, E.; Gündüz, U.; Yücel, M.; Eroglu, I. Hydrogen Production by Hup(–) Mutant and Wild–Type Strains of *Rhodobacter capsulatus* from Dark Fermentation Effluent of Sugar Beet Thick Juice in Batch and Continuous Photobioreactors. *Bioprocess Biosyst. Eng.* **2015**, *38* (10), 1935–1942.

Valdivia, M.; Galan, J. L.; Laffarga, J.; Ramos, J. L. Biofuels 2020: Biorefineries Based on Lignocellulosic Materials. *Microb. Biotechnol.* **2016**, *9* (5), 585–594.

Vendruscolo, T. P.; Barelli, M. A.; Castrillon, M. A.; da Silva, R. S.; de Oliveira, F. T.; Corrêa, C. L.; Zago, B. W.; Tardin, F. D. Correlation and Path Analysis of Biomass Sorghum Production. *Genet. Mol. Res.* **2016**, *15* (4).

Wang, T.; Zhai, Y.; Li, H.; Zhu, Y.; Li, S.; Peng, C.; Wang, B.; Wang, Z.; Xi, Y.; Wang, S.; Li, C. Co–Hydrothermal Carbonization of Food Waste–Woody Biomass Blend towards Biofuel Pellets Production. *Bioresour. Technol.* **2018**, *267*, 371–377.

Wang, X.; Feng, Y.; Wang, H.; Qu, Y.; Yu, Y.; Ren, N.; Li, N.; Wang, E.; Lee, H.; Logan, B. E. Bioaugmentation for Electricity Generation from Corn Stover Biomass Using Microbial Fuel Cells. *Environ. Sci. Technol.* **2009**, *43* (15), 6088–6093.

Wang, Y.; Ho, S. H.; Yen, H. W.; Nagarajan, D.; Ren, N. Q.; Li, S.; Hu, Z.; Lee, D. J.; Kondo, A.; Chang, J. S. Current Advances on Fermentative Biobutanol Production Using Third Generation Feedstock. *Biotechnol. Adv.* **2017**, *35* (8), 1049–1059.

CHAPTER 8

Intrusion of Biotechnology for Degradation of Organic Wastes

RUBIYA DAR[1*], BABA UQAB[2*], SHAH ISHFAQ[2], SALEEM FAROOQ[2], RIASA ZAFFAR[2], and HINA MUSHTAQ[2]

[1]Centre of Research for Development, University of Kashmir, Srinagar, India

[2]Department of Environmental Science, University of Kashmir, Srinagar, India

*Corresponding author. E-mail: rubi07@rediffmail.com; uqabaalibaba@gmail.com

ABSTRACT

In the recent years, quality of the agricultural products have been improved with biotechnological interventions and this intervention has emerged as powerful tool for improving productivity, nutrition, and physicochemical attributes. With such interventions, the economy of various agro-based products has shown gradual increase. From the day of creation of earth, the main obligation of man is to improve the quality of life for better living standards. Agriculture and technology combined together not only have helped to achieve this goal but have also helped for achieving many of anthropogenic goals. With the use of modern technology and other tools, man has been able to cultivate crops of his choice and rear animals for his needs. Technological intervention is good until it does not hamper our environment, and major fallout of all these activities is waste generation. Generation of wastes demand that measures must be taken in order to avert the damage caused by the unpleasant consequences of their accumulation. The only solution to this problem is waste management that too in a scientific and proper way. Modern concept of waste management does not consider waste as waste but treat it as resource that can be judiciously used for betterment of life. Agro wastes are the best examples to fit in this concept as major portion of biomass of agro wastes are not being utilized in a proper and sustainable way. The resources that find their way in the waste bins are bioconvertible into useful products. This chapter will focus on creation of wealth using biotechnological tools from wastes with special emphasis on recoverable agrowastes. Biotechnological methods, such as composting, biodegradation, and bioremediation can be implemented for solid waste management as these techniques are efficient biotechniques.

8.1   INTRODUCTION

Agricultural wastes when defined scientifically are residues of raw agricultural products which comprises of both natural and artificial wastes produced from variety of activities. The activities mainly comprise "dairy farming, horticulture, seed growing, livestock breeding, grazing land, market gardens, nursery plots, and even woodlands." These wastes are classified into solid, liquid, or slurries depending upon their agricultural activities. In order to avoid interference of the wastes with other agricultural activities, wastes should be removed from the fields (Sarmah, 2005). They are typically nonproducts of agricultural products and include mostly materials that can support humans and are also economically advantageous in all terms, such as collection, transport, and processing costs. On the other hand, animal waste, food waste, crop waste, and fruit and vegetable waste constitute the agro waste. Although Agricultural waste is thought to be a major constituent of total waste matter in developed world, but recent estimates have shown that agricultural waste arising are rare. Expanding agricultural production has resulted in increased livestock waste, agro-industrial by-products and agricultural residues. If the developing countries continue to intensify the farming system in the same pace, there is an apprehension that there will be global increase in agricultural wastes. Yearly statistics have shown that at least 998 million tons of wastes from agriculture are being generated and almost 80% constitutes the organic waste out of the total solid waste generated in any farm (Overcash, 1973).

8.2   BIOTECHNOLOGY AND AGRICULTURAL WASTE

Biotechnology is often defined as the technology of hope as it promises sustainability in food, health, and environment and for this reason it is also called far-reaching technology. New advances in life sciences are biotechnology-driven as this technology has energized and unfolded many new scenarios. Biotech drugs and vaccines are benefiting over 100 million people worldwide, this has not only provided the health benefits but on the other hand accounts for a US$40 billion market. Large number of products especially agri-biotech and industrial biotech have helped mankind, and Biotechnology finds its applications in diverse fields like agriculture, industry, and environment. Techniques like tissue culture, genetic engineering, and molecular breeding have opened new opportunities and have improved the quality of life. This demonstrates that involvement of biotechnology in diverse fields. Biotechnology has no doubt answered a lot of queries related to many mysteries, but one biological and technical domain where it needs intervention is the agricultural waste recycling and application on agro-food industry. Serious environmental injure is being caused by adding enormous quantity of vineyard and winery wastes annually.

Lignocelluloses have the property of creating stern environmental effects if they are allowed to accumulate in the system. At the same time, significant proportion of cereal by-products are being produced all over the globe (Moser, 1994; Verstraete and Top, 1992). Such environmental problems can be easily tackled by biological interventions with the aim to recycle them (Smith, 1998). The cultivation of both edible and medicinal mushrooms was implemented using both solid-state cultivation and operated submerged fermentation of various natural agro-food by-products (Petre and Teodorescu, 2009; Stamets, 2000). No doubt that technological innovations have led to rising standards of living but have also resulted in elevated levels in terms of quantity and variety of solid waste generated by various sectors. The condition is further aggravated by the increasing urbanization and growth of populations. The definition of solid wastes as per U.S. Environmental Protection Agency (EPA) is "any garbage or refuse, sludge from a wastewater treatment plant, water supply treatment plant or air pollution control facility and other discarded material, including solid, liquid, semi-solid, or contained gaseous material resulting from industrial, commercial, mining and agricultural operations and from community activities." India approximately produces about 36.5 million tons of waste annually and it has been calculated that India will generate 400,000–125,000 metric tons of waste by 2030. Industrial solid waste encompasses a wide variety of materials including paper, packaging materials etc.

Biotechnology has attained more attention attributed to its new technologies that have appeared since 1970. Agricultural biotechnology when defined in terms of biotechnology is involving living organisms for decomposition and control of agricultural wastes in the environment. Around 100,000 metric tons of wastes are being generated in India per day and when it comes to Delhi and Mumbai the generation is 8300 and 9000 metric tons per day. MSW generation has reached figures of 1.5 billion tons per day and if such trends continued, then the projected waste generation in 2025 will be 2.5 billion tons. Inefficient and insufficient waste infrastructure development in India is the reason behind the increase in agricultural waste and to convert solid waste into a resource management method aided with sustainability is the need of the hour, and biotechnology offers a suitable answer to this problem.

8.3   WASTES PRODUCED BY CULTIVATION ACTIVITIES

Tropical climates are often considered as optimum for growing crops and the fact that insects and weeds also prefer the same climate is also true. Hence, in order to eradicate pests, high demand of pesticides are being used to prevent the spread of disease. In order to get high yields, a lot of misuse occurs of these pesticides and the containers in which they have been packed (Dien and Vong, 2006).

Thanks to their highly strong and toxic chemicals, these contaminants have the credibility to cause unforeseen environmental effects, such as food poisoning, unsafe food safety and polluted farmland. No doubt, fertilizers play a vital role in the management of plant productivity and quality in agricultural production. Most farmers, however, put in extra fertilizer to their crops than the plants need (Hai and Tuyet, 2010). The stern outcome of such undue use of the application of fertilizer is that it is used to increase the annual production to the point of abuse. When it comes to absorption rate, it varies and depends on the soil characteristics, plant species, and fertilizer compounds (Thao, 2003). Soil possesses the property to retain excess fertilizers, and from soil, it may enter any other ecosystem and may result in pesticide pollution in various ecosystems.

8.3.1   PRODUCTION OF WASTES FROM LIVESTOCK

A slaughter house contains solid waste in the form of organic materials, manure, and also the pollutants like H_2S and CH_4, and odors. Hence, the pollution caused by the livestock production is a matter of concern as most of them are constructed in residential areas. The proportion of ammonia, hydrogen sulfide, and methane varies with the digestion process and also depends on the health status of organic materials, food ingredients, micro-organisms, and animals.

Untreated and nonreusable waste sources may produce emissions, especially greenhouse gases, thus having a negative impact on soil fertility and causing water pollution. Many species of microorganisms and parasitic eggs exist (Hai and Tuyet, 2010), and these germs and substances can spread disease to humans and can cause many environmental hazards.

8.3.2 PRODUCTION OF AQUACULTURE WASTE

Aquaculture growth has demanded increased use of feed, and the quality of feed used determines the quality of waste generated. The main type of waste generated through the aquaculture is the metabolic waste and approximately (30%) of the feed used IN farms become toxic wastes.

Temperature increase results in increased feeding which results in increased waste generated. Water flow patterns are critical for waste management in production units, as proper flow would decrease the depletion of fish populations and allow settable solids to settle and concentrate rapidly. This can be important because it is easy to collect a high percentage of nonfragmented feces to significantly reduce the dissolved organic waste (Mathieu and Timmons, 1995).

8.4 ROUTES/WAYS FOR UTILIZATION OF WASTE

Technology that is being implemented in use of agricultural waste must have the features that it must use the residues efficiently or it should store residues in such conditions that do not induce spoilage. There are a number of applications which include:

8.4.1 APPLICATIONS AS FERTILIZER

The implemention of animal engravings for fertilizer definitely affects the input energy requirements at the farm level (Timbers and Downing,

1977). Manure could supply nitrogen, phosphorus, and potassium in chemical fertilizers at 19%, 38%, and 61%. However, the use of large containment fertilizers is linked to high energy costs for transport, allocation, processing time, odor problems, and the potential for groundwater contamination. It has been reported by Mokwunye (2000) that manure from poultry contains phosphorus and this phosphorus has shown positive effects on crop productivity. Phosphorous addition results nutrient enrichment and improves cation exchange efficiency there by improves physical condition and water retention capacity.

8.4.2 ANAEROBIC DIGESTION

Methane gas, especially manure, can be produced from agricultural waste and it is best suited for heating purposes, such as grilling, water heating, and food drying. Microbial fermentation takes part in the anaerobic digestion of agricultural waste in two stages and the final product is methane-rich gas. Acid forming bacteria especially methanogenic species break down solids into organic acid and final product is methane gas. The composition of the standard gas produced is methane, 50%–70%; CO_2, 25%–45%; N_2, 0.5%–3%; H_2, 1%–10% with traces of H_2S; and the calorific value of the gas is 18–25 MJ/m^3 (Timbers and Downing, 1977). Methane gas being highly volatile and the high capital expenses for the production of this gas are some of the main drawbacks of the digestion system. Nevertheless, the benefits far outweigh the aforesaid drawbacks.

8.4.3 ELIMINATION OF HEAVY METALS BY ADSORBENTS FROM AGRICULTURAL WASTES

Due to industrialization and urbanization, the unregulated introduction of heavy metals into the atmosphere has presented a major problem worldwide. In healthy end products, including synthetic contaminants, heavy metal ions, such as copper, cadmium, mercury, arsenic, chromium, and lead ions do not degrade, the majority of which are prone to biological degradation (Gupta et al., 2001). Studies on the handling of heavy metal effluent have shown that adsorption is a highly efficient process for the extraction of heavy metal from waste streams and that activated charcoal is widely used (Chand et al., 1994). Agricultural waste has confirmed to be a low-cost alternative in recent years for the management of effluents that contain heavy metals through the adsorption method. Different researchers have explored low-cost agricultural waste, such as rice husk and coconut husk (Ayub and Khan, 2002; Tan et al., 1993), sawdust, oil palm shell (Ajmal, et al., 1996; Khan et al., 2003), neem bark (Ayub et al., 2006), etc., for the removal of contaminants from waste water, particularly heavy metals.

8.4.4 PYROLYSIS

Pyrolysis is the process that takes place in the absence of oxygen and the process vaporize a portion of the material and the temperature rises to 400°C–600°C, leaving char. This is known to be a higher technology method for the use of agricultural waste. Hydrogasification and hydrolysis are others methods used both for the treatment of farm waste chemicals and for energy recovery. In the preparation of fuel alcohol, ammonia for fertilizers, glucose for food and feed, agriculture is of particular importance.

8.4.5 ANIMAL FODDER

Agricultural waste can be utilized as feed for animals, fish, etc. Many agricultural wastes need treatment prior to consumption as certain wastes cannot be used directly by animals. These wastes can be treated mechanically or chemically in order to make them edible. Wastes like fiber residues and roughage need enrichment due to their low nutritive value. Use of agro-industrial waste – corn barn for feeding fish like Tilapa fish (Otubusin, 2001) has yielded increased fish weight, best feed conversion rate and showed specific growth rate. Mechanically, the waste is shredded and subjected to high temperature and pressure in order to increase its digestibility. Chemicals are applied to the agricultural wastes in which chemical treatment is provided to the finely shredded agricultural waste, and the chemical used to treat waste is either ammonia or urea. Addition of chemical not only increases the digestibility but also enhances the nutritional value by increased protein content. Biological process involves the burial of waste under soil without aeration for 2 or 3 months and making it feasible for animal feeds.

8.4.6 DIRECT COMBUSTION

This method is one of the oldest ways to process the agricultural waste by burning the waste as fuel for heating, producing coal, cooking, etc. The energy in the form of radiation and thermal energy can be put to best use by manufacturing the waste in solid form before using it for the production of electricity. Combustion of agro waste means "Rapid chemical reaction of biomass and oxygen (oxidation), release of energy and the simultaneous formation of organic matter-CO_2 and water ultimate oxidation products" (Klass, 2004).

8.5 SYSTEM FOR AGRICULTURAL WASTE MANAGEMENT

Development of agricultural management systems for organic farming and sustainable development is of great concern from policy point of view (Hai and Tuyet, 2010). The management of waste without causing the environmental pollution is not possible without better technologies and best use of resources, behavioral, and psychological drifts toward sustainable management. Organic wastes, if poorly handled can result in air, water, and soil pollution and can become a threat to public health. It can also lead to volatilization of ammonia which may result in acid rains (Wright, 1998). Practices of livestock processing arises questions related to degradation of water quality, pathogens, odor problems, release of ammonia, methane and other noxious emissions and increased potassium and phosphorus loadings in soil (Fabian et al., 1993).

USDA (2012) defines "Agricultural Waste Management Systems" (AWMS) as a well-designed system in which all necessary components are constructed and learned to control and use by-products of agricultural production in a way that maintains or improves the quality of air, water, soil, plant, and animal husbandry. it also defines basic functions of AWMS which are "produce", "set", "transport," "refining," "transition," and "use." Output is a feature of the quantity and quality of the produced agricultural waste. If the volumes generated are necessary to become a resource issue, the pollution needs control. A full output review includes the form, quality, quantity, position, and timing of the waste generated. An AWMS Program includes collection process with position of collection points, collection schedule, manpower specifications, necessary equipment or construction infrastructure, product management and installation costs, storage process with facility design and capacity and treatment of waste including physical, chemical, and biological treatments including pretreatment and posttreatment.

8.5.1 THE 3RS IN AGRICULTURAL WASTE MANAGEMENT (AWM)-A HIERARCHICAL APPROACH

The 3R strategy is stated by USDA (2012) as a conventionally articulated Hierarchical pyramid in which the position of each method arises in the environmental benefits from bottom to top. In AWM, the 3R approach refers to the removal of waste by reducing the volume of waste, reusing the waste products, and finally recycling the waste to generate the same or changed products. Many waste products can be used as tools to produce different items or the same commodity, implying that the same material can be recovered. If waste is regularly collected, it prevents processing of fresh similar or identical items. This 3R concept preserves fresh energy, adds value to the capital already used and most critically minimizes waste quantity and its adverse effects. Reduce, Reuse, and Recycle is aimed to achieve an effective minimization of waste production by:

- Choice to use an item carefully in order to lessen amount of generated waste.
- Use of usable items or parts of items repeatedly.
- Use of waste as resource.

8.5.2 POULTRY WASTE MANAGEMENT OPTIONS

The AWMS provides strategies to manage poultry wastes by applying each function of AWMS.

8.5.2.1 PRODUCTION

Poultry waste contains compost and dead chicken. In addition, it includes garbage, flush water, and feed waste depending upon the system.

8.5.2.2   COLLECTION

The poultry waste is collected on the floor where it is combined with feces and creates a "pan" which is typically extracted between flocks. This litter pack is collected either in deep stacks or frequently isolated using shallow pits or belt scrappers places under the cages.

8.5.2.3   STORAGE

Poultry waste is deposited outside the dwelling or building. It can be directly used to fields or compacted for later use. For longer collection, the poultry litter should be kept in a roofed facility where it can be processed under dry conditions.

8.5.2.4   TREATMENT

Poultry litter is compostable. Composting stabilizes the litter into odorless and disease-free mass which can be used as bedding or supplementary feed stock.

8.5.2.5   TRANSFER

The waste can be transferred via trucks or pipes, tank wagons or gutters depending upon the state of the litter, that is, solid or liquid.

8.5.2.6   UTILIZATION

Poultry waste can be used for agricultural lands as manure or marketed as biofertilizers. It can also be used as fuel, bedding for animals or supplementary feed for livestock.

8.6   BIOLOGICAL TECHNIQUES FOR THE MANAGEMENT OF AGRICULTURAL WASTE

The primary objective of this chapter is to study the various biotechnology processes used in the treatment of agricultural waste, and also to recognize the efficacy of the different biotechnological methods used in the treatment of agricultural waste. There was an effort to research the appropriate and creative biotechnologies for the fluctuating intensity of agricultural waste production.

8.6.1   COMPOSTING

Composting is an antiquated method (King, 1911). It is a process primarily designed for the use in agriculture of solid wastes of animal and plant origin. Recently the idea of farm composting has been extended to treat industrial garbage, water, and night soil of human origin. Much of the early work on composting was related to evolving measures to achieve quick composting by the addition of chemicals and determining the factors involved in optimum composting. Artificial farmyard manure fortified with chemicals was known earlier (Russel and Richards, 1917). Similarly, the role of nitrogenous starters (0.7–0.8 g mineral nitrogen per 100 g composting materials) and the importance of C/N ratio of composting material as actors in decomposition were also recognized early (Hutchinson and Richards, 1921). At the same time, it was soon realized that fungi were involved in cellulose breakdown (Rege, 1927; Tenney and Waksman, 1929; Waksman and Stevens, 1928). Some of the important points which are relevant in scientific composting are as follows:

- Shredding of materials to smaller pieces.
- The initial C/N ratio of the material because a ratio of 30–35 is ideal for composting. A low C/N ratio encourages the ammonia to escape while a high C/N ratio is not congenial for microbial equilibrium.
- Frequent turning of compost material for proper decomposition and controlling fly breeding and lessening noxious odors.
- Maintenance of optimum temperature at 50°C–60°C which helps in destroying harmful pathogens and Ascaris eggs (Acharya, 1940; Yadav and Subba Rao, 1980).

In aerobic method, a pit is dug near the cattle house on a site free from waterlogging. Plant residues are shredded and piled up in the pit with the excreta of cattle in layers. The layering of plant residues and cattle excreta is done alternately so as to fill the pit completely. The residues are turned once in a fortnight and a good quality compost is made in about 3 months. In the anaerobic method, the layers in the pit are sealed with plaster of mud.

Due to the anaerobic conditions, the temperature inside the pit rises. Variations of this process involve the opening of the mud plaster periodically, followed by the turning over of the contents and the resealing of the pit by mud plaster. This leads to an alteration of aerobic and anaerobic conditions. Individual preferences, practices, and needs dictate the adoption of any of these methods. In china (FAO report), various devices are adopted to combine the anaerobic and aerobic methods in circular or rectangular pits filled with aquatic weeds, rice straw, animal dung, silt, shredded crop residues, night soil, urine, etc. The turning over of the residues at intervals and the resealing of the mud plaster of the pits are essential ingredients of the Chinese methods. Hollow bamboo poles are driven into the sealed mud plaster to provide aeration which can easily be removed and the holes plastered again with mud to provide anaerobic conditions.

8.6.2 ACCELERATION OF COMPOSTING

The composting process is essentially a microbiological process. This cycle can be enabled by raising microbial activity by introducing nitrogen-rich materials, such as dried blood, cow dung, oil-cakes, ammonium sulfate, sodium nitrate, urea, or ammonia. The composting cycle can be speeded up by adding chosen cellulolytic microorganisms to the composting pit.

A biological mechanism where microbes convert the plant material into soil-like material is called composting and the material and the material formed is called compost. The frequently used processing method is that of given by Geisel (2001) and El Hagga et al. (2004). Environmental conditions play a significant role in decomposing organic waste and microbial behavior is also affected by these conditions. Certain factors influence the process of composting which ultimately influence the decomposition rate also, and these factors include particle size, moisture content, aeration, temperature, C/N ratio, and pH. Subsequently, application of moisture is likely to have an effect on the N fertilizer content of the mature compost. In order to be more efficient and to dominate the indigenous microbiota, the microbial inoculations of bacterial species like *Streptomyces aurefaciens, Trichoderma viridie, T. harzianum, Bacillus subtilis,* and *B. licheniformis* can be used for successful degradation (Badr El Din et al., 2000). Composting has vital benefits in eliminating waste, killing plant seeds, and pathogenic microorganisms (Bernal et al., 2009). Additional benefits of composting as a waste management tool include generating useful soil changes, low operating costs, simple implementation in most developing countries and facilitating environmentally friendly activities, such as greenhouse gas emission mitigation, supporting fertilizer application performance (Hoornweg et al., 2000). One benefit of sustainable agriculture is its lower reliance on chemical fertilizers and on-farm residue recycling to preserve and/or boost soil fertility. Agricultural waste recycling is worthy of providing high-quality organic fertilizers which can be used to fertilize farmland. Management of the residue plays a vital role in nutrient cycle and plant pathogen dynamics. Microorganisms because of their powerful enzymatic mechanisms play significant role in recycling (Blaine Metting, 1993). Microbial technologies for agriculture and waste management are receiving considerable notice to meet the special need of developing countries. Among the technologies available, composting is a fundamental method that suits different scales of application, from individual to community level. The process of waste decomposition and wastewater treatment is carried out under controlled conditions with the help of microorganisms and Flea present in nature. This approach contains the following basic steps to complete the procedure.

- Site planning.
- Waste separation (food and vegetable waste).
- Organic matter (dry) collection.
- Waste layer-by-layer inclusion.
- Cover the plastic covered container to sustain the required humidity and temperature.

In this process, as a material, the organic matter is transformed into humus, the final product looks like soil rich in carbon and nitrogen, and the plants are optimally grown. The final product developed after the composting process increases the sensitivity of the soil to retain water and makes it easier to cultivate the soil. As the breaking of solid waste starts in the presence of worms and increases the richness of the nutrients in compost, vermicomposting has recently become more effective. Such procedures are time-consuming and human monitoring is required. Hence in recent years, rapid composting has been implemented. Fast composting increases the degradation rate of chemicals or waste by maintaining a high nutrient content. The main goal for this process is to maintain the optimum condition of all vital parameters and to increase the composting rate. Rapid composting of solid waste by incorporating chemicals is a process by which organic compounds, such as bauxite, phosphogysum, and pesticides are applied to increase the amount of waste biodegradation. The application of bauxite raises the heat, pH, and serves as a mechanism for composting mixture aeration. Glucose introduction as instant carbon often increases the rate of composting mixture decomposition. Another approach – The cycle of Berkley rapid composting (BRC) includes regular composting content turning and shredding. This system is ideal for the scale of 1/2 to 1 1/2 inches composting waste. It takes about 2–3 weeks for this process.

8.6.3   VERMICOMPOSTING

Vermicomposting is the process of the production of compost through earthworm action. It is an eco-biotechnological process that converts organic substances rich in energy into stabilized vermin-compost-like humus products. Vermi compost preparation is an effective as well as easily adopted compost preparation technique. Not only can this composting system decompose a huge amount of organic waste but it also helps to maintain a higher nutrient status in composted materials (Bajsa et al., 2004; Lazcano and Domínguez, 2011; Hema and Rajkumar, 2012). An environmentally safe way to convert waste into nutritious compost for crop production is the earthworm vermicomposting technology (Edward et al., 1985; Yadav and Garg, 2011). In addition, this technology transforms the problem into a resource through the processing of waste and provides good manure that can be used to improve the soil quality (Azarmi et al., 2008).

One value of vermicomposting is the obvious advantage of holding out of landfills rotting waste. Furthermore, the passing through the digestive tracts of the worms alters the very structure of the substance, turning certain nutrients into a shape that is much more bioavailable to the plants to which it is added. Bins are inexpensive to make and retain, and the fertilizer produced by the worms is completely organic and also much healthier than store-bought fertilizer.

Vermicomposting is a biooxidant cycle in which earthworms in the decomposer culture intensively associate with microorganisms, speeding up organic matter through stabilization with modified physical and biochemical properties. Vermicomposting varies from traditional composting, because the digestive systems of earthworms absorb the organic material. The digested casts may be used to improve soil fertility and physical properties. In this cycle, the earthworms actively participate through physical and biochemical intervention in the destruction of organic matter. Physical involvement in the oxidation of organic substrates results in separation, to improve activity and aeration of the surface area. Alternatively, metabolic shifts in organic matter oxidation are achieved by microorganisms by enzymatic fermentation, nitrogen excretion accumulation, and inorganic and organic content transport. The earthworms contribute significantly to the recycling of organic waste and the development of high humic organic manure which helps to maintain the soil structure, aeration, and fertility. The bioactive substances in the humic acid fertilizer will increase the physiological metabolism, development, production, germination of seeds, etc., while in ordinary fertilizers those characteristics are absent. The use of humic acid fertilizer can also increase crop anti-dry and ant frigidity potential effectively and prevent underground plant disease, insect pests, and pathogenic bacteria. The present reviews describe various aspects that include the composting of organic waste by different worm organisms with vermin.

8.6.4 BIOREMEDIATION

Environmental threats and risks arising from residual lethal chemicals or other waste and toxins could be reduced or eliminated through introducing biotechnology in the process of (bio) treatment/(bio)remediation of historical emissions, and by mitigating contamination resulting from existing manufacturing processes by emission prevention and control methods. The US Environmental Protection Agency (USEPA) describes bioremediation as "a controlled or random activity in which microbiological processes are used to degrade or turn pollutants into less harmful or nontoxic materials, thus remediating or removing environmental contamination" (USEPA, 1994; Talley, 2005). Bioremediation involves four processes that act on the contaminant (Asante- Duah, 1996; FRTR, 1999; Khan et al., 2004; Doble and Kumar, 2005; Gavrilescu, 2006).

1. Elimination of contaminant from site without need for separation from host medium.
2. Separation from host medium.
3. Destruction or degradation of contaminants to less toxic compounds.
4. Retention of subsurface movement pollutants.

The first three are processes meant to reduce the contaminant while containment is a monitoring technique that checks the movement of the contaminant to different susceptible receptors (Watson, 1999; Khan et al., 2004; Gavrilescu, 2006).Two routes may be used to eliminate any toxins from the atmosphere; degradation and immobilization through a process that makes it biologically unavailable for degradation and thus effectively eliminates it (Evans and Furlong, 2003). Immobilization may be achieved by chemicals released by animals or introduced in the neighboring atmosphere that capture or chelate the contaminant, rendering it insoluble and therefore inaccessible as an individual in the ecosystem. Immobilization may also be a major remediation issue, as it can contribute to aging pollution and a lot of research effort needs to be applied to identify solutions for changing the cycle back. Destruction involves biotransformation and biodegradation by an individual or mixture of species. It forms an important process of environmental cleanup methods and is referred to as cornerstone of environmental biotechnology. Valuable microbial reactions, including oxidation and detoxification of toxic organic compounds, inorganic materials, metal transformation, gaseous, liquid and solid waste, rely on biological processes (Eglit, 2002; Evans and Furlong, 2003; Gavrilescu, 2004). Full biodegradation helps in detoxification of carbon dioxide, soil, and innocuous inorganic salts by mineralizing the toxins. Incomplete biodegradation may result in breakdown products that may or may not be less harmful than the initial pollutant, so solutions, such as dispersion, dilution, biosorption, volatilization, and/or chemical or biochemical must be addressed. (Lloyd, 2002; Gavrilescu, 2004).

Bioaugmentation often includes the systematic introduction of microorganisms that have been bred, modified, and optimized at the site for particular pathogens and environments. Biorefining involves microbes being used in mineral-processing systems. It is an environmentally friendly procedure and in some situations, makes mineral recovery and resource use that would otherwise not be feasible.

Microbes and plants are considered vital in environmental biotechnology as both have potential to remediate the contaminants from the environmental (Evans and Furlong, 2003). The generic term "microbe" includes prokaryotes such as bacteria (or arcaea) and eukaryotes like yeasts, fungi, protozoa, and unicellular plants, rotifers.

Some of these species have the potential to kill some of the most toxic and recalcitrant compounds, as they have been found in unfriendly conditions where their composition and metabolic capability is impaired by survival needs. Microorganisms may exist in mixed cultures (consortia) as free individuals or as groups that are of particular interest in many specific environmental technologies, such as activated sludge (Gavrilescu and Macoveanu, 1999; Gavrilescu and Macoveanu, 2000; Metcalf and Eddy, 1999). Microbial community structures in activated sludge, made up of activated sludge flocks containing various types of microorganisms, are one of the most important elements in the design of biological wastewater treatment systems (Wagner and Amann, 1997; Wagner and Loy, 2002). The role of plants in the cleanup of the environment is exercised during the oxygenation, filtration, and conversion of solid to gas or removal of contaminants. The use of animals to eliminate contaminants is based on the concept that every organism should extract substances from the atmosphere for its own development (Hamer, 1997; Saval, 1999; Wagner and Loy, 2002; Doble et al., 2004; Gavrilescu, 2004; Gavrilescu and Chisti, 2005). Both bacteria as well as fungi are very effective in complex chemicals to safer end products. Fungi will ingest complex organic substances that are not typically processed by other organisms.Protozoa–algae and plants have proven ideal for the removal of nutrients like nitrogen, phosphorus, sulfur and many minerals and metals from the atmosphere.

8.6.5 BIOFILTERATION

Biofilteration can be used as pollution combating of air and water with partition coefficient less than 1. A biofilter consists of a filter bed composed of organic matter, for example, peat, sawdust, compost, etc. acting as carrier for biomass as well as nutrient source. Biofilteration is recommended for poorly water soluble pollutants. Various benefits of biofiltering are provided in agricultural sector applications. Anaerobic and aerobic biofilters have been successfully utilized for the treatment of dilute piggery wastes. A variety of anaerobic and aerobic biofilters are postulated to be capable of providing a fast detention period system for heavy agricultural effluents capable of achieving high effluent efficiency from operations where historically suspended growth systems have been implemented.

8.6.6 BIOSORPTION

Biosorption is a metabolically passive process, implying that it does not involve energy, and the amount of pollutants that a sorbent may eliminate depends on the kinetic equilibrium and the cellular surface composition of the sorbents. Pollutants adsorb onto the surface of the tissue. Bioaccumulation is an ongoing biochemical mechanism powered by energy from a living organism which involves respiration. The approach may be an alternative to traditional waste treatment plants (Yurtsever and Engil, 2009). A number of microbial and other forms of biomass have shown good potential for biosorption, and several have been suggested as the basis for the treatment of metal-bearing waste (Orhan et al., 2006). As a polishing or adjunct process, biosorption provides low-cost benefits, better efficiency especially in comparison to techniques, such as precipitation and ion exchange, and does not produce high-metal sludge without high-metal sludge (Pino et al., 2006). The metals are not only extracted from waste in biosorption but are also recycled to be reused for various purposes. One of the more common questions that biosorption processes pose includes the fate of the after-process biosorbent. The fate of the condensed substances produced after the phase of elution is still largely unanswered. A solution can be extracted from these highly concentrated substances using another method, such as precipitation or electrowinning. The ultimate treatment of the substance should be discussed even if the biosorbent can be effectively recovered over several cycles (Vijayaraghavan and Yun, 2008)

8.6.7 ANAEROBIC TREATMENT

By holding temperature, moisture content, and pH close to their optimum values, anaerobic degradation of organic waste accelerates the natural decomposition of organic material without oxygen. The produced CH_4 can be used for generating heat and/or electricity. (Mata-Alvarez et al., 2000; Salminen and Rintala, 2002). Solid waste biotreatment is one of the most popular applications (TBV GmbH, 2000).

There are a variety of interdependent steps in the anaerobic cycle. In order to make organic compounds simpler, complex organic compounds, such as lipids, proteins, and carbohydrates are originally hydrolyzed. Acidogens then ferment these into volatile fatty acids (VFAs).. Of these fatty acids, the most common is ethanoic acid. However, propanoic, butanoic, and pentanoic acids may also be present in varying amounts depending on the process's stability. Despite the production of acids through the process, the device must be sufficiently buffered to prevent pH declines that can adversely affect the further progress of the method.

The acidogens contain both optional and mandatory anaerobic bacteria. The total amount of organic material in the wastewater would not have changed significantly until this point in the process, although the type and complexity of organic compounds could have significantly changed. The carbon dioxide is the gaseous by-product of the acid reactions. The methanogenic process is preceded by the acidogenic period. The methanogens are mandatory anaerobes that turn the fatty acids into methane and carbon dioxide from acidogenesis. This results in a substantial reduction of the wastewater's organic content. The generated methane offers an avenue for recovering energy. The anaerobic cycle is a complex process, and it has a significant risk of becoming dysfunctional and ultimately collapsing. The lack of molecular oxygen is among the important environmental factors that ought to be present. This is particularly so for high-rate systems. These anaerobic structures should be equipped with reactors that impose positive pressure inside the vessels to exclude oxygen. In an anaerobic reactor, the microbial consortium often requires an adequate macronutrient and micronutrient balance to ensure that microbial growth may occur. Though anaerobes have a relatively slow rate of growth, this is a good trait for sludge digestion as it means low solid production (Wun Jern, 2008).

The benefits of anaerobic biotechnology are summarized as shown below:

i)   Stability of processes.
ii)   Cutback of costs for waste biomass disposal.
iii) Decrease of supplementation costs for nitrogen and phosphorus.
iv)   Reduction of requirements for installation space.
v)   Energy efficiency, promising ecological and economic benefits.
vi)   Elimination of pollution by off-gas air.
vii) Prevention of foaming of excess water from surfactants.
viii) Biodegradation of nonbiodegradables from aerobic sources.
ix)   Significant decrease in levels of chlorinated organic toxicity.
x)   Seasonal treatment provision.

While these positive features are listed above in anaerobic biotechnology, there are also some negative conditions for this technology as follows:

i)   Long start-up prerequisite for biomass inventory production.
ii)   Inadequate potential for inherent alkalinity generation in dissolved or carbohydrate wastewater.
iii) Inadequate effluent quality for surface water discharge in some cases.

iv) Inadequate methane production from dilute wastewaters to provide for heating to the 35°C optimal temperature.

v) The generation of sulfide and odor from sulfate feed stocks.

vi) Nitrification not possible.

vii) Significantly greater toxicity of chlorinated aliphatic agents than on aerobic heterotrophs to methanogens.

viii) Low rates of kinetics at low temperatures.

ix) High levels of NH_4 (40–70 mg/L) required to maximize biomass activity (Speece, 1996).

8.7 CONCLUSION

Agricultural wastes are residues from the harvesting and processing of raw agricultural products, are nonproduct manufacturing and processing sources and may include materials that can benefit people. Such residues are generated from a variety of agricultural operations, including grains, livestock, and aquaculture. When properly managed by applying knowledge of agricultural waste management systems such as the 3Rs, this type of waste can be transformed into beneficial materials for human and agricultural use. From the results, it is important to note that adequate selection, storage, care, transition, and reuse of waste represent the panacea to a healthy environment. Proper waste management would help develop our agricultural sector and provide many with a viable biofuel option. Developing environmentally friendly, energy-efficient, and cost-effective waste solutions with market potential to meet the needs of citizens in rural and urban areas is today's need. The specific challenge is to arrive at a sustainable and scalable solution for value-added product development. Solid waste production is increasing exponentially around the globe. Biotechnologies are proposed as effective techniques for safe management of solid wastes. For simple scaling of organic waste from individual to community level, the composting method is most appropriate. Bioremediation is the most appropriate method for various types of solid waste, ranging from agricultural waste to industrial heavy metals. It also poses future potential research in converting applying bioremediation for chemical and petroleum industrial solid wastes. Nonetheless, waste generation remains a major consequence of almost all human activity, including agro-food production and processing operations, and these have far-reaching environmental and health consequences in general, calling for measures to be taken to control them if the adverse results of their accumulation are to be prevented. However, the bulk of waste management solutions introduced over time continued to see waste as a worthless thing that has to be securely disposed of. A close look at most natural systems, however, gives a verdict that omnipotent and omniscient creators abhor waste of any kind, since nature has a way to use every bit of its resources. Biotechnological concepts see waste from a different perspective, because what is considered to be waste may not truly be entirely so. In agro-food wastes, this is especially evident because most of what is often regarded as wastes consigned to the waste bins contain a reasonable percentage of bioconvertible reserves that are bioconvertible into useful products trapped in them. In this regard, the role of modern technology particularly biotechnology, in the exploitation of resources is difficult to overestimate, as is evident from a wide range of useful resources: animal feed and feed, biofertilizers, industrial chemicals/raw materials, biofuels, biogas, and other alternatives to renewable energy, etc. derived from so-called waste, using a biotechnology tool.

KEYWORDS

- **biotechnology**
- **agricultural wastes**
- **energy-efficient**
- **vermicomposting**
- **organic wastes**

REFERENCES

Acharya, C. N. Composts and Soil Fertility. *Indian Farm.* **1940,** *1*, 66–68, 121–125.

Acharya, C. N. Relation between Nitrogen Conservation and Quantity of Humus Obtained in Manure Preparation. *Indian Farm.* **1946,** 7, 66–67.

Acharya, C. N. Studies on the Hot Fermentation Processes for the Composting of Town Refuse and Other Waste Material. *Ind. J. Agri.Sci.* **1939,** *9*, 817–833.

Agamuthu, P. Challenges and Opportunities in Agro-Waste Management: An Asian Perspective. In *Inaugural Meeting of First Regional 3R Forum in Asia,* 2009; pp 11–12.

Ahmed, Z.; Lim, B.R.; Cho, J.; Song, K. G.; Kim. K.P.; Ahn, K. H. Biological Nitrogen and Phosphorus Removal and Changes in Microbial Community Structure in a Membrane Bioreactor: Effect of Different Carbon Sources. *Water Res.* **2008**, *42*, 198–210.

Ajmal, M.; Rao, R. A. K.; Siddiqui, B. A. Studies on Removal and Recovery of Cr (VI) from Electroplating Wastes. *Water Res.* **1996**, *30*, 1478–1482.

Aksu, Z.; Yener, J. Investigation of the Biosorption of Phenol and Monochlorinated Phenols on the Dried Activated Sludge. *Process Biochem.* **1998**, *33*, 649–655.

Aktan, G. Treatment and Evaluation of Wastes via Microorganisms. *Indust. Microbiol.* **1983**, 404–410.

Al-Malack, M. H. Determination of Biokinetic Coefficients of an Immersed Membrane Bioreactor. *J. Membr. Sci.* **2006**, *271*, 47–58.

Antizar-Ladislao, B.; Galil, N. I. Biosorption of Phenol and Chlorophenols by Acclimated Residential Biomass under Bioremediation Conditions in a Sandy Aquifer. *Water Res.* **2006**, *38*, 267–276.

Ayub, S.; Ali, S. I.; Khan, N. A. Adsorption Studies on the Low Cost Adsorbent for the Removal of Cr(VI) from Electroplating Wastewater. *Environ. Pollut. Cont. J.* **2006**, *5*, 10–20.

Azarmi, R.; Giglou, M. T.; Taleshmikail, R. D. Influence of Vermicompost on Soil Chemical and Physical Properties in Tomato (*Lycopersicum esculentum*) Field. *Afr. J. Biotech.* **2008**, *7*, 2397–2401.

Bahadir, T.; Bakan, G.; Altas, L.; Buyukgungor, H. The Investigation of Lead Removal by Biosorption: An Application at Storage Battery Industry Wastewaters. *Enzyme Microb. Technol.* **2007**, *41*, 98–102.

Bajsa, O.; Nair, J.; Mathew, K. and Ho, G. E. Vermiculture as a Tool for Domestic Waste Water Management. *Water Sci. Tech.* **2004**, *48*, 125–132.

Barbosa, V. L.; Tandlich, R.; Burgess, J. E. Bioremediation of Trace Organic Compounds Found in Precious Metal Refineries' Wastewaters: A Review of Potential Options. *Chemosphere* **2007**, *68*, 1195–1203.

Bermek, E. Importance of Biotechnology and Sciences. *Tubitak Bull.* **1989**, *6*, 16–17.

Betmann, H.; Ehrhardt, H. M.; Rehm, H. J. Degradation of Phenol by Immobilized Microorganisms. *3rd Eur. Biotechnol. Congress.* **1984**, 27–33.

Brown and Root Environmental Consultancy Group. Environmental Review of National Solid Waste Management Plan. Interim Report Submitted to the Government of Mauritius, 1997.

Buyukgungor, H. Biotechnology and Environment. First International Scientific Congress between Turkish States, 1992; pp 168–179.

Chand, S.; Aggarwal, V. K.; Kumar, P. Removal of Hexavalent Chromium from the Wastewater by Adsorption. *Indian J. Environ. Health* **1994**, *36*, 151–158.

Chang, Y; Hudson, H. J. The Fungi of Wheat Straw Compost. I. Ecological Studies. *Trans. Br. Myco. Soc.* **1967**, *50*, 649–666.

Charcosset, C. Membrane Processes in Biotechnology: An Overview. *Biotechnol. Adv.* **2006**, *24*, 482–492.

Chavan, A.; Mukherji, S. Treatment of Hydrocarbon-Rich Wastewater Using Oil Degrading Bacteria and Phototrophic Microorganisms in Rotating Biological Contactor: Effect of N:P ratio. *J. Hazard. Mater.* **2008**, *154*, 63–72.

Council for Agricultural Science and Technology Utilization of Animal Manures and Sewage Sludge in Food and Fiber Production. Report No. 41, 1975.

DeCarolis, J. F.; Adham, S. Performance Investigation of Membrane Bioreactor Systems during Municipal Wastewater Reclamation. *Water Environ. Res.* **2007,** *79,* 2536–2550.

Department of Environment. National 3R Strategy for Waste Management. Ministry of Environment and Forests, Government of the People's Republic of Bangladesh, 2010.

Dermou, E.; Vayenas, D. V. A Kinetic Study of Biological Cr(VI) Reduction in Trickling Filters with Different Filter Media Types. *J. Hazard. Mater.* **2007,** *145,* 256–262.

Dien, B. V.; Vong, V. D. Analysis of Pesticide Compound Residues in Some Water Sources in the Province of Gia Lai and DakLak. Vietnam Food Administrator, 2006.

DiPalma, L.; Verdone, N. The Effect of Disc Rotational Speed on Oxygen Transfer in Rotating Biological Contactors. *Bioresour. Technol.* **2009,** *100,* 1467–1470.

Doan, H. D.; Wu, J.; Jedari, E. M. Effect of Liquid Distribution on the Organic Removals in a Trickle Bed Filter. *Chem. Eng. J.* **2008,** *139,* 495–502.

Dobson, R. S.; Burgess, J. E. Biological Treatment of Precious Metal Refinery Wastewater: A Review. *Miner. Eng.* **2007,** *20,* 519–532.

Eastwood, D. J. The Fungus Flora of Composts. *Trans. Br. Myco. Soc.* **1952,** *35,* 215–220.

Edwards, C. A.; Burrows, I.; Fletcher, K. E.; Jones, B. A. The Use of Earthworms for Composting Farm Wastes. In *Composting Agricultural and Other Wastes*; Gasser, J. K. R., Ed.; Elsevier: London and New York, 1985; pp 229–241.

EFB (European Federation of Biotechnology). Environmental Biotechnology. Briefing Paper 4, 1999; pp 1–4.

Elliot, L. F.; Travis, T. A. Detection of Carbonyl Sulfide and Other Gases Emanating from Beef Cattle Manure. *Soil Sci. Soc. Am. Proc.* **1973,** *37,* 700–702.

Environmental Engineers' Handbook, 2nd ed.; CRC: USA, 2007.

Fan, F.; Zhou, H.; Husain, H. Use of Chemical Coagulants to Control Fouling Potential for Wastewater Membrane Bioreactor Processes. *Water Environ. Res.* **2007,** *79,* 952–957.

Gadd, G. M. Bioremedial Potential of Microbial Mechanisms of Metal Mobilization and Immobilization. *Curr. Opin. Biotechnol.* **2000,** *11,* 271–279.

Gannoun, H.; Khelifi, E.; Bouallagui, H.; Touhami, Y.; Hamdi, M. Ecological Clarification of Cheese Whey Prior to Anaerobic Digestion in Upflow Anaerobic Filter. *Bioresour. Technol.* **2008,** *99,* 6105–6111.

Gavrilescu, M.; Chisti, Y. Biotechnology—A Sustainable Alternative for Chemical Industry. *Biotechnol. Adv.* **2005,** *23,* 471–499.

Golucke, C. G. Composting: A Review of Rationale, Principles and Public Health. *Compost Sci.* **1976,** *17,* 20–28.

Gray, K. R.; Sherman, K.; Biddlestone, A. J. A Review of Composting-Part I. *Proc. Biochem.* **1971,** 32–36.

Guler I.; Gurel, L.; Bahadir, T.; Buyukgungor, H. Biosorption of Nickel (II) Ions from Aqueous Solutions by Rhizopus Arrhizus Attached on Rice Bran. *J. Biotechnol.* **2007,** 131–139.

Guler, I.; Buyukgungor, H. The Treatment of Wastewater Containing Phenolic Compounds Using Biological Methods. In *5th IWA Leading-Edge Conference on Water and Wastewater Technologies,* 2008.

Guo, W.; Vigneswaran, S.; Ngo, H. H.; Xing, W.; Goteti, P. Comparison of the Performance of Submerged Membrane Bioreactor (SMBR) and Submerged Membrane Adsorption Bioreactor (SMABR). *Bioresour. Technol.* **2008,** *99,* 1012–1017.

Gupta, V. K.; Gupta, M.; Sharma, S. Process Development for the Removal of Lead and Chromium from Aqueous Solution Using Red Mud—An Aluminum Industry Waste. *Water Res.* **2001,** *35,* 1125–1134.

Gurel, A. C.; Buyukgungor, H. Phenol Removal Using Phenobac Bacteria. Bioremediation of Soils Contaminated with Aromatic Compounds; NATO ARW; 2004; p 27.

Hai, H. T.; Tuyet, N. T. A. Benefits of the 3R Approach for Agricultural Waste Management (AWM) in Vietnam. Under the Framework of Joint Project on Asia Resource Circulation Policy Research Working Paper Series. Institute for Global Environmental Strategies Supported by the Ministry of Environment, Japan, 2010.

Hema, S.; Rajkumar, N. An assessment of Vermicomposting Technology for Disposal of Vegetable Waste Along with Industrial Effluents. *J. Environ. Sci. Comp. Sci Engi. Technol.* **2012,** *1,* 5–8.

Hussein, S. D. A.; Sawan, O. M. The Utilization of Agricultural Waste as One of the Environmental Issues in Egypt (A Case Study). *J. Appl. Sci. Res.* **2010,** *6,* 1116–1124.

Hutchinson, H. B.; Richards, E. H. Artificial Farm Yard Manure. *J. Min. Agric.* **1921,** *28,* 398–411.

Iqbal, M.; Edyvean, R. G. J. Biosorption of Lead, Copper and Zinc Ions on Loofa Sponge Immobilized Biomass of *Phanerochaete chrysosporium. Miner. Eng.* **2004,** *17,* 217–223.

Judd, S. The Status of Membrane Bioreactor Technology. *Trends Biotechnol.* **2008,** *26,* 109–116.

Khan, N. A.; Shaaban, M. G.; Hassan, M. H. A. Removal of Heavy Metal Using an Inexpensive Adsorbent. *Proc. UM Research Seminar 2003 organized by Institute of Research Management and Consultancy (IPPP),* University of Malaya, Kuala Lumpur, 2003.

King, F. H. Farmers of Forty Centuries on permanent Agriculture in China, Korea and Japan, Jonathan Cape, London, 1911.

Klass, D. L. Biomass for Renewable Energy and Fuels. In *Encyclopedia of Energy;* Cleveland, C. J., Ed.; 2004; pp 193–212.

Kochtitzky, O. W.; Seaman, W. K.; Wiley, J. S. Municipal Composting Research at Johnson City, Tennessee. *Comp. Sci.* **1969,** *9,* 5–16.

Krishna, M. *A Manual on Compost and Other Organic Manures;* Today and Tomorrow's Printers and Publishers, 1978.

Krishna, G.; Kumar, P. Treatment of Low-Strength Soluble Wastewater Using an Anaerobic Baffled Reactor (ABR), 2003.

Lazcano, C.; Domínguez, J. The Use of Vermicompost in Sustainable Agriculture: Impact on Plant Growth and Soil Fertility. In *Soil Nutrients;* Mohammad, M., Ed.; Nova Science Publishers, Inc, 2011.

Lefkosa, Buyukgungor, H.; Wilk M.; Schubert, H. Biosorption of Lead by *C. freundii* Immobilized on Hazelnut Shells. *Proc. V. AIChE World Con.* **2006,** *II,* 437–442.

Leng, R. A.; Choo, B. S.; Arreaze, C. Practical Technologies to Optimize Feed Utilization by Ruminants. In *Legume Trees and Other Fodder Trees as Protein Sources for Livestock;* Speedy, A.; Pugliese, P. L., Eds.; FAO: Rome, Italy, 1992; pp 145120.

Mba, D.; Bannister, R. H. Ensuring Effluent Standards by Improving the Design of Rotating Biological Contactors, 2007.

McAdam, E. J.; Judd, S. J.; Cartmell, E.; Jefferson, B. Influence of Substrate on Fouling in Anoxic Immersed Membrane Bioreactors. *Water Res.* **2007,** *41,* 3859–3867.

Melin, T.; Jefferson, B.; Bixio, D.; Thoeye, C.; De Wilde, W.; De Koning, J.; Van der Graaf, J.; Wintgens, T. Membrane Bioreactor Technology for Wastewater Treatment and Reuse. *Desalination.* **2006,** *187,* 271–282.

Miller, D.; Semmens, K. Waste Management in Aquaculture. Aquaculture in Form at Ion Series, Extension Service, West Virginia University, 2002.

Mohan, D.; Singh, K. P. Multi Component Adsorption of Cadmium and Zinc using Activated Carbon Derived from Bagasse—An Agricultural Waste. *Water Res.* **2002,** *36,* 2304–2318.

Mokwunye, U. Meeting the Phosphorus Needs of the Soils and Crops of West Africa: The Role of Indigenous Phosphate Rocks. Paper presented on Balanced Nutrition Management Systems for the Moist Savanna and Humid Forest Zones of Africa at a Symposium Organized by IITA at Ku Leuva at Cotonun, Benin Republic, Oct 9–12, 2000.

Molin, G.; Nilsson, I. Attached Fermentation Applied to Phenol Degradation with R. Putida. In *3rd European Biotechnology Congress,* 1984; pp 69–74.

Munich, B. H. Using of Agricultural Wastes as Energy Sources. In *2nd International Agricultural Mechanization and Energy Symposium,* 1993; pp 129–136. Ankara (in Turkish).

Munich, M. Anaerobic Biosorption for the Removal of Heavy Metals from Wastes. In *3rd European Biotechnology Congress,* 1984; pp 41–50.

Munich, M. G.; Mohseni, M. The Treatment of Waste Air Containing Phenol Vapors in Biotrickling Filter. *Chemosphere* **2008,** *72,* 1649–1654.

Nakhla, G.; Lugowski, A.; Patel, J.; Rivest, V. Combined Biological and Membrane Treatment of Food-Processing Wastewater to Achieve Dry-Ditch Criteria: Pilot and Full-Scale Performance. *Bioresour. Technol.* **2006,** *97,* 1–14.

Navarro, A. E.; Portales, R. F.; Sun-Kou M. R.; Llanos, B. P. Effect of pH on Phenol Biosorption by Marine Seaweeds. *J. Hazard. Mater.* **2008,** *156,* 405–411.

OECD (Organisation for Economic Cooperation and Development) Report. *A Framework for Biotechnology Statistics;* OECD Publications, 2005; pp 1–52.

Oliveira, E. A.; Montanher, S. F.; Andrade, A. D.; Nobrega, J. A.; Rollemberg, M. C. Equilibrium Studies for the Sorption of Chromium and Nickel from Aqueous Solutions Using Raw Rice Bran. *Process Biochem.* **2003,** *40,* 3485–3490.

Orhan, Y; Hrenovic, J; Buyukgungor, H. Biosorption of Heavy Metals from Wastewater by Biosolids. *Eng. Life Sci.* **2006,** *4,* 399–402.

Overcash, M. R. *Livestock Waste Management;* Humenik, F. J., Miner, R. J., Eds.; CRC Press: Boca Raton, 1973.

Pino, G. H.; Souza de Mesquita, L. M.; Torem, M. L.; Pinto G. A. S. Biosorption of Cadmium by Green Coconut Shell Powder. *Miner. Eng.* **2006,** *19,* 380–387.

Plaza, C.; Xing, B.; Fernandez, J. M.; Senesi N.; Polo, A. Binding of Polycyclic Aromatic Hydrocarbons by Humic Acids Formed During Composting. *Environ. Pollut.* **2009,** *157,* 257–263.

Poincelot, R. P. A Scientific Examination of the Principles and Practices of Composting. *Comp. Sci.* **1975,** *15,* 24–31.

Rege, R. D. Biochemical Decomposition of Cellulosic Material with Special Reference to the Action of Fungi. *Anna. App. Bio.* **1927,** *14,* 1–44.

Roberts, J. A.; Sutton, P. M.; Mishra, P. N. Application of the Membrane Biological Reactor System for Combined Sanitary and Industrial Wastewater Treatment. *Int. Biodeter. Biodegr.* **2000,** *46,* 37–42.

Russel, E. J.; Richards, E. H. The Changes Taking Place during the Storage of Farmyard Manure. *J. Agric. Sci.* **1917,** *8,* 495–563.

San Diego, Chandrasekeran, P.; Urgun-Demirtas, M.; Pagilla, K. R. Aerobic Membrane Bioreactor for Ammonium-Rich Wastewater Treatment. *Water Environ. Res.* **2004,** *79,* 2352–2362.

Sarmah, A. K. Potential Risk and Environmental Benefits of Waste Derived from Animal Agriculture. In *Agriculture Issues and Policies Series—Agricultural Wastes*; Ashworth, G. S.; Azevedo, P., Eds.; Nova Science Publishers, Inc.: New York, 2005. ISBN 978-1-60741-305-9,

Speece, R. E. *Anaerobic Biotechnology for Industrial Wastewaters*; Archae Press: USA, 1996.

Spellman, F. R. *Handbook of Water and Wastewater Treatment Plant Operations*; Lewis Publishers: USA, 2003.

Stutzenberger, F. J.; Kaufman, A. J.; Lossin, R. D. Cellulolytic Activity in Municipal Solid Waste Composting. *Can. J. Micro.* **1970,** *16,* 553–560.

Tan, W. T.; Ooi, S. T.; Lee, C. K. Removal of Chromium (VI) from Solution by Coconut Husk and Palm Pressed Fibre. *Environ. Technol.* **1993,** *14,* 277–282.

Tenney, F. G.; Waksman, S. A. Composition of Natural Organic Material and Their Decomposition in the Soil, IV. The Nature and Rapidity of Decomposition of the Various Organic Complexes in Different Plant Materials under Aerobic Condition. *Soil Sci.* **1929,** *28,* 55–84.

Thao, L. T. H. Nitrogen and Phosphorus in the Environment. *J. Surv. Res.* **2003,** *15,* 56–62.

Thawornchaisit, U.; Pakulanon, K. Application of Dried Sewage Sludge as Phenol Biosorbent. *Bioresour. Technol.* **2007,** *98,* 140–144.

Timbers, G. E.; Downing, C. G. E. Agricultural Biomass Wastes: Utilization Routes. *Can. Agric. Eng.* **1977,** *19,* 84–87.

Totova, N. A. Iron Humus Complexes in Some Soils. *Pochvovedenie* **1962,** *12,* 38–43.

Ünlü, A.; Hasar, H.; Kınac, C.; Çakmakc, M.; Koçer, N. N. Real Role of an Ultrafiltration Hollow-Fiber Membrane Module in a Submerged Membrane Bioreactor. *Desalination* **2005,** *181,* 185–191.

USDA. Agricultural Waste Management Field Handbook; United States Department of Agriculture, Soil Conservation Service, 2012. http://www.info.usda.gov/ viewerFS. aspx?hid=21430 (accessed 10 June 2016).

Ustun, S.; Buyukgungor, H. Removal of Phenol from Aqueous Solutions Using Various Biomass. *J. Biotechnol.* 2007; pp 131:135.

Vijayaraghavan, K.; Yun, Y. S. Bacterial Biosorbents and Biosorption. *Biotechnol. Adv.* **2008,** *26,* 266–291.

Wagner, M.; Loy, A. Bacterial Community Composition and Function in Sewage Treatment Systems. *Curr. Opin. Biotechnol.* **2002,** *13,* 218–227.

Waksman, S. A.; Stevens, R. K. Contribution to the Chemical Composition of Peat, 1. Chemical Nature of Organic Complexes in Peat and Method of Analysis. *Soil Sci.* **1928,** *26,* 113–137.

Waksman, S. A.; Cordon, T.; Hulpoi, H. Influence of Temperature Upon the Microbiological Population and Decomposition Process in Composts of Stable Manures. *Soil Sci.* **1939,** *47,* 83–114.

Webley, D. M. The Microbiology of Composting, I. The Behavior of the Changes Taking Place during Composting. *Proc. Soc. Appl. Bacteriol.* **1947**, *2*, 83–89.

Weiner, R. F.; Matthews, R. A. *Environmental Engineering*, 4th ed.; Butterworth-Heinemann: USA, 2002.

Woodard, F. *Industrial Waste Treatment Handbook*; Butter worth Heinemann: USA, 2001.

Wright, R. J. Executive Summary, 1998. www.ars.usda.gov/is/np/agbyproducts/agbyexec-summary.pdf (accessed on 25 April 2016).

Wu, J; Yu, H. Q. Biosorption of Phenol and Chlorophenols from Aqueous Solutions by Fungal Mycelia. *Process Biochem.* **2006**, *41*, 44–49.

Wun, J. N. G. *Industrial Wastewater Treatment*; Imperial College Press: Singapore, 2006.

Yadav, A.; Garg, V. K. Recycling of Organic Wastes by Employing *Eisenia fetida. Biol. Technol.* **2011**, *102*, 2874–2880.

Yadav, A.; Subba Rao, N. S. Use of Cellulolytic Microorganisms in Composting. In *Recycling Residues of Agriculture and Industry. Proceedings of a Symposium at the Punjab Agricultural University*, Ludhiana, India (Edited by M. S. Kalra), 1980; pp 267–273.

Yuan, L. M.; Zhang, C. Y.; Zhang, Y. Q.; Ding, Y.; Xi, D. L. Biological Nutrient Removal Using an Alternative of Anoxic and Anaerobic Membrane Bioreactor (AAAM) Process. *Desalination* **2008**, *221*, 566–575.

Yurtsever, M.; Engil, A. Biosorption of Pb(II) Ions by Modified Quebracho Tannin Resin. *J. Hazard. Mater.* **2009**, *163*, 58–64.

CHAPTER 9

Biotechnological Intervention for the Management of Agricultural Waste

TIJJANI SABIU IMAM[1], HADIZA ABDULLAHI ARI[2], and
ADAMU YUNUSA UGYA[3*]

[1]Department of Biological Sciences, Bayero University Kano,
Kano State, Nigeria

[2]Faculty of Sciences, National Open University of Nigeria, Nigeria

[3]Department of Environmental Management, Kaduna State University,
Kaduna, Nigeria

*Corresponding author. E-mail: ugya88@kasu.edu.ng

ABSTRACT

The unintended presence of agricultural waste in the aquatic environment results to changes in the water characteristics, causing contamination and producing severe damages to the freshwater bodies. Different agricultural waste management technologies have been applied in the treatment of agricultural waste before discharge, however, the use of biotechnology will surely help in the management of these agricultural waste because most agricultural waste management technologies are expensive, difficult to access, or inefficient. Biotechnology is employed in the reduction of risks associated with the pollution of soil, water, and air due to agricultural processes, thereby serving as an effective tool for the achievement of agricultural sustainability as it can be used for the monitoring of polluted sites. Some tools employed in biotechnology include genetic engineering, cell tissue culture, molecular techniques, bioinformatics, etc. Aquaculture, livestock production and cultivation activities have been identified as sources of agricultural waste. Different methods of agricultural waste management include compositing, recycling separation and sorting and incineration. The intervention of biotechnology in the treatment of agricultural waste has led to the conversion of the waste to bioethanol and biogas, which are eco-friendly, thus preventing the consequences of greenhouse effect. The utilization of agricultural waste as biomass for bioethanol and biogas production is advantageous because these biomasses are renewable substances which will reduce the rate of consumption of fossil fuel, for instance, the fuel needed by the entire transportation sector will not be satisfied by ethanol, but its combination with fossil fuels will reduce the pressure on fossil fuel and also reduce the environmental impact.

9.1   INTRODUCTION

The desire to improve the living standard of humans has led to an unintentional massive production of waste, which unfortunately has negative consequences on the living environment (Ugya et al., 2019a). Activities from agro-based industries like rubber processing, palm oil production, wood processing, and other wastes from agricultural events are collectively classified as agricultural waste. The unplanned presence of these wastes in the aquatic environment invariably changes the water characteristics, causing contamination and thereby making water unsafe for drinking, recreational activities, irrigation, and sometimes uninhabitable for aquatic organisms due to the damaging effects it exerts on the freshwater bodies (Hasan et al., 2019; Ugya et al., 2019c). The effects of agricultural waste on the soil have led to a decrease in soil fertility as a result of an alteration of soil PH, and an increase in environmental problems such as soil erosion, flooding, and soil degradation. Although soil degradation may occur naturally, agricultural activities are major contributors to soil degradation (Moss, 2008; Ugya, 2015).

The world as it is today is being affected by improper management of agricultural waste, which is mostly being discharged into the environment without any treatment. Although different agricultural waste management systems (AWMSs) have been developed for the treatment of agricultural waste before discharge, the involvement of biotechnology will surely help in the management of these agricultural waste because most AWMS technologies are expensive, difficult to access, or inefficient (Rao and Rathod, 2019; Ugya and Imam, 2017).

Biotechnology is a system involving the use of organisms to develop products according to human purpose (Singhania et al., 2017; Ugya et al.,

2019b). The tools employed in biotechnology include genetic engineering, cell tissue culture, molecular techniques, bioinformatics, etc. Recently, biotechnology is widely used in the field of medicine, an advancement which is beyond its known application in the genetic modification of crops (Pakshirajan et al., 2015; Ugya and Aziz, 2016). Biotechnology is employed in the reduction of risks associated with the pollution of soil, water, and air due to agricultural processes, thereby serving as an effective tool for the achievement of agricultural sustainability as it can be used for the monitoring of polluted sites. The knowledge of biotechnology has been adopted in the treatments of wastewater and solid waste; this knowledge could also be effective in the eradication of the menace associated with agricultural waste management. This chapter is aimed at accessing the benefit of biotechnological intervention in the management of agricultural waste (Pakshirajan et al., 2014).

9.1.1 AGRICULTURAL WASTE GENERATION

The activities in agriculture that involves the production and processing of agricultural products are usually accompanied by waste generation, which is as a result of chemical application, irrational farming method, etc. These wastes generated tend to depend on the type of agricultural activity in practice. The different agricultural practices that lead to the generation of agricultural waste include aquaculture, livestock production, and cultivation activities (Nigussie et al., 2015).

9.1.2 AQUACULTURE AS A SOURCE OF AGRICULTURAL WASTE

The unceasing reliant on aquaculture due to the effect of the over-exploitation of wild fish has led to the growing demand for the basic needs of aquaculture, which include water, land, and natural resources. This rivalry has necessitated the need of more inputs such as feeds per unit area of water and land leading to waste generation (Loehr, 1978; Sundström et al., 2014). The generated waste is in the form of unused inputs or by-products with little or no economic value and hence regarded as nuisance in the environment. Other sources of waste in aquaculture apart from feeds include chemical and pathogens. The chemicals used in aquaculture are in the form of medicine, disinfectants, and antifoulants (Bostock et al., 2010).

9.1.3   *AGRICULTURAL WASTE FROM LIVESTOCK PRODUCTION*

The livestock industries produce large volumes of wastewater and solid waste, which are unfavorable to the environment. These wastes tend to contain pollutants such as animal urine, unused milk, animal dung, poultry mess, left-over feeds, poultry litter, etc. Air pollutants produce as a result of the processing of livestock include methane, hydrogen sulphide, and odor. The odor is a result of putrefaction of the mixture of the livestock dung and urine while the production of hydrogen sulphide and methane is as a result of the component of the animal feeds and the activities of microorganisms (Grossi et al., 2018; Hutchison et al., 2005).

9.1.4   *AGRICULTURAL WASTE GENERATION FROM CULTIVATION ACTIVITIES*

The cultivation of agricultural products is associated with some damaging practices, such as the application of pesticide and fertilizer. The never-ending application of pesticide is due to the mounting number of insects and weed predominantly in the tropic areas. To prevent the loss of crop yield to these insects and pest, pesticides are applied to kill the weeds and insects (Cerda et al., 2019). Meanwhile, the incessant use of fertilizer is to maintain the soil fertility and increase productivity of agricultural products. The excess application of fertilizer and pesticide causes chemical leaching into the soil and runoff water bodies thereby exerting toxic effects on aquatic flora and fauna and also affect soil biodiversity. The presence of pollutants pesticides and fertilizer in the aquatic environment tend to encourage algae blooms and subsequent death of aquatic flora and fauna (Danewalia et al., 2016).

9.2   ENVIRONMENTAL IMPACT OF AGRICULTURAL WASTE

One of the world's largest contributors of agricultural waste is from livestock and these have become a major environmental problem, which has resulted to greenhouse gases, ammonia, and nitrous oxide. Studies indicate that livestock activity is responsible for the emission of about 18%, 37%, 65%, and 64%, of world CO_2, CH_4, N_2O, and NH_3, respectively. The emission of these greenhouse gases has contributed to global warming thereby causing the changes in rainfall trends, average temperature, weather conditions, change in atmospheric carbon dioxide, depletion of the ozone layer, etc. Inorganic pollutants are pollutants including sulfur and its compounds—ammonia, nitrate, phosphate, heavy metals, etc.—which release into freshwater as a result of agricultural runoff that tend to facilitate the depleting of oxygen in freshwater habitat and also lead to the change of the physical chemistry of freshwater bodies (Camargo and Alonso, 2006; Zhou et al., 2014). These pollutants tend to be washed away as runoff into nearby stream where they cause degradation of aquatic ecosystem due to their persistence's nature leading to the death of aquatic lives. The pollutants can also leach into the soil thereby affecting biodiversity of the soil and this can decrease soil fertility. Once present in the aquatic environment, these pollutants tend to encourage eutrophication leading to algae growth. It is a well-known fact that microalgae can be very dangerous to freshwater habitat because of the formation of algae blooms and production and release of ROS (Mohamed, 2018). Freshwater microalgae blooms vary in color from green, yellow, or brown depending on the species of microalgae involved (Wells et al., 2015). The blooms are a result of excessive utilization of nutrients by freshwater microalgae, particularly in slow-moving water bodies such as lakes and ponds (Mohamed, 2018).

These blooms produced by freshwater microalgae are extremely toxic to freshwater biota because they tend to block the gills of fishes and other freshwater micro-invertebrate and also deplete the existing dissolved oxygen in freshwater habitat leading to the death of these freshwater organisms (Dorantes-Aranda et al., 2015). They also make an ecosystem to be an unsustainable one because they tend to be poisonous to the smaller fishes and other filter-feeder organisms that feed on them, and this toxicity is passed across the food web leading to the death of macro-organisms in the freshwater environment (Paerl, 2018; Paerl et al., 2016; Schmale et al., 2019). Freshwater microalgae can also disturb the aquatic productivity as they prevent the penetration of sunlight energy to reach deep into other freshwater zones thereby leading to the reduction in the diversity of organisms in the freshwater habitat (Wu et al., 2015). The prevention of sunlight by freshwater microalgae could be due to the decoloration of freshwater habitat by freshwater microalgae or due to the formation of thick film on the surface of freshwater by freshwater microalgae (Wang et al., 2017). These also affect the mobility of freshwater fauna such as coral and other freshwater flora from getting sunlight for photosynthesis because they submerged the freshwater flora and also serve as a shade thereby preventing sunlight energy from reaching these organisms (Havens, 2008). In addition, freshwater microalgae are able to produce microtoxins that have detrimental effects on freshwater biota, particularly fishes and other micro- and macro-invertebrate (Ferrão-Filho and Kozlowsky-Suzuki, 2011; Fox, 2012; Lowenstine, 2008).

9.3 METHODS OF AGRICULTURAL WASTE MANAGEMENT

9.3.1 COMPOSTING AS A METHOD OF AGRICULTURAL WASTE MANAGEMENT

Composting can simply be denoted as a method where organic matter can be converted into a fine structure called compost. This is the easiest method used in the treatment of agricultural waste because the organic materials in the waste are allowed to form compost, which is also called the soil conditioner. The composting method aid in the increase in soil fertility because the compost, formed as a result of the recycling of numerous organic materials in agricultural waste, is rich in nutrients. Compost is usually used as a soil conditioner, natural pesticide, and fertilizer because it increases the amount of humus in the soil. Compost is suitable for the control of soil erosion, land reclamation, wetland construction, and filling of landfill because the process of its formation involves the accumulation of large mass of organic mass that degrades into humus after a period of time. The process of the decay of these organic matters is assisted by the presence of water and detritivores such as fungi, earthworm, and anaerobic organisms. The essential elements that aid the functioning of an organism in compost formation include carbon, nitrogen, oxygen, and water (Nsimbe et al., 2018). The organisms utilized the nitrogen and water for growth and other metabolic activities and also utilized oxygen for the oxidation of the carbon presences for the formation of energy. Both carbon and nitrogen are the major constituents of plants and animal, although the amount tend to vary across the different organisms (Pergola et al., 2020). Microorganisms also make tremendous contribution in the formation of compost because they release enzymes that bring about the decaying of the organic matter into compost. The ability of a microorganism to bring about the decomposition of organic substances into compost is subject to the appropriate mixture of nitrogen, oxygen, carbon, and water. The types of microorganisms that play an active role in compost include bacteria, actinobacteria, and rotifer; some invertebrates, such as earthworm, can also help in the decomposition of organic waste material in agricultural waste into compost (Sanchez et al., 2017).

9.3.2 STAGES OF COMPOSTING

Several studies have tried to modify the traditional method of agricultural-waste composting, the modification ranges from the control of the carbon–nitrogen ratio to the water content of the compost. However, most of the studies show that there are no significant differences in the compost produced from agricultural waste resulting from the alteration of parameters. The improvement in the microorganisms shows significant differences between compost resulting from traditional and modern methods of agricultural composting. The stages of composting as a result of microbial activities occur in three stages that includes the mesophilic, thermophilic, and maturation stage. The mesophilic stage is the stage by which mesophilic organism causes the decomposition of agricultural waste under moderate temperature. The thermophilic stage of composting results due to the activities of the thermophilic organisms under high temperature. The maturation stage of composting of agricultural waste is the last stage of composting in which mesophilic organisms lead to the formation of compost and also a decrease in temperature is experienced (Kumar, 2011).

9.3.3 USES OF COMPOST

Composts are added to the soil to increase the humus and nutrient content of the soil. They perform as an absorbent that encompass moisture and soluble minerals because they are porous due to the presence of absorbent materials. The porous nature of compost supports the bonding of nutrients. However, compost is infrequently used alone, plants can successfully grow from mixed soil, sand, grit, bark chips, vermiculite, perlite, or clay granules to yield loam soil. Compost is usually ploughed straight into the soil or growth medium to enhance the level of organic matter and the total fertility of the soil (Kumar, 2011). Generally, direct growing of seedlings of plants in compost is not endorsed due to the presence of toxins in the compost that could hinder the germination of plant seedling. It is best to use mixtures of 20–30% compost for the transfer of seedlings at cotyledon stage or later. The immunity of plants can also be increased using compost because it protects plants from diseases and pests. The ability of compost to prevent pathogens and weeds from attacking plants have been revealed to be due to the presence of microbial pesticide present in compost. This microbial pesticidal activity of compost is due to the thermophilic and mesophilic temperatures produced by compost (Kumar, 2011).The process of composting involves the conversion of organic compounds in agricultural waste into useful material via the process of decomposition. The process of composting prevents the creation of landfill due to the fact that the material to be disposed can be converted into useful compost rather than dumping in a landfill; these processes prevent the formation of landfill leachate, which tend to contaminate underground water. The process of o-composting involves the processing of organic compounds with other non-organic in agricultural waste for compost formation. The process of sorting of agricultural waste stream using in-vessel composting method is referred to as mechanical sorting. This mechanical biological treatment method can be used for the regulation of the amount of organic matter from agricultural that can be deposited in a landfill to prevent climatic effect. This is because this organic matter in agricultural waste, once in landfill, can biodegrade forming methane which can lead to global warming (Pan et al., 2012).

9.3.4 RECYCLING AS A METHOD OF AGRICULTURAL WASTE MANAGEMENT

The process of converting agricultural waste materials into new resources for utilization is referred to as recycling. The ability of an agricultural waste to be recycled is dependent on the capability of the agricultural waste to regain the characteristics it has lost. These methods of agricultural waste management are a substitute to conventional way of disposing agricultural waste because they aid in sustainability and prevent greenhouse emission. These processes of recycling conserves agricultural raw material because they slow the rate of consumption of new agricultural raw materials, and thereby are preventing the occurrences of air pollution and water pollution resulting from incineration and landfilling, respectively. These processes of agricultural waste management are regarded as an important part of modern waste reduction. The process of agricultural waste recycling supports environmental sustainability because the amount of usage of agricultural raw material is minimized. Agricultural waste is gaining more acceptance due to the adoption of recycling policy by more farms and companies leading to recycling of agricultural waste rather than transferring them to landfill. The continuous increase in the recycling of agricultural waste worldwide has prevented the loss of money for agricultural waste disposal leading to an increase in agricultural production (Xue et al., 2016).

9.3.5 SEPARATION AND SORTING

With the right recycling equipment and processing machinery, agricultural waste can easily be segregated, processed, and sold for profit. All of the following waste can be processed and recycled using small mobile recycling equipment:

- Green waste such as old trees, branches, and hedges can be easily separated and shredded for reuse.
- Mounds of used soil or waste from the ground can be crushed using a cone crusher, and passed through a trammel to provide you fresh top soil and clean rocks that are useful for landscaping.
- Plastics from your farm, such as bale wraps, feed bags, and buckets, and any kind of plastic container, can all be separated and then baled together easily for sale.
- Plastics—silage and domestic
- Hazardous substances and their containers—chemicals
- Packaging—cardboard
- Treated and untreated timber—old fencing
- Metals—redundant machinery
- Tyres—ex-silage weighting down
- Construction and demolition waste (Alvarenga et al., 2017).

9.3.6 INCINERATION

Incineration is a waste-treatment process that involves the combustion of organic substances contained in waste materials, and just like other high-temperature waste treatment systems, incineration is described as "thermal treatment" in the sense that waste materials are converted into ash, flue gas, and heat. The ash is mostly formed by the inorganic constituents of the waste and may take the form of solid lumps or particulates carried by the flue gas, which must be cleaned of gaseous and particulate pollutants before they are dispersed into the atmosphere. In some cases, the heat generated by incineration can be used to generate electric power. Incineration with energy recovery is one of several waste-to-energy technologies such as gasification, pyrolysis, and anaerobic digestion. While incineration and gasification technologies are similar in principle, the energy produced from incineration is high-temperature heat; whereas, combustible gas is often the main energy product from gasification. Incineration and gasification may also be employed without energy and materials recovery (Zhang et al., 2001). Experts and local communities from several countries are concerned about the environmental effect of incinerators. In some countries, for instance, incinerators built just a few decades ago often did not include materials separation to remove hazardous, bulky, or recyclable materials before combustion. These tend to pose risk to the health of the plant workers and the local environment due to inadequate levels of gas cleaning and combustion process control. Most of these facilities did not generate electricity. Incinerators reduce the solid mass of the original waste by 80–85% and the volume (already compressed somewhat in garbage trucks) by 95–96%, depending on the composition and degree of recovery of materials such as metals from the ash for recycling. This implies that while incineration does not entirely replace landfilling, it nevertheless, significantly reduces the necessary volume for disposal (Santiago-De la Rosa et al., 2017). Garbage trucks often reduce the volume of waste in a built-in compressor before delivery to the incinerator. Alternatively, at landfills, the volume of the uncompressed garbage can be reduced by approximately 70% by using a stationary steel compressor, although with a significant energy cost. Simpler waste compaction is a common practice for compaction at landfills in several countries. Incineration has particularly strong benefits for the treatment of certain waste types in niche areas such as clinical wastes and certain hazardous wastes where pathogens and toxins can be destroyed by high temperatures. Examples include chemical multiproduct plants with diverse toxic or very toxic wastewater streams, which cannot be routed to a conventional wastewater treatment plant. In countries such as Japan, Singapore, and the Netherlands where land is a scarce resource, waste combustion is commonly practiced. For more than a century, Denmark and Sweden have been using the energy generated from incineration in localized combined heat and power facilities supporting district-heating schemes, for instance, in 2005, waste incineration produced 4.8% of the electricity consumption and 13.7% of the total domestic heat consumption in Denmark. Other European countries such as Luxembourg, the Netherlands, Germany, and France rely heavily on incineration for handling municipal waste (Paul et al., 2019).

9.4   THE INTERVENTION OF BIOTECHNOLOGY IN AGRICULTURAL WASTE MANAGEMENT

The constituent of agricultural waste is rich in starch and nitrogen and as such is useful for the production of energy and feeds. The starchy composition of agricultural waste varies according to the biomass composition but is said to always contain at least 40% cellulose, 30% hemicellulose, and 25% lignin (Singhania et al., 2017). The waste is used in energy production and also in the production of other metabolites such as enzymes and organic acids.

9.4.1 AGRICULTURAL WASTE AS A RAW MATERIAL FOR ENERGY

In developing countries with low technological know-how, agricultural waste in the form of sugar cane and rice husk are either allowed to rot on the farm or are used as an energy source via direct combustion. The increasing awareness of the air pollution resulting from the direct combustion of agricultural waste has discouraged the use of this method as an option for agricultural waste management due to the climatic effect in impact of man and its environment. The intervention of biotechnology has provided solution to these problems because agricultural waste is converted to bioethanol and biogas, which are eco-friendly, thus preventing the consequences of greenhouse effect. The utilization of agricultural waste as biomass for bioethanol and biogas production is advantageous because these biomasses are renewable substances and, as such, will reduce the rate of consumption of fossil fuel, which are the product of nonrenewable substances. These methods are effective in the management of agricultural waste because farmers tend to save these waste substances properly for sale to companies for extra sources of livelihood (Owusu and Asumadu-Sarkodie, 2016; Singhania et al., 2017). In the current state, the fuel needed by the entire transportation sector will not be satisfied by ethanol, but its combination with fossil fuels will reduce the pressure on fossil fuel and also reduce the environmental impact (Owusu and Asumadu-Sarkodie, 2016).

9.4.2 THE USE OF AGRICULTURAL WASTE FOR BIOPROCESSES

Agricultural waste is utilized in the production of different types of metabolites, this biotechnological intervention involves the use of carbon sources present in agricultural waste for various bioprocesses, for example, during fermentation process, the starch component of agricultural wastes such as cassava bagasse, brown rice bran, wheat bran, etc. are used for the production of metabolites such as enzyme and carboxylic acids.

Other components of agricultural waste utilized for bioprocesses include cellulose, protein, and crude nitrogen. Cellulose component of agricultural waste is used for the production of enzymes of cellulose origin whereas protein and crude nitrogen are used for production of therapeutic products and products used for the control and remediation of environmental pollution (Bharathiraja et al., 2017).

9.5 CONCLUSION

Undoubtedly, the production of agricultural products from different wastes materials depends on the type of technology involved. This chapter shows that biotechnology will help in solving the problems associated with the generation of this waste. This waste can be utilized as a raw material for renewable energy production especially biogas and bioethanol thereby serving as a cost-effective source of energy. Biotechnology plays a role in the reduction of landfill due to the utilization of the biomass resulting from agricultural activities. Biotechnology will change the perception of farmers from agricultural waste burning thereby preventing the negative effect on the environment.

KEYWORDS

- agricultural waste
- biotechnology
- greenhouse gases
- algae blooms

REFERENCES

Alvarenga, P.; Palma, P.; Mourinha, C.; Farto, M.; Dores, J.; Patanita, M.; Cunha-Queda, C.; Natal-da-Luz, T.; Renaud, M.; Sousa, J. P. Recycling Organic Wastes to Agricultural Land as a Way to Improve Its Quality: A Field Study to Evaluate Benefits and Risks. *Waste Manag.* **2017,** *61,* 582–592.

Bharathiraja, S.; Suriya, J.; Krishnan, M.; Manivasagan, P.; Kim, S. K. Production of Enzymes from Agricultural Wastes and Their Potential Industrial Applications. *Adv. Food Nutr. Res.* **2017,** *80,* 125–148.

Bostock, J.; McAndrew, B.; Richards, R.; Jauncey, K..; Telfer, T.; Lorenzen, K.; Little, D.; Ross, L.; Handisyde, N.; Gatward, I.; Corner, R. Aquaculture: Global Status and Trends. *Phil. Trans. R. Soc. Lond. Ser. B, Biol. Sci.* **2010**, *365*, 2897–2912.

Camargo, J. A.; Alonso, A. Ecological and Toxicological Effects of Inorganic Nitrogen Pollution in Aquatic Ecosystems: A Global Assessment. *Environ. Int.* **2006**, *32*, 831–849.

Cerda, A.; Artola, A.; Barrena, R.; Font, X.; Gea, T.; Sánchez, A. Innovative Production of Bioproducts from Organic Waste Through Solid-State Fermentation **2019**, *3*.

Danewalia, S. S.; Sharma, G.; Thakur, S.; Singh, K. Agricultural Wastes as a Resource of Raw Materials for Developing Low-Dielectric Glass-Ceramics. *Sci. Rep.* **2016**, *6*, 24617.

Dorantes-Aranda, J. J.; Seger, A.; Mardones, J. I.; Nichols, P. D.; Hallegraeff, G. M. Progress in Understanding Algal Bloom-Mediated Fish Kills: The Role of Superoxide Radicals, Phycotoxins and Fatty Acids. *PloS One* **2015**, *10*, e0133549–e0133549.

Ferrão-Filho, A. D. S.; Kozlowsky-Suzuki, B. Cyanotoxins: Bioaccumulation and Effects on Aquatic Animals. *Marine Drugs* **2011**, *9*, 2729–2772.

Fox, J. W. 114 - Venoms and Poisons from Marine Organisms. In *Goldman's Cecil Medicine*; Goldman, L., Schafer, A. I., Eds.), 24th ed; W. B. Saunders: Philadelphia, 2012; pp 697–700.

Grossi, G.; Goglio, P.; Vitali, A.; Williams, A. G. Livestock and Climate Change: Impact of Livestock on Climate and Mitigation Strategies. *Animal Front.* **2018**, *9*, 69–76.

Hasan, M. K.; Shahriar, A.; Jim, K. U. Water Pollution in Bangladesh and Its Impact on Public Health. *Heliyon.* **2019**, *5*, e02145–e02145.

Havens, K. E. Cyanobacteria Blooms: Effects on Aquatic Ecosystems. *Adv. Exp. Med. Biol.* **2008**, *619*, 733–747.

Hutchison, M. L.; Walters, L. D.; Avery, S. M.; Munro, F.; Moore, A. Analyses of Livestock Production, Waste Storage, and Pathogen Levels and Prevalences in Farm Manures. *Appl. Environ. Microbiol.* **2005**, *71*, 1231–1236.

Kumar, S. Composting of Municipal Solid Waste. *Crit. Rev. Biotechnol.* **2011**, *31*, 112–136.

Loehr, R. C. Hazardous Solid Waste from Agriculture. *Environ. Health Perspect.* **1978**, *27*, 261–273.

Lowenstine, L. J. Algal Bloom Toxicity in Marine Animals. In *Zoo and Wild Animal Medicine*; Fowler, M. E.; Miller, R. E., Eds.; W. B. Saunders: Saint Louis, 2008; pp 341–348.

Mohamed, Z. A. Potentially Harmful Microalgae and Algal Blooms in the Red Sea: Current Knowledge and Research Needs. *Mar. Environ. Res.* **2018**, *140*, 234–242.

Moss, B. Water Pollution by Agriculture. *Phil. Trans. R. Soc. Lond. Ser. B Biol. Sci.* **2008**, *363*, 659–666.

Nigussie, A., Kuyper, T. W., and de Neergaard, A. Agricultural Waste Utilisation Strategies and Demand for Urban Waste Compost: Evidence from Smallholder Farmers in Ethiopia. *Waste Manag.* **2015**, *44*, 82–93.

Nsimbe, P.; Mendoza, H.; Wafula, S. T.; Ndejjo, R. Factors Associated with Composting of Solid Waste at Household Level in Masaka Municipality, Central Uganda. *J. Environ. Public Health* **2018**, 1284234–1284234.

Owusu, P. A.; Asumadu-Sarkodie, S. A Review of Renewable Energy Sources, Sustainability Issues and Climate Change Mitigation. *Cogent. Eng.* **2016**, *3*, 1167990.

Paerl, H. W. Mitigating Toxic Planktonic Cyanobacterial Blooms in Aquatic Ecosystems Facing Increasing Anthropogenic and Climatic Pressures. *Toxins* **2018**, *10*, 76.

Paerl, H. W.; Gardner, W. S.; Havens, K. E.; Joyner, A. R.; McCarthy, M. J.; Newell, S. E.; Qin, B.; Scott, J. T. Mitigating Cyanobacterial Harmful Algal Blooms in Aquatic Ecosystems Impacted by Climate Change and Anthropogenic Nutrients. *Harmful Algae* **2016**, *54*, 213–222.

Pakshirajan, K.; Rene, E. R.; Ramesh, A. Biotechnology in Environmental Monitoring and Pollution Abatement. *BioMed. Res. Int.* **2014**, 235472–235472.

Pakshirajan, K.; Rene, E. R.; Ramesh, A. Biotechnology in Environmental Monitoring and Pollution Abatement. *BioMed. Res. Int.* **2015**, 963803–963803.

Pan, I.; Dam, B.; Sen, S. K. Composting of Common Organic Wastes Using Microbial Inoculants. *3 Biotech.* **2012**, *2*, 127–134.

Paul, S.; Mbewe, P.; Kong, S.; Šavija, B. Agricultural Solid Waste as Source of Supplementary Cementitious Materials in Developing Countries. *Materials* **2019**, *12*, 1112.

Pergola, M.; Persiani, A.; Pastore, V.; Palese, M. A.; D'Adamo, C.; De Falco, E.; Celano, G. Sustainability Assessment of the Green Compost Production Chain from Agricultural Waste: A Case Study in Southern Italy. *Agronomy* **2020**, *10*.

Rao, P.; Rathod, V. Valorization of Food and Agricultural Waste: A Step towards Greener Future. *Chem Rec.* **2019**, *19*, 1858–1871.

Sanchez, O. J.; Ospina, D. A.; Montoya, S. Compost Supplementation with Nutrients and Microorganisms in Composting Process. *Waste Manag.* **2017**, *69*, 136–153.

Santiago-De la Rosa, N.; Mugica-Alvarez, V.; Cereceda-Balic, F.; Guerrero, F.; Yanez, K.; Lapuerta, M. Emission Factors from Different Burning Stages of Agriculture Wastes in Mexico. *Environ. Sci. Pollut. Res. Int.* **2017**, *24*, 24297–24310.

Schmale, D. G.; Ault, A. P.; Saad, W.; Scott, D. T.; Westrick, J. A. Perspectives on Harmful Algal Blooms (HABs) and the Cyberbiosecurity of Freshwater Systems. *Front. Bioeng. Biotechnol.* **2019**, *7*, 128–128.

Singhania, R. R.; Patel, A. K.; Pandey, A. Biotechnology for Agricultural Waste Recycling. In *Current Developments in Biotechnology and Bioengineering*; Wong, J. W. C.; Tyagi, R. D.; Pandey, A., Eds.; Elsevier, 2017; pp 223–240.

Sundström, J. F.; Albihn, A.; Boqvist, S.; Ljungvall, K.; Marstorp, H.; Martiin, C.; Nyberg, K.; Vågsholm, I.; Yuen, J.; Magnusson, U. Future Threats to Agricultural Food Production Posed by Environmental Degradation, Climate Change, and Animal and Plant Diseases—A Risk Analysis in Three Economic and Climate Settings. *Food Sec.* **2014**, *6*, 201–215.

Ugya, A. Y. The Efficiency of Lemna minor L. in the Phytoremediation of Romi Stream: A Case Study of Kaduna Refinery and Petrochemical Company Polluted Stream. *J. Appl. Biol. Biotechnol.* **2015**, *3*, 011–014.

Ugya, A. Y.; Aziz, A. A concise Review on the Effect of Tannery Wastewater on Aquatic Fauna. *Merit Res. J. Med. Med. Sci.* **2016**, *4*, 476–479.

Ugya, A. Y.; Hua, X.; Ma, J. Phytoremediation as a Tool for the Remediation of Wastewater Resulting from Dyeing Activities. *Appl. Ecol. Environ. Res.* **2019a**, *17*, 3723–3735.

Ugya, A. Y.; Imam, T. S.; Hua, X.; Ma, J. Efficacy of Eicchornia Crassipes, Pistia Stratiotes and Nymphaea Lotus in the Biosorption of Nickel from Refinery Wastewater. *Appl. Ecol. Environ. Res.* **2019b**, *17*.

Ugya, A. Y.; Imam, T. S.; Li, A., Ma, J.; Hua, X. Antioxidant Response Mechanism of Freshwater Microalgae Species to Reactive Oxygen Species Production: A Mini Review. *Chem. Ecol.* **2019c**, 1–20.

Ugya, A. Y.; Imam, T. S. Temporal Heavy Metals Variation in Vegetables Sampled at Kasuwan Mata, Kaduna, Etropolis, Nigeria. *Malaysia J. Sci.* **2017**, *36*, 63–73.

Wang, L.; Wang, X.; Jin, X.; Xu, J.; Zhang, H.; Yu, J.; Sun, Q.; Gao, C.; Wang, L. Analysis of Algae Growth Mechanism and Water Bloom Prediction under the Effect of Multi-Affecting Factor. *Saudi J. Biol. Sci.* **2017**, *24*, 556–562.

Wells, M. L.; Trainer, V. L.; Smayda, T. J.; Karlson, B. S. O.; Trick, C. G.; Kudela, R. M.; Ishikawa, A.; Bernard, S.; Wulff, A.; Anderson, D. M.; Cochlan, W. P. Harmful Algal Blooms and Climate Change: Learning from the Past and Present to Forecast the Future. *Harmful Algae* **2015**, *49*, 68–93.

Wu, T. T.; Liu, G. F.; Han, S. Q.; Zhou, Q.; Tang, W. Y. Impacts of Algal Blooms Accumulation on Physiological Ecology of Water Hyacinth. *Huan Jing Ke Xue* **2015**, *36*, 114–120.

Xue, L.; Zhang, P.; Shu, H.; Chang, C. C.; Wang, R.; Zhang, S. Agricultural Waste. *Water Environ. Res.* **2016**, *88*, 1334–1373.

Zhang, F. S.; Yamasaki, S., Nanzyo, M. Application of Waste Ashes to Agricultural Land–Effect of Incineration Temperature on Chemical Characteristics. *Sci. Total Environ.* **2001**, *264*, 205–214.

Zhou, G. J.; Ying, G. G.; Liu, S.; Zhou, L. J.; Chen, Z. F.; Peng, F. Q. Simultaneous Removal of Inorganic and Organic Compounds in Wastewater by Freshwater Green Microalgae. *Environ. Sci. Process Impacts.* **2014**, *16*, 2018–2027.

CHAPTER 10

Biofuels: Current Challenges and Possible Solutions

ADEYEMI ADESINA

Department of Civil and Environmental Engineering, University of Windsor, Windsor, Canada

Corresponding author. E-mail: adesina1@uwindsor.ca

ABSTRACT

Advances in recent years have seen the rapid evolution of biofuels to replace conventional fossil fuels. Biofuels are not only sustainable and renewable, but they also create a supplement for the other reserves of fuels. However, despite the advancement in the use and development of biofuels, some limitations exist that have limited its large-scale applications, especially in certain parts of the world. This chapter presents the current challenges facing the production and use of biofuels alongside the possible solutions to these challenges. Limitations with the use of biofuel that might ensue with the production and use of biofuel in the future were also discussed.

10.1 INTRODUCTION

From ancient times, energy has been demanded by everyday activities. All over the world, fossil fuels are the major sources of energy used for various applications ranging from the manufacturing sector to the transportation sector. However, with the continuous and increasing use of fossils fuels comes various detrimental threats to the environment. Fossil fuels are one of the major contributors to the world's anthropogenic emissions, which have resulted in various side effects such as climate change. Climate change has resulted in various catastrophic effects such as extreme temperatures, a rise in sea levels resulting in flooding, etc. The high demand for energy has also resulted in a continuous increase in the price of fossil fuels, which consequentially affects economies all over the world. The world's energy demand in the next 10 years has been reported to increase by 50% compared to the current energy demand (Lin, 2013; Viana et al., 2014). With the current depletion of the reserve of fossil fuels, a shortage in the energy supply in the future might occur due to the limited source of energy. On the contrary, biofuels are sustainable alternatives to conventional fossil fuels. Biofuels are obtained from bioresources, which makes them sustainable and renewable. Also, their use results in a net-zero carbon-dioxide emission on the environment. In addition to the sustainable advantage of biofuels, its use will supplement the decrease in the production of fossil fuels due to its plummet in the available reserves. It can be seen that a decrease in the availability of fossil fuel is imminent. Therefore, there is a need for rapid research, development, and application of biofuels so as to meet the future demand of energy in a sustainable way. However, as the generation and use of biofuels are still emerging and relatively new compared to fossil fuels, it is faced with many challenges that impede both its production and use. This chapter presents the current challenges facing the development and use of different types of biofuels. It is anticipated that the discussion in this chapter will gear more research and developments in solving the major challenges associated with the use and application of biofuels. It is also hoped that this chapter will create more awareness of the need to diversify our source of energy and explore the numerous benefits of biofuels.

10.2 BIOFUELS

As mentioned earlier, biofuels are promising sustainable alternatives to replace fossil fuels. Biofuels are sustainable because they are biodegradable, and their use will result in a net-zero carbon dioxide emission. Also, since they are obtained from bioresources, their production process is less energy-intensive and more environmentally friendly compared to the production of fossil fuels. Biofuels can exist as solids, liquids, or gases. Although biofuels are mostly used for liquid and gas fuels from biological sources, its use in this chapter is for all types of fuels (i.e., solid, liquid and gas fuels from biological sources). Biofuels can also be classified as first, second, or third generation depending on the production process involved.

The first-generation biofuels are produced using the transesterification process while the second-generation fuels are produced by cellulose hydrolysis and sugar fermentation method. Biofuels produced from algae are referred to as the third-generation biofuels. Biofuels can be obtained from both plants and animals. Table 10.1 presents a few examples of the sources and products that can be used for biofuel generation. Although biofuels are deemed more environmentally friendly compared to fossil fuels, several challenges need to be overcome in order for the full sustainability and cost advantages of biofuels to be evident.

TABLE 10.1   Examples of Biofuel Sources and Products.

Source	Products
Crop residue	Rice husk
	Palm oil
	Castor oil
	Cassava
	Coconut shells
	Rapeseed
	Jatropha
	Sugarcane bagasse
	Roots
	Corn
	Canola
	Soybeans
	Algae
Animal	Cow dung
	Buffalo dung
	Chicken fat
	Pig dung
	Poultry dung

10.3   CURRENT CHALLENGES AND POSSIBLE SOLUTIONS

Despite the promising benefits of using biofuels, several challenges are limiting its large-scale and cost-effective applications. Some of these challenges are associated with a certain type of biofuel or/and some part of the world, while some are global challenges affecting the use of biofuels.

10.3.1   SUSTAINABLE PRODUCTION PROCESS

The primary objective of developing biofuels is to replace fossil fuels in order to reduce its carbon footprint and provide an alternative source of fuel for energy demand. Therefore, it is critical to ensure that the production process of these biofuels is sustainable in order to conserve the sustainability advantages of biofuels. This sustainable production process can be achieved by the involvement of the government and energy stakeholders giving support to only companies producing biofuels in an environmentally friendly way. Also, strict policies should be put in place to guide the biofuels producers on the sustainability levels required for their production activities. Countries such as the United States and Brazil have put in place such frameworks (IEA, 2020). However, there is an imminent need for other countries and/or regions to follow suit in order to ensure the global production of sustainable biofuels. Some other initiatives that can be taken are restricting the production of certain types of biofuels. For example, the European Union has banned the production of first-generation biofuels due to its negative impact on food production, land use, and the ecosystem. In general, sustainable production of biofuel will ensure that neither the food supply nor the environment is placed under any threat and will also create an avenue to manage different wastes. The sustainable production process of biofuels should also create more employment opportunities in order to improve social quality.

10.3.2   LIMITED KNOWLEDGE AND TECHNOLOGY

The limited knowledge and technology in the field of biofuels have also resulted in the non-efficient of production of biofuels from sources such as algae. More research and development in this area will result in significant improvement in the efficiency of growing and harvesting algae. Also, there is an imminent need to develop low-cost systems to main the growth conditions which will result in a corresponding increase in productivity.

10.3.3 UNAVAILABILITY OF BIORESOURCES

The limited availability of some types of bioresources to produce certain biofuels has limited the use of biofuels in industries such as the aviation and marine industries. Also, most sources used for the production of biofuels are seasonal due to the yield of plants at different times of the year. This varying yield of plants also creates a corresponding different type of crop residue all through the year for biofuel generation. Especially in Northern countries such as Canada and the United States, limited farming and harvesting go on in the winter months, which creates a gap in the supply of residues for biofuel generation during those months. The effect of this challenge on the production of biofuels can be reduced or eliminated by putting in place a framework to manage the storage of biomass all through the year so as to ensure a minimal impact on the supply of residue for biofuel generation.

10.3.4 INEFFICIENT USE

Billions of people in the rural areas of the world use biomass for cooking. However, these typical uses of biofuels in these rural areas is insufficient as a lot of time is taken in gathering the biomass used for cooking and the combustion process is not controlled resulting in not maximizing the energy output from the biomass. In addition, the traditional use of biomass results in significant pollution to the users and the environment where these biomasses are used for cooking. To ensure the effective use of these solid biofuels, technological processes and products need to be developed to convert the energy from the biomass to electrical energy, then the electrical energy can be used for cooking. Some of these technological processes or products can involve the development of charcoal stoves, producer gas generators, etc. The incorporation of this technological advancement will ensure that the energy outputs from the biomass are maximized and will eliminate or reduce the pollution from the conventional burning of the biomass.

10.3.5 LIMITED ENERGY CONTENT

Compared to petroleum diesel, biofuels such as biodiesel has been found to have lower energy content (Kaplan et al., 2006; Murillo et al., 2007).

This lower energy content will result in more fuel consumption when biofuels are used for various applications. An increase of up to 10% in biofuels consumption has been reported when biodiesel is used to replace petroleum diesel (Atabani et al., 2012). Therefore, finding ways to enhance the energy content of these biofuels or the use of alternative sources to obtain biofuels with higher energy content will gear more interest in the application of biofuels.

10.3.6 COMPATIBILITY

With the production of alternative sources of fuels comes a major challenge of compatibility of biofuels with current systems in place that uses fuels. For example, cars, electric generators, airplanes, etc., have an energy conversion system in place to convert fossil fuels to energy. However, as fossil fuels have a different composition, biofuels cannot be used for conventional systems. Therefore, it is imperative for future development that the manufacturing energy conversion system should be versatile so that different types of fuels can be used in the systems for energy generation.

10.3.7 LACK OF MARKET AND POLICY FRAMEWORK

Like the survival of any other commodity, a market and policy framework need to be put into place in order to control the quality and cost of biofuels. In contrast to fossil fuel, there exists no universal market that oversees the supply and demand of biofuels. Development of a region and universal market and policy framework to control the cost and quality of different types of biofuels will go a long way in increasing the production and use of biofuels.

10.3.8 STORAGE

Although the storage of biofuels and biomass has been found to be less expensive there are other challenges associated with its storage. (Rajesh et al., 2008). The storage of biomass used for biofuel generation at room temperature can result in a reduction in the heating value and biomass degradation (Rentizelas et al., 2009). The storage of these biomasses can also have a detrimental impact on the health and safety of humans and animals. To prevent degradation or reduction in the heating value, these biomasses can be stored in a controlled heated environment. However, the storage of biomass in such an environment will result in a corresponding increase in the cost of storage of the biomass. Several other types of methods can be used to treat the biomass before storage in order to conserve its heating value (Uslu et al., 2008).

10.3.9   HIGH PRODUCTION COST

The production cost of most biofuels is still relatively expensive. The high production cost of biofuel ranges from the high cost of the catalyst used in the production of these biofuels to the advanced technological equipment required. The production of biofuels involves the use of enzymes. However, the enzymes used are expensive and this has limited the possibility to produce biofuels in large quantities (He et al., 2014). Also, the current production of biofuel from animal wastes is more expensive than that of fossil fuels. Even for biofuels obtained from crop residues, the collection and transportation process are very expensive resulting in a corresponding increase in the cost of biofuels. Therefore, the production of cheaper catalysts and the use of locally sourced crop residues will result in a significant reduction in the cost of these biofuels. Also, the government can provide incentives such as a reduction in taxes and duties for producers of biofuels in order to reduce their overall operational costs.

10.3.10   FUEL FOOD COMPETITION

The increasing production and interest in biofuels are expected to have a negative influence on the food and water supply as the raw materials used for the production of biofuels are in one way or the other related to food production. In addition to the production and use of biofuels limiting the supply of food, it can also result in an increase in the price of the food if most of the resources for food production are geared toward the production of biofuels. The study by Koizumi (2015) and Ajanovic (2011) showed that this fuel food competition challenge might occur if resources such as lands meant for food production is directed toward the production of biofuels. Therefore, in order to produce biofuels without harming the food supply, the development of biofuels should gear toward using plants such as Jatropha that are not consumed as food. Also, more incentives should be put in place for farmers to produce more food products as farmers now tend to produce non-food plants for biofuels generation due to its higher economic advantage. More land should also be dedicated to the production of food plants only as most landowners might prefer to use their land spaces to grow only plants for biofuel production which might limit the available land space for growing food products.

10.3.11   WASTE MANAGEMENT

The production of biofuels from various sources results in a corresponding generation of wastes. These wastes range from solid, liquid to gasses. Since biofuels are considered carbon-neutral due to the equivalent absorption of carbon dioxide during their life cycle, the waste in the form of greenhouse gases can be neglected. However, there still exist wastes in the form of solids and liquids, which is a major challenge for small producers of biofuels. The liquid wastes generated can be eliminated by treating the liquid wastes and reusing the water in the biofuel production process. The solid wastes can be recycled as binders or aggregates in cementitious composites depending on their physical and chemical properties.

10.3.12   POOR ACCEPTANCE

Anything new takes time before it is universally accepted by the people. As the use of biofuels on a large scale is relatively new compared to fossil fuels, its acceptance by the public is expected to take time. The currently limited acceptance of biofuel is a major challenge to its universal applications. Incentives such as the elimination of purchase cost given to buyers of electric cars can be put in place for buyers buying products using biofuels. This type of initiative among others will encourage more use and acceptance of biofuel as a sustainable source of energy.

10.4   CONCLUSIONS

The depleting deposits of fossil fuels coupled with their high carbon footprint have called for an imminent need to find alternative renewable and sustainable sources of energy. One of the promising renewable energy that will serve the future energy demand is biofuels. However, the use and development of biofuels are still emerging and faced with various challenges. This chapter discussed the major issues associated with the development and use of biofuels alongside possible solutions. Discussion in this chapter shows that one of the major challenges that need to be solved is the production of economically sustainable biofuels. Generally, there is an imminent need to find ways to reduce the cost and carbon footprint of the production of biofuels. Also, effective economical storage methods need to be put in place in order to conserve the properties of biomass used for biofuel production. Putting in place a market and policy framework will also help to control the pricing and quality of biofuels.

KEYWORDS

- **bioresources**
- **biofuel**
- **renewable energy**
- **biotechnology**
- **fossil fuels**

REFERENCES

Ajanovic, A. Biofuels versus Food Production: Does Biofuels Production Increase Food Prices? *Energy*, 2011. https://doi.org/10.1016/j.energy.2010.05.019.

Atabani, A. E.; Silitonga, A. S.; Badruddin, I. A; T. Mahlia, M.I.; Masjuki, H. H.; Mekhilef, S. A Comprehensive Review on Biodiesel as an Alternative Energy Resource and Its Characteristics. *Renew. Sustain. Energy Rev.* **2012**. https://doi.org/10.1016/j.rser.2012.01.003.

Demirbas, A. *Biodiesel: A Realistic Fuel Alternative for Diesel Engines. Biodiesel: A Realistic Fuel Alternative for Diesel Engines*, 2008. https://doi.org/10.1007/978-1-84628-995-8.

He, J.; Ai M. W.; Daiwen, C.; Bing, Y.; Xiangbing, M.; Ping, Z.; Jie, Y.; Gang, T. Cost-Effective Lignocellulolytic Enzyme Production by Trichoderma Reesei on a Cane Molasses Medium. *Biotechnol. Biofuels* **2014**. https://doi.org/10.1186/1754-6834-7-43.

Kaplan, C.; Arslan, R.; Sürmen, Ali. Performance Characteristics of Sunflower Methyl Esters as Biodiesel. *Energy Sources, Part A Recov., Utilization Environ. Effects* **2006**. https://doi.org/10.1080/009083190523415.

Koizumi, T. Biofuels and Food Security. *Renew. Sustain. Energy Rev* **2015**. https://doi.org/10.1016/j.rser.2015.06.041.

Lin, C. Y. Effects of Biodiesel Blend on Marine Fuel Characteristics for Marine Vessels. *Energies* **2013**. https://doi.org/10.3390/en6094945.

Murillo, S.; Míguez, J. L.; Porteiro, J.; Granada, E.; Morán, J. C. Performance and Exhaust Emissions in the Use of Biodiesel in Outboard Diesel Engines. *Fuel.* **2007**. https://doi.org/10.1016/j.fuel.2006.11.031.

Rajesh; Arthe R.R.; Rajesh, E. M.; Rajendran, R.; Jeyachandran, S. Production of Bio-Ethanol from Cellulosic Cotton Waste through Microbial Extracellular Enzymatic Hydrolysis and Fermentation. *Electron. J. Environ. Agric. Food Chem.* **2008**.

Rentizelas, A. A.; Tolis, A. J.; Tatsiopoulos, I. P. Logistics Issues of Biomass: The Storage Problem and the Multi-Biomass Supply Chain. *Renew. Sustain. Energy Rev.* **2009**. https://doi.org/10.1016/j.rser.2008.01.003.

Transport Biofuels—Tracking Transport—Analysis—IEA. n.d. https://www.iea.org/reports/tracking-transport-2019/transport-biofuels (accessed April 19, 2020).

Uslu, A.; Faaij, A. P. C.; Bergman, P. C.A. Pre-Treatment Technologies, and Their Effect on International Bioenergy Supply Chain Logistics. Techno-Economic Evaluation of Torrefaction, Fast Pyrolysis and Pelletisation. *Energy* **2008**. https://doi.org/10.1016/j.energy.2008.03.007.

Viana, M.; Pieter H.; Augustin C.; Xavier Q.; Bart D.; Ina de V.; John van A. Impact of Maritime Transport Emissions on Coastal Air Quality in Europe. *Atmospheric Environ.* **2014**. https://doi.org/10.1016/j.atmosenv.2014.03.046.

CHAPTER 11

Biotechnological Intercession in Biofuel Production

WASIA SHOWKAT[1], MOONISA ASLAM DERVASH[2*],
KHALID Z. MASOODI[1], JAVEED A. MUGLOO[3], S. A. GANGOO[3], and
SABA MIR[1]

[1]*Division of Plant Biotechnology, SKUAST-K, Shalimar, Srinagar, J&K, India*

[2]*Division of Environmental Sciences, SKUAST-K, Shalimar, Srinagar, J&K, India*

[3]*Faculty of Forestry, SKUAST-K, Benihama, Ganderbal, J&K, India*

Corresponding author. E-mail: moonisadervash757@gmail.com

ABSTRACT

Fossil fuels in large quantities are being used despite their role in increasing environmental problems. Worldwide usages of biomass-based biofuels are necessary for environment preservation and development of sustainable communities. Therefore, to move from an economy zone that is dependent on nonrenewable sources of energy to one with advanced multiplicity of renewable sources will definitely not be an effortless progression. It will rather require investment of imperative exploration efforts to get adapted to dynamics of a varying energy market, make cost-effective progressions, and evade overlap with shared significance markets, such as, agriculture, food, and livestock production. Research and development projects working toward production of biofuels are mainly targeted at conversion of plant/microbial biomass to renewable fluidized fuels. Foremost challenges encountered for the production of biofuels include lack of domestication of biofuel crops, low oil/fatty acid produced from biofuel plants/microbes, as well as of lignocellulose recalcitrance to chemical and enzymatic breakdown. Day by day, research on genetic, genomic resources available for plant/microbial enhancement, elucidation of lipid metabolism to assist exploitation of fatty acid biosynthetic approach and phyto cell wall biosynthesis and assemblage all over the world are broadening. The knowledge so brought to the forefront will be used for the production of next generation biofuels by mounting fatty acid concentration and by optimizing hydrolysis of phyto cell walls to liberate fermentable sugars. This chapter briefly describes the biotechnological interventions that can make major advancements in strain improvement for commercial scale production of biofuels.

11.1 INTRODUCTION

Man has started to use biomass as an energy source decades ago in order to protect himself from vagaries of changing environment to achieve food security goals. Globally, increasing population, unrelenting energy crisis, rapid degradation of nonrenewable energy sources, excessive use of automobiles, pollution troubles from fuel emissions and associated health diseases demand immediate need for alternative sources of fuels (Wellmer et al., 2019). Each solution is quite critical for transition toward future with sustainable energy resources. As such, completely new technologies need to be utilized for production of efficient sustainable energy resources (Ciardelli et al., 2019). The biotechnological tools with efficient and reliable techniques can prove very useful for the production of biobased biofuels. Also, due to heavy deterioration of the fossilized reserves, for example, coal, natural gas, or petroleum, increasing trends in global warming, climatic changes, and increasing future energy demand due to high emission of fossil fuels, there is an urgent need to search for substitute basis of energy. Among various alternative sources available today, biofuels remain one of the best substitutes for decreasing dependency on fossil fuels by replacing them completely or partially, thereby, also providing global energy security (Purdy, 2019).

Generally, biofuels are derivative of renewable organic products and are frequently considered as green substitutes to fossil energy repository owing to their enormous contribution to carbon dioxide emissions reduction (Alco-forado, 2019). Vegetation undoubtedly represents unmatched resources for biofuel production as they globally produce about 200×10^9 tons of biomass annually, for example, polysaccharides, sugars, oils, and supplementary biopolymers. However, regardless of their cornucopia and impending environmental benefits, their proficient and sustainable consumption for energy compendium relics an exigent commission necessitates foremost investments in scientific technology (Ogunwole et al., 2019). None of the prospective crops has so far been cultured or bred for enhanced polysaccharide or fatty acid content as far as biofuel production is concerned. For this rationale, biofuel exploration is mainly directed toward deciphering the strategies for phytometabolism and characteristics that require to be tailored to optimize yield for biofuel fabrication. In model crop systems, such as rice, "*Arabidopsis* and *Brachypodium,*" genetic and genomic resources available are being utilized to respond basic scientific queries that cannot be tackled unswervingly using impending biofuel crops (Ghag et al., 2019). Therefore, successful drift toward biofuel energy-based future requires use of the biotechnological tools and as such cost of conversion from biomass to biofuel can be reduced (Ghag et al., 2019).

Among other biofuels, biodiesel has been recipient of suitable consideration, because to its resemblance with traditional diesel in requisites of chemical configuration and energy magnitude. Also, no new changes need to be made in diesel engine, as biodiesel finds compatibility with present available locomotive models and has been profitably mixed with diesel as haulage fuel in various nations including "*Italy, Germany* and *Malaysia*" (Kehrein et al., 2020). Diverse onset of the biotechnology companies and research centers have taken up microbial fermentation-based biodiesel production, aimed to produce more resourceful, cost-effective and sustainable biodiesel. Genetic engineering plays a pivotal role in transforming microbial strains into recommendable cell factories with towering competence of biodiesel fabrication (Zuccaro et al., 2020).

Biotechnology uses reliable and efficient tools for producing biofuels which increase yield, without consuming much energy needed for production. Notable improvements in the activity of organisms and enzymes have been made with the help of molecular biology in the past few decades (Amaniampong et al., 2020). Apart from sugarcane ethanol, biofuel industry still remains backward globally. Although some products of biofuel industry including bioethanol derivative of corn starch and biodiesel gained from vegetation, such as, soybean, canola, and sunflower have been marketed (Soltanian et al., 2020), these plants still do not get dues/credits and find their place as any other food crop. Plus, production of starch and plant oil from such plants are moderate to meet mammoth prerequisite of transportation fuels, which in turn has elicited production of substitutive biofuel production based on lignocellulosic biomass. Lignocellulose, mainly comprised of polysaccharides including *"cellulose, hemicellulose,* and *lignin,"* is the foremost ample biomaterial on the globe (Liu et al., 2017). With the help of the biotechnological strategies, cell wall architecture and lignocellulosic constitution in plant cells can be changed for enhancement of ethanol yield. Biotechnology can enhance yield density by manipulating plant physiology, architecture, photosynthetic efficiency and decrease agronomic inputs, for example, herbicides and pesticides (Suganya et al., 2016). Progress is rapidly being made on traits which permit crops to consume nutrients more efficiently, thus providing them to be grown with less amount of fertilizers. Generating biomass crops on lands that are extremely dry or with deprived soil features, can raise the scale of biofuel production without any influence on food production (Enamala et al., 2018).

Biotechnology has been focusing on the engineering of saline, drought, cold, and heat stress tolerant plants as well as plants that can thrive on an extensive range of soil habitats. Biomass stock plants with high quantities of cellulose and hemicellulose content would give better fermentation yields and thus, higher amounts of ethanol per ton of biomass (Unkefer et al., 2017). Research has already been executed successfully for cloning of genes that code for *"cellulases* and *polygalacturonase"* enzymes to produce cost-effective biorefinery approach for attaining maximum biomass conversion (Faraji and Voit, 2017). Difficulties faced by biofuel producers in processes, such as microbial digestion and fermentation can effectively be reduced through enhancement of the biotechnological strategies.

11.2   BIOFUEL SOURCES (CROPS/MICROORGANISMS)

Plants that mostly find use in research and development of biofuels include *"Arabidopsis thaliana, Oryza sativa* (rice), *Zea mays* (corn), *Glycine max* (soybean), *Brassica napus* (canola/oilseed rape), *Panicum virgatum* (switchgrass,) *Populus trichocarpa* (poplar), *Eucalyptus globules, Jatropha curcas, Elaeis guineensis* (oil palm), *Saccharum officinarum* (sugarcane)" (Shih et al., 2016). Crops for production of biofuels have been chosen on the account of elevated production capability in economical input agri-designs. This strategy is very important as it results in decrease of spatial extent required for the generation of biofuel crops and subsequently bypasses the necessity to apply soil amendments to escalate the production of food crops (Miao et al., 2017). Molecular, genetic, genomic, and biotechnological resources (Confalonieri and Sparvoli, 2019) available for contender biofuel crops are restricted at a particular temporal instant, but are showing an increase quantitatively. Genetic engineering strategies for crop improvement via "*Agrobacterium*"-mediated genetic alteration have been devised for "switchgrass, *Jatropha*, poplar, and *Brachypodium*" (Confalonieri and Sparvoli, 2019). For the manufacture of polyhydroxybutyrate (PHB) in transgenic switchgrass, triumphant engineering of a well-designed metabolic conduit has been documented, recommending multifaceted traits that can be tailored in biofuel crop production (Muthamilarasan et al., 2019). For exploitation in transient gene expression studies, a protocol for producing switchgrass protoplasts has also been reported, thereby, permitting for swift testing of contestant genes for purposeful analysis (Srivastava et al., 2020). For the multiplication of "*Miscanthus* and *Jatropha* explants," tissue culture strategies (in vitro) have also been highly recommended (Liang et al., 2019). While working on genome schemes of switchgrass and oil palm, "*Brachypodium*, poplar, sorghum and maize" genome sequencing have been recently studied during a growing phase (Liang et al., 2019). As such, these compendiums will be relatively supportive in devising the paraphernalia for well-designed genomic and proteomic assays and eventually permit relative genomics between model species and biofuel crops turn into an actuality.

Some autotrophic microalgae having competence of oil accumulation have been found, such as "*Navicula pelliculosa, Scenedesmus acutus, Crypthecodinium cohnii, Chlorella vulgaris, Botryococcus braunii, Dunaliella primolecta, Neochloris oleoabundans, Monallanthus salina, Phaeodactylum tricornutum,* and *Tetraselmis sueica*" (Abdel et al., 2019). On an average, the microalgae produce 1%–70% oil content, which remarkably gets affected by cultivation conditions as well as algae species. While enhanced TAG amount in "*Monodus subterraneus*" under phosphate-limited settings was reported, "microalga *Scenedesmus* sp. LX1" produced maximum quantity of lipids (35.7%) at cultivation temperature 20°C (Favaro et al., 2019). Frequently exploited microorganisms for ethyl alcohol production are yeasts, among which "*Saccharomyces cerevisiae*" is mostly preferred as concentration of ethyl alcohol produced by it (as elevated as 18% of the fermentation broth). "*Zymomonas mobilis*" is an additional well-deliberated ethyl alcohol generating strain with published genome sequence (Huang et al., 2020). "*Z. mobilis*" has been engineered for transformation of wood hydrolysate possessing glucose, mannose, and xylose as major sugar constituents to ethyl alcohol by introducing genes encoding mannose and xylose catabolic enzymes (Thanapimmetha et al., 2019). As such, ethyl alcohol production reached up to 89.8% of the theoretical yield within 72 h (Thanapimmetha et al., 2019). Genetic engineering of "*Escherichia coli*" strains has also proved quite useful for the ethanol generation and it represents a one of the classical victorious relevance of metabolic engineering (Thanapimmetha et al., 2019).

The biotechnological tools will majorly harmonize basic research, currently designed at understanding features of sources of biofuels (plants/ microorganisms) that can be besieged for blueprint of improved yield and more proficiently relevant sources via genomic/transcriptomic/metabolic engineering (Lapeña et al., 2020).

11.3   BIOFUEL SOURCE IMPROVEMENT

Altogether renewed focus on crop/microbial improvement for enhancing biofuel productivity is highly needed (Temme et al., 2019). Generation of renewable resources by genetic enhancement of plants/microbes can be accomplished through altering genetic composition for improvement of the dispensation characteristics, plant stature, structural design, resistance to biotic and abiotic stresses, and assembling potential overall biomass yield a target for crop/microbial enhancement (Damude et al., 2018) and generalized view for the same is shown in Figure 11.1. Biofuel sources that have so far been located are at diverse degrees of cultivation and cultivar selection, whereas genetic, genomic resources for these species together with genome sequences, and transformation protocols are presently under development process (Meyer et al., 2017). Major break-throughs to understand lipid metabolism in order to triumph over the low oil/fat production and recalcitrance of lignocellulose for competent and economical alteration to biodiesel is briefly discussed below.

11.3.1   *TARGETING PLANT OIL METABOLISM*

Augmentation of oil concentration in seeds has been attained through exploitation of expression levels of enzymes implicated in the production of triacylglycerols (TAGs) (Li and Zheng, 2017). Approximately, 1.5% enhancement in oil quantity in soybean seeds due to overexpression of fungal DAGT2 enzyme has been achieved. Overexpression of DGAT cDNA led to a resemblant phenotype in "*Arabidopsis.*" Overexpression of two soybean transcription factors *Dof4* and *Dof11* led to enhancement in total fatty acid and lipid seed quantity in "*Arabidopsis*" by commencement of fatty acid biosynthetic cascade. It seemed that TF's *Dof4* and *Dof11* activated lipid biological production in *Arabidopsis* by stimulating acetyl-CoA carboxylase and long acyl-CoA synthase, enzymes caught up in fatty acid production (Zabed et al., 2016).

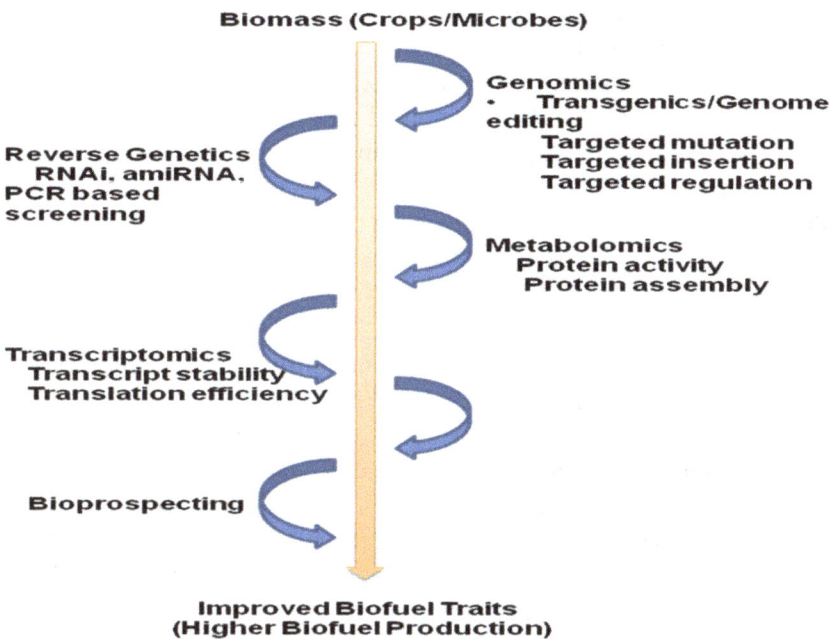

FIGURE 11.1 Genetic enhancement of Biomass (crops/microbes) for increased biofuel production with biotechnology tools.

Enrichment in lipid content by 40% in canola seeds has been observed by overexpression of yeast glycerol 3-phosphate dehydrogenase (ghpd1) gene (Melati et al., 2019). Enhancement in overall production of oils for biodiesel fabrication can be accomplished by engineering oil accretion in vegetative parts, for example, leaves (Qin et al., 2017). Fatty acid breakdown and overexpression of seed development/growth transcription factor LEC2A in aging leaves of "*Arabidopsis*" mutants led to notable oil accumulation. Piling up of high quantity of TAGs (up to 2% leaf dry weight) in aging leaves of "*Arabidopsis*" mutants contrasted to wild-type was reported, while senescent leaves overexpressing LEC2 stored TAGs around 1% leaf dry weight (Damm et al., 2017). Therefore, senescence-inducing promoters that either subdue fatty acid breakdown or induce seed improvement/growth process in foliage play critical role in enhancing prospect of engineering biofuel crops (Damm et al., 2017). Expansion in lipid metabolism research can thus prove useful for better understanding of plant wax biosynthetic mechanism. Among other compounds, the phytowaxes store alkanes derived from fatty acids, as such better understanding of alkane manufacturing process can result in the fabrication of hydrocarbons as subsequent generation biofuels (Kim, 2018).

Overexpression of *SUB1A-1* resulted into a phenotype, where starch storage can be two-folded, particularly in pre-flowering stages due to an effect of ethylene response factors (ERFs) are naturally caught up with plant reply to flooding stress (Montella et al., 2017). As such, it is assumed that SUB1A-1 sturdily inhibits the expression of classical flowering genes "*CONSTANS* and *FLOWERING LOCUS T.*" through correlated effect of flowering time inhibition (Montella et al., 2017).

Flowering modification has also been reported with TFs of "*SQUA-MOSA PROMOTER BINDING PROTEIN-LIKE (SPL)*" family. It has been established that *SPLs* take part as final resort flowering mechanisms stimulated by plant age and naturally inhibited in juvenile plants by miRNA156 (Li et al., 2018). In this manner, the overexpression of miRNA156 in natural mutants or genetically engineered plants leads to an extension of a juvenile phase, thereby increasing starch as well as cell wall saccharification (Li et al., 2018).

11.3.2 TARGETING LIGNOCELLULOSIC BIOMASS

The recalcitrant character of multifarious combination of polysaccharides and lignin remains representative of the major difficulties for competent lignocellulose alteration to biofuels. Exhaustive studies on lignocellulosic characteristics in forage and pulping industries have shown lignin as possessing a foremost role in cell wall recalcitrance (Krasznai et al., 2018). Discourse on the linkage between lignification and forage digestibility has improved over the years due to genetic and biochemical analyses of brown midrib mutants of maize, sorghum, and resemblant grasses. As such, current genetic engineering approaches aim at producing crops with altered lignin, with the anticipation that these approaches will increase forage digestibility and/or pulping effectiveness (Zabed et al., 2016). Knowledge acquired from bioengineering research has largely broadened our comprehension of finest lignin features essential for diverse purposes of lignocellulosic materials while also adding to our knowledge of lignin biosynthetic cascade. Recent growth of significance in cellulosic biofuel generation has gained attention of lignin engineering with *"Populus trichocarpa* and *Brachypodium distachyon"* novel emerging model crops for energy production. As such, lignin study on these model crops, as well as on diversity of other energy crop species, will elucidate lignin biosynthesis and its directive in energy crops, which would escort to logical genetic engineering strategies to alter lignin for enhanced biofuel synthesis (Damm et al., 2017).

Notable progression in fermentable sugars released from lignocelluloses by down-regulating certain monolignol biosynthetic enzymes was accomplished in a transgenic *"alfa alfa."* Silencing of cinnamate 4- hydroxylase (C4H), hydroxycinnamoyl CoA shikimate hydroxycinnamoyl transferase (HCT), and coumaroyl shikimate 3-hydroxylase (C3H) resulted into the lower lignin quantities that correlated with improved cell wall enzymatic hydrolysis. Overexpression of xyloglucanase of *"Aspergillus"* led to decrease of xyloglucan in a poplar, thereby improving saccharification of wood, apparently by producing cellulose readily available to enzymatic hydrolysis (Tiwari and Verma, 2019).

In poplar, silencing of *"PoGT47C,"* a glycosyltransferase homologous to *"Arabidopsis FRA8"* and implicated in glucuronoxylan production, resulted in the enhancement of glucose manufacture following enzymatic hydrolysis, demonstrating that dipping of xylan amount leads to enhanced saccharification competence (Sánchez-Rodríguez et al., 2017). Switching over to cellulose biosynthesis is an optional target approach for lignocellulose upgradation. Cellulose synthase complexes and other proteins involved in the methods have been recognized in *"Arabidopsis"* and other higher plants. In recent times, identification of transcriptional regulation machinery via small RNAs shows that cellulose synthase and several hemicellulosic biosynthetic genes are uniformly down-regulated during leaf growth stages (Nething, 2017). Refinement of this miniature

RNA-directed biocascade has enormous influence for metabolic engineering of cellulose biosynthetic mechanism (Nething, 2017).

11.3.3 TARGETING MICROBIAL FATTY ACID METABOLISM

Approximately, 40% enhancement of the total fatty acid amount in yeast "*Hansenula polymorpha*" was observed due to an overexpression of "*Mucor rouxii*" Acetyl-CoA (ACC). 100-fold enhancement of malonyl-CoA pool and six-fold enhancement in the fatty acid generation were observed in *E. coli* due to an overproduction of ACC (Silva, 2017). Notable enhancement in the incorporation of acetate from the medium due to overexpression of acyl-CoA synthetase (*ACS*) gene in "*E. coli*" contributed to an amplification in the rate of fatty acid generation (Silva, 2017).

An essential technique to increase lipid storage in some strains, such as, "*Mucor circinelloides* and *S. cerevisiae*" is due to the overexpression of ME (an enzyme that catalyzes conversion of malate into pyruvate and concurrently trims down a NADP+ molecule into NADPH, is indicated to be a reducing power contractor for lipogenic enzymes. such as, ACC, FAS, and ACL) (Bellou et al., 2016).

It has been demonstrated that thioesterases are responsible for the liberation of fatty acid chains from acyl carrier protein and generation of free fatty acids. Considerable enhanced fatty acid fabrication in many strains such as "*E. coli* and cyanobacteria" has also been reported due to overexpression of thioesterases. The obstruction of lipid degradation cascade and other metabolic cascades will eventually escort to an accretion of fatty acid compendium (Dourou et al., 2018).

"*Chlorella pyrenoidosa*," a starchless mutant exhibits a strategy to promote polyunsaturated fatty acid content. Overexpression of zinc finger protein TFs that link a DNA sequence within the promoter in microalgae leads to amplified lipid production (You et al., 2017).

11.4 PRODUCTION OF BIOFUELS FROM GENETICALLY MODIFIED SOURCES (CROPS/MICROBES)

Biofuels mainly bioethanol, biobutanol, biodiesel, biogas have so far been produced from genetically engineered crops/microbes. The production processes for each have been discussed briefly in the following sections.

11.4.1 PRODUCTION OF BIOETHANOL

As much as 24% reduction in lignin content was observed in transgenic switchgrass lines, down-regulated with COMT by RNAi (Bardi, 2018). Down-regulated with COMT, switchgrass produced ethanol 38% more than wild-type (Bardi, 2018). Similar quantity of ethyl alcohol was generated from this transgenic switchgrass with less stern pretreatment, much lower cellulose dosages as the control. A 2 year old field grown "COMT RNAi" switchgrass plants shows a same sugar release and ethyl alcohol produce as those grown under controlled conditions. Also, plants did not exhibit any unconstructive consequences on biomass produce or pathological vulnerability, specifying that biofuel generated from COMT RNAi switchgrass can be cost-effective. Nearly, 80% decline of plant protein extract 4CL commotion in switchgrass generated by RNAi silencing of 4CL gene (Korkhovoy et al., 2016) leads to decrease of lignin amount up to 32% in a severe knockdown line. Around 57.2% more fermentable sugar was produced by dilute acid pretreated transgenic biomass than the pretreated wild-type wood (Korkhovoy et al., 2016).

Under fermentation conditions, "CAD1 and COMT" down-regulated "*Brachypodium distachyon*" plants produced 9% and 17% more ethyl alcohol than control (Papp et al., 2016). About 46% and 44% enhancement in saccharification efficiencies can be demonstrated by two "*B. distachyon*" mutants that exhibit point mutations at diverse locations in the "CAD gene." Lignin reduction in these species proves to be quite fruitful for saccharification competence. Five brown midrib (bm) maize mutants and four bm sorghum mutants possessing decreased lignin, have been characterized as "CAD or COMT mutants" which resulted in 22% and 21% amplification in the alteration of cellulose to ethyl alcohol, respectively, thereby increasing digestibility (Bansal et al., 2018). Saccharification efficiency shows three-fold increase due to overexpression of polygalacturonase or pectin methylesterase in transgenic "*Arabidopsis.*" tobacco and wheat leaves, with the alteration of pectin composition or architecture (Papanikolaou and Aggelis, 2019). Additionally, transgenic stems also show increased efficiency of enzymatic saccharification. Pectin degradation in poplar by overexpression of pectate lyase (that degrades HG) leads to upgradation in wood saccharification, as a consequence of the higher liberation of pentoses and hexoses (Ghildiyal et al., 2017).

Cell surface engineering technology enables to equip yeast cells with capability to break down cellulose. Following the same approach, EG II and CBH II enzymes from *"Trichoderma reesei"* and BGL1 from *"Aspergillus aculeatus"* were used in the study and were subsequently attached on the yeast cell surface (Passoth, 2017). The surface engineered yeast cells so obtained were perched and developed in anaerobic condition with phosphoric acid swollen cellulose and amorphous cellulose after aerobic growth, directly yielded ethanol from amorphous cellulose (Passoth, 2017).

Sweet potato flour medium (SPFM) containing 80% moisture, ammonium sulfate 0.2%, pH 5.0, co-cultured strain of *"Trichoderma* sp. and *Saccharomyces cerevisiae"* (1:4 ratio), 10% inoculum size and fermented at 30°C for 72 h produced utmost ethyl alcohol (172 g/kg substrate) (Rodionova et al., 2017). The highest ethyl alcohol concentration, maximum ethyl alcohol productivity (2.8 g/kg substrate/h), microbial biomass (23 × 108 CFU/ g substrate), ethanol yield (47 g/100 g sugar consumed) and fermentation efficiency (72%) were also achieved under these settings (Soares et al., 2019). Between competent cells of *"Trichoderma* sp. and *S. cerevisiae* when co-cultured, familiar cell interaction was observed. Ethanol generating capability by co-culture was 65% greater than single culture of *"S. cerevisiae"* from unsaccharified SPFM (Wargacki et al., 2012).

11.4.2 PRODUCTION OF BIOBUTANOL

Due to an already known genome sequence of *"Clostridia acetobutylicum,"* a significant perceptive of its metabolic conduits, cellular regulation, and genetics is possible (Weyens et al., 2009). On employing the genome and metabolic repository, *"C. acetobutylicum"* has been shown to enhance butanol yield by metabolic engineering (Martien and Amador-Noguez, 2017).

Utilizing antisense RNA (asRNA) approach, improvement in selectivity for butanol generation has been shown (Ong et al., 2016). Although, only asRNA against *ctfB* (the second CoA transferase gene in the polycistronic *aad-ctfA-ctfB* message) in *"C. acetobutylicum"* considerably causes decline in acetone and butanol levels compared with control (Ong et al., 2016), asRNA against *ctfB* combined with overexpression of alcohol–aldehyde dehydrogenase gene (*aad*) results in increased produce of butanol/acetone ratio (Ong et al., 2016). An improvement in concentration of butanol yield by mutant strain was by about 230% of control strain and maximum ever documented so far in *"C. acetobutylicum"* (Leong et al., 2018). As such, the investigation verified that asRNA against *ctfB* degraded entire "sol operon (*aadctfA- ctfB*) transcript." Likewise, butanol/acetone ratio of mutant strain doubled compared with control strain (Leong et al., 2018). Thus, asRNA approach proved successful in increasing selectivity of butanol yield (George et al., 2017).

Strain SolRH produced by an inactivation of *solH* in "*C. acetobutylicum*" leads to an increase in glucose exploitation rate and yields higher butanol in contrast to wild-type. Gene product of *solH* is an acknowledged inhibitor of solvent formation gene (Bankar et al., 2013). As such, strain SolRH (pTAAD) produces even more concentrations of solvents than strain SolRH (Bankar et al., 2013).

Cloning and subsequent expression of n-butanol biosynthesizing genes of "*C. acetobutylicum*" including "ATCC 824 (*thl, hbd, crt, bcd, etfA, etfB*, and *adhE2*)" in "*E.coli*" under an aerobic environment yields *n*-butanol and additional *n*-butanol by obliterating the pathways contending with *n*-butanol production when bacteria is cultivated in enriched media (Liu and Khosla, 2010). In the similar fashion, when diverse genes of "*C. acetobutylicum* ATCC 824 (*thl, hbd, crt, bcd-etfB-etfA*, and *adhe1* or *adhe*)" were cloned and expressed in "*E. coli*," the biotailored strain with *adhe* produced four times as much n-butanol as engineered strain with *adhe1* (Bhatia et al., 2017).

Among various organisms, "*Lactobacillus brevis*" unsurprisingly possess the utmost tolerance to butanol (3.0% of butanol), thus it has been subjected to metabolic engineering to yield butanol (Kour et al., 2019). "Recombinant *L. brevis* strains" were proficient to yield *n*-butanol on glucose medium only compared with "*E. coli*" strain that produced butanol and other alcohols due to overexpression of KDC and ADH and addition of specific 2-keto acids to growth medium (Singh et al., 2020).

An increment in intracellular levels of threonine was gained through overexpression of "2-ketovalerate, *thrABC, ilvA*, and *leuABCD*" in "*E. coli* strain" with competing pathways obliterated. Product of "*thrABC* gene" is involved in threonine generation, whereas the product of "*ilvA* gene" is threonine dehydrogenase, which catalyzes the reaction of threonine to 2-ketobutyrate and protein product of "*leuABCD* gene" catalyzes the conversion of 2-ketobutyrate to 2-ketovalerate (Srivastava et al., 2020). As such, engineered strain yields 0.9 g/L *n*-butanol in initial shake flask experiments lacking much optimization. Overexpressed "*alsS, ilvC*, and *ilvD* in *E.coli*" strain leads to carbon reserve being transformed to

2-ketoisovalerate, and in turn, 2-ketoisovalerate is transformed to isobutanol using KDC and ADH. Deletion of genes of competing pathways results in the increments in the levels of pyruvate existent for synthetic isobutanol pathway (Yan et al., 2020), which inturn yielded 20 g/L isobutanol in engineered strain with 86% of maximum yield (Yan et al., 2020).

CRISPR analysis for gene *spo0A* deletion in "*Clostridium beijerinckii*" proved useful for acetone–butanol–ethanol generation. "*C. acetobutylicum*" acetone–butanol–ethanol fermentation pathway employed in the generation of butanol was first constructed in "*E.coli*" to lay a foundation for relationship to other hosts (Tang et al., 2020). Co-expression of "*S.cerevisiae*" formate dehydrogenase and overexpression of "*E.coli*" glyceraldehyde 3-phosphate dehydrogenase to elevate glycolytic flux enhanced titers to 580 mg/L and butanol generation to 200 mg/L (Saumya et al., 2019). Recently, testing of "*E.coli*" for native promoter of hydrogenase I cluster Phya Bw2V which carries plasmid pCNA-PHC and pENA-TA in anaerobic fermentation with extra glucose, results in butanol production up to 2.8 g/L in batch culture bioreactor (Saumya et al., 2019).

"*Thermoanaero bacterium saccharolyticum*" strain already has been exemplified and biotailored for the generation of biohydrogen, ethanol, and butanol. Different gene clusters used for the study were "*hbd, crt, bcd, eftA,* and *eftB*" from "*Thermoanaerobacterium thermosaccharolyticum,* DSM571 and *adhE2* from *C. acetobutylicum*" (Huang et al., 2020). "*Klebsiella pneumonia*" has also been engineered for the generation of 2-butanol and 1-butanol from crude glycerol as a single carbon source (Huang et al., 2020). 1-butanol production from "*Klebsiella*" was carried out by altering the CoA-dependent pathway, and 2-2-keto acid pathway was developed by expressing genes for *ter-bdhB-bdhA* and *kivd* (Huang et al., 2020). Butanol titer and produce were found to be 15.03 mg/L and 27.79 mg butanol/g-cell and 28.7 mg/L and 51.58 mg butanol/g-cell (Huang et al., 2020). Native products can be inhibited through antisense RNA approach. 1-butanol was yielded by biotailoring a-ketoisovalerate decarboxylase (*kivd*) and alcohol dehydrogenase (*adh*) from "*Lactococcus lactis* into *Klebsiella pneumonia,*" thereby bypassing the pathway for generation of 2,3-butandiol. Production was 320 mg/L which displayed two-fold increment (Huang et al., 2020). Engineering of "*Geobacillus thermoglucosidasius*" with acetohydroxy acid synthase gene, 2-ketoisovalerate dehydrogenase gene from "*Bacillus subtilis* and *Lactococcus lactis*" and promoter region of lactate dehydrogenase gene from "*Geobacillus thermodenitrificans*" yields 3.3 g/L isobutanol from glucose as substrate (Guirimand et al., 2019).

For butanol production, "*Pyrococcus furiosus*" has been genetically tailored by the expression of lactate dehydrogenase gene from "*Caldicellulosiruptor bescii*" for the production of 3-hydroxypropionate (additionally employed as electrofuel) (Hu et al., 2019) employing hydrogen as substrate and by developing 1-butanol, 2-butanol production pathway (Hu et al., 2019). In the study, genes responsible for the activity of enzymes involved in first three reactions were isolated from different sources, that is, acetyl-CoA to crontyl CoA from "*Thermoanaerobacter tengcongensis,*" trans-2-enoyl-CoA reductase (*ter*) from "*Spirochaete thermophila*" and butyraldehyde dehydrogenase (*Bad*) and butanol dehydrogenase (*Bdh*) from "*Thermoanaerobacter* sp. *X514*" (Creutz, 2019). 1-butanol and 2-butanol production were 70 and 15 mg/L after 48 h from genetically engineered "*Pyrococcus furiosus*" at 60°C (Creutz, 2019). Synthetic pathway of isobutanol through keto acid pathway is shown in Figure 11.2.

11.4.3 PRODUCTION OF BIODIESEL

For fatty acid ethyl esters (FAEE) production from "*E. coli,*" ethanol cascade from "*Zymomonas mobilis,*" pyruvate decarboxylase (*pdc*) and alcohol dehydrogenase (*adhB*) and unspecific acyltransferase (*atfA*) from "*Acinetobacter*" is required (Stamenković et al., 2011). The process of ethyl alcohol manufacture was merged with the esterification of ethyl alcohol with the acyl moieties of coenzyme A, thioesters from fatty acids in the supplied glucose and oleic acid (Demirbas, 2008), and as such 1.28 g/L of FAEE under aerobic environment in the presence of glucose and oleic acid by fed-batch fermentation was generated by the engineered strain (Demirbas, 2008). Additionally, there are supplementary reports on biotailored "*E.coli*" with capability to yield FAEE (Stamenković et al., 2011). It has been reported that engineered "*E.coli*" strain has yielded FAEE directly from glucose and ethanol, from only glucose and from glucose and xylan (Sivasamy et al., 2009). This strategy of synthetic biology will make vast improvements in engineering endeavors to manufacture multiplicity of biodiesels (Sivasamy et al., 2009).

In similar fashion, heterologous generation of bifunctional wax ester synthase/acyl-coenzyme A: diacylglycerol acyltransferase (WS/DGAT) from "*A.calcoaceticus* ADP1 in *S. cerevisiae*" can lead to manufacture of FAEEs and fatty acid isoamyl ester (Demirbaş, 2002). Investigations regarding the generation of microbial lipids have also been carried out in oleaginous yeasts, such as, *"Rhodosporidium toruloides"* (Demirbaş, 2002). Different genetic engineering strategies applied in microalgae for biodiesel application are shown in Figure 11.3.

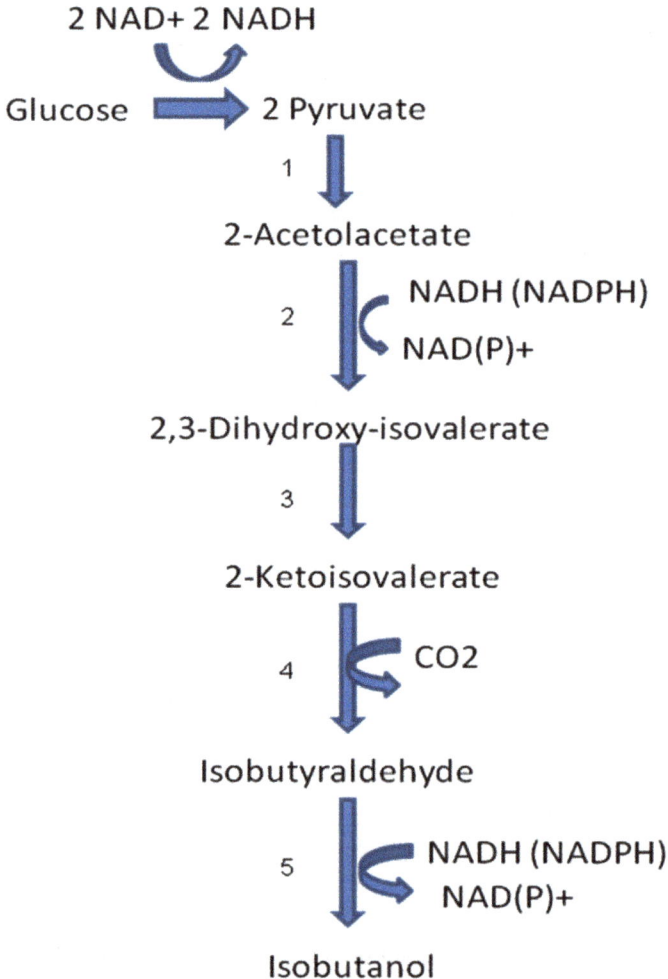

FIGURE 11.2 Synthetic pathway of isobutanol through keto acid pathway. Numbers refer to the enzymes: 1: acetolactate synthase (AIsS), 2: acetohydroxy acid isomeroreductase (IIvC), 3: dihydroxyacid dehydratase (IIvD), 4: 2-keto acid decarboxylase (KDC), 5: alcohol dehydrogenase (ADH).

Regulon Engineering

Transgenic Cells

⬇

| Optimising light harvesting | **Photosynthesis** | Enhancing carbon capturec |

⬇

| Manipulating precursor building pathways | **Respiration** | Blocking starch synthesis |

⬇

| Modification of FA profile | **Fatty Acid Biosynthesis** | Modulating FA synthesis |

⬇

| Blocking lipolysis | **TAG Assembly** | Stimulating TAG synthesis |

Stress alebation for optimal growth

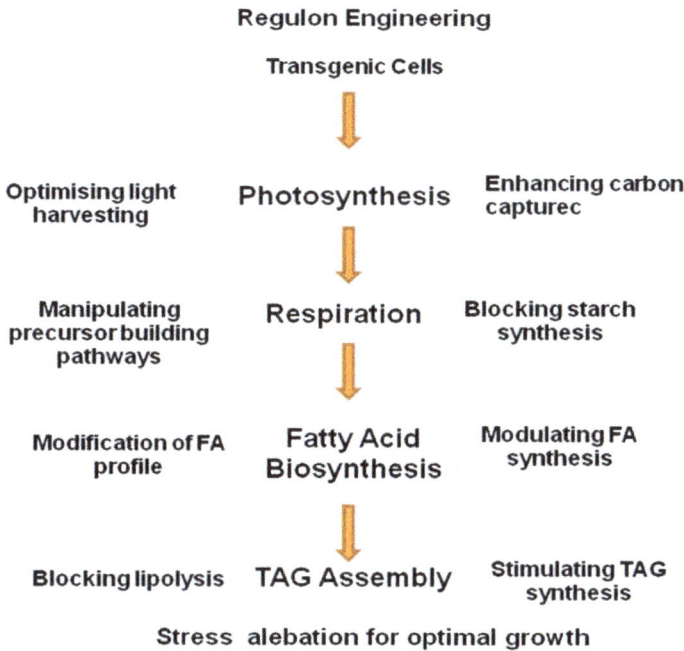

FIGURE 11.3 Schematic illustration of different genetic engineering strategies applied in microalgae for biodiesel application.

11.4.4 PRODUCTION OF BIOGAS

A study has been carried out where laboratory equipment, purposely designed for biogas generation was established and tested (Demirbaş, 2002). The equipment is a specifically designed batch fermentation system and is an appropriate model of incessantly operated mass scale bioindustrial biogas fermentor (Demirbaş, 2002). "*C. saccharolyticus*" and its flexible hydrogen yielding activity increase the biogas productivity from diverse substrates (from Banana, 2007). In incessantly operated mass scale bioindustrial biogas technology, imperative factor regulating the worth of this biotechnological enhancement of performance was the persistence of the advantageous consequence in time (Demirbaş, 2002). "*C. saccharolyticus*" is competent for integration into natural biogas producing consortia under suitable environment. An imperative feature in this relevance is the rate at which fermentor is supposed to be encumbered with organic substrates (from Banana, 2007).

11.5   CONCLUSIONS

For sustainable society to persist, reliance on biofuels would decrease dependence on fossil fuels as energy sources. Advancement in phytobiomass features for biofuel generation is in the inaugural phases. Biofuel crops have already been recognized and are at diverse stages of breeding and cultivar selection, while genetic, genomic resources for these species including genome sequences and transformation protocols are being fabricated. Understanding the lipid metabolism and phyto cell wall biosynthesis are required to triumph over low oil/fat yields and recalcitrance of lignocelluloses for resourceful and economical alteration to biodiesel and other biofuels. Even though few targets for genetic engineering in oil seeds and monolignol replacement in cell walls have been determined, the general effect of modification of these characteristics in committed biofuel crops is still lacking. Although plant biotechnology plays key role in triumphant production of energy crops, it still should go collectively with breeding strategies besieged at sustaining or augmenting significant agronomic characteristics that convert these plants into striking domains for biofuel production, such as, escalated resistance to abiotic and biotic factors, low soil amendment requirements and perpetual life cycle. Additionally, amalgamation of cell surface biotailoring and metabolic biotailoring will definitely generate additional profitable microorganisms for economical and less exhaustive manufacturing of biofuels. It is only in the presence of appropriate profitable, scientific, and social policies that biofuels will contribute to energy safekeeping without threatening food security of the underprivileged. And that needs to be worked out in diverse framework of food and agriculture system of the world.

KEYWORDS

- **biotechnology**
- **microbes**
- **biofuels**
- **lignocellulose**
- **biogas**

REFERENCES

Abdel, H.; Luo, Z.; Ray, D. Genetic Improvement of Guayule (Parthenium argentatum A. Gray): An Alternative Rubber Crop. In *Advances in Plant Breeding Strategies: Industrial and Food Crops*; Springer, 2019; pp 151–178.

Alcoforado, F. Global Climate Change and Its Solutions. *J. Atmos. Earth Sci.* **2019**, *2* (007).

Amaniampong, P.; Asiedu, N.; Fletcher, E.; Dodoo, D.; Olatunji, O.; Trinh, Q. Conversion of Lignocellulosic Biomass to Fuels and Value-Added Chemicals Using Emerging Technologies and State-of-the-Art Density Functional Theory Simulations Approach. In *Valorization of Biomass to Value-Added Commodities*; Springer, 2020; pp 193–220.

from Banana, B. Production of Biogas from Banana and Plantain Peels. *Adv. Environ. Biol.* **2007**, *1* (1), 33–38.

Bankar, S.; Survase, S.; Ojamo, H.; Granström, T. Biobutanol: The Outlook of an Academic and Industrialist. *RSC Adv.* **2013**, *3* (47), 24734–24757.

Bansal, N.; Khot, M.; Jana, A.; Nautiyal, A.; Sharma, T.; Dasgupta, D.; Mohapatra, S.; Yadav, S. K.; Hazra, S.; Ghosh, D. Oleaginous Yeasts: Lignocellulosic Biomass Derived Single Cell Oil as Biofuel Feedstock. *Principles App. Ferment. Technol.* **2018**, 263–306.

Bardi, L. Production of Bio-Oils from Microbial Biomasses. In *Mycoremediation and Environmental Sustainability*; Springer, 2018; pp 61–89.

Bellou, S.; Triantaphyllidou, I.; Aggeli, D.; Elazzazy, A.; Baeshen, M.; Aggelis, G. Microbial Oils as Food Additives: Recent Approaches for Improving Microbial Oil Production and Its Polyunsaturated Fatty Acid Content. *Curr. Opin. Biotechnol.* **2016**, *37*, 24–35.

Bhatia, S.; Kim, S.; Yoon, J.; Yang, Y. Current Status and Strategies for Second Generation Biofuel Production Using Microbial Systems. *Energy Conv. Manage.* **2017**, *148*, 1142–1156.

Ciardelli, F.; Bertoldo, M.; Bronco, S.; Passaglia, E. Environmental Impact. In *Polymers from Fossil and Renewable Resources*; Springer, 2019; pp 161–187.

Confalonieri, M.; Sparvoli, F. Recent Advances in Medicago spp. Genetic Engineering Strategies. In *The Model Legume Medicago Truncatula*, 2019; pp 1149–1161.

Creutz, C. Expression of Metazoan Annexins in Yeast Provides protection against Deleterious Effects of the Biofuel Isobutanol. *Sci. Rep.* **2019**, *9* (1), 1–6.

Damm, T.; Grande, P.; Jablonowski, N.; Thiele, B.; Disko, U.; Mann, U.; Schurr, U.; Leitner, W.; Usadel, B.; de María, P. OrganoCat Pretreatment of Perennial Plants: Synergies between a Biogenic Fractionation and Valuable Feedstocks. *Bioresour. Technol.* **2017**, *244*, 889–896.

Damude, H.; Hitz, W.; Meyer, K.; Yadav, N. DGAT genes from Yarrowia Lipolytica for Increased Seed Storage Lipid Production and Altered Fatty Acid Profiles in Soybean. Google Patents, 2018.

Demirbaş, A. Biodiesel from Vegetable Oils via Transesterification in Supercritical Methanol. *Energy Conv. Manage.* **2002**, *43* (17), 2349–2356.

Demirbas, A. Comparison of Transesterification Methods for Production of Biodiesel from Vegetable Oils and Fats. *Energy Conv. Manage.* **2008**, *49* (1), 125–130.

Dourou, M.; Aggeli, D.; Papanikolaou, S.; Aggelis, G. Critical Steps in Carbon Metabolism Affecting Lipid Accumulation and Their Regulation in Oleaginous Microorganisms. *Appl. Microbiol. Biotechnol.* **2018**, *102* (6), 2509–2523.

Enamala, M.; Enamala, S.; Chavali, M.; Donepudi, J.; Yadavalli, R.; Kolapalli, B.; Aradhyula, T. V.; Velpuri, J.; Kuppam, C. Production of Biofuels from Microalgae-A Review on Cultivation, Harvesting, Lipid Extraction, and Numerous Applications of Microalgae. *Renew. Sustain. Energy Rev.* **2018**, *94*, 49–68.

Faraji, M.; Voit, E. Improving Bioenergy Crops through Dynamic Metabolic Modeling. *Processes* **2017**, *5* (4), 61.

Favaro, L.; Jansen, T.; van Zyl, W. Exploring industrial and Natural Saccharomyces Cerevisiae Strains for the Bio-Based Economy from Biomass: The Case of Bioethanol. *Crit. Rev. Biotechnol.* **2019**, *39* (6), 800–816.

George, I.; Bogaerts, P.; Gilis, D.; Rooman, M.; Flot, J. New tools for bioprocess analysis and optimization of microbial fuel production. *Microbial Fuels: Technologies and Applications* **2017**, 427–494.

Ghag, S.; Vavilala, S.; D'Souza, J. Metabolic Engineering and Genetic Manipulation of Novel Biomass Species for Biofuel Production. In *Advanced Bioprocessing for Alternative Fuels, Biobased Chemicals, and Bioproducts*; Elsevier, 2019; pp 13–34.

Ghildiyal, R.; Bisht, G.; Shrivastava, R. Engineering Yeast as Cellular Factory. In *Metabolic Engineering for Bioactive Compounds*; Springer, 2017; pp 173–208.

Guirimand, G.; Bamba, T.; Matsuda, M.; Inokuma, K.; Morita, K.; Kitada, Y.; Kobayashi, Y.; Yukawa, T.; Sasaki, K.; Ogino, C. Combined Cell Surface Display of β-d-Glucosidase (BGL), Maltose Transporter (MAL11), and Overexpression of Cytosolic Xylose Reductase (XR) in *Saccharomyces cerevisiae* Enhance Cellobiose/Xylose Coutilization for Xylitol Bioproduction from Lignocellulosic Biomass. *Biotechnol. J.* **2019**, *14* (9), 1800704.

Hu, Y.; Zhu, Z.; Nielsen, J.; Siewers, V. Engineering *Saccharomyces cerevisiae* Cells for Production of Fatty Acid-Derived Biofuels and Chemicals. *Open Biol.* **2019**, *9* (5), 190049.

Huang, X.; Bai, S.; Liu, Z.; Hasunuma, T.; Kondo, A.; Ho, S. Fermentation of Pigment-Extracted Microalgal Residue Using Yeast Cell-Surface Display: Direct High-Density Ethanol Production with Competitive Life Cycle Impacts. *Green Chem.* **2020**, *22* (1), 153–162.

Kehrein, P.; van Loosdrecht, M.; Osseweijer, P.; Dewulf, J.; Garfi, M.; Duque, J. A Critical Review of Resource Recovery from Municipal Wastewater Treatment Plants–Market Supply Potentials, Technologies and Bottlenecks. *Environ. Sci.: Water Res. Technol.* **2020**.

Kim, D. Physico-Chemical Conversion of Lignocellulose: Inhibitor Effects and Detoxification Strategies: A Mini Review. *Molecules* **2018**, *23* (2), 309.

Korkhovoy, V.; Tsarenko, P.; Blume, Y. Genetically Engineered Microalgae for Enhanced Biofuel Production. *Curr. Biotechnol.* **2016**, *5* (4), 256–265.

Kour, D.; Rana, K.; Yadav, N.; Yadav, A.; Rastegari, A.; Singh, C.; Negi, P.; Singh, K.; Saxena, A. Technologies for Biofuel Production: Current Development, Challenges, and Future Prospects. In *Prospects of Renewable Bioprocessing in Future Energy Systems*; Springer, 2019; pp 1–50.

Krasznai, D.; Champagne, R.; Roy, H.; Champagne, P.; Cunningham, M. Compositional Analysis of Lignocellulosic Biomass: Conventional Methodologies and Future Outlook. *Crit. Rev. Biotechnol.* **2018**, *38* (2), 199–217.

Lapeña, D.; Kosa, G.; Hansen, L.; Mydland, L.; Passoth, V.; Horn, S.; Eijsink, V. Production and Characterization of Yeasts Grown on Media Composed of Spruce-Derived Sugars and Protein Hydrolysates from Chicken By-products. *Microb. Cell Factories* **2020**, *19* (1), 19.

Leong, W.; Lim, J.; Lam, M.; Uemura, Y.; Ho, Y., Third Generation Biofuels: A Nutritional Perspective in Enhancing Microbial Lipid Production. *Renew. Sustain. Energy Rev.* **2018**, *91*, 950–961.

Li, X.; Chang, S.; Liu, R. Industrial Applications of Cellulases and Hemicellulases. In *Fungal Cellulolytic Enzymes*; Springer, 2018; 267–282.

Li, X.; Zheng, Y. Lignin-Enzyme Interaction: Mechanism, Mitigation Approach, Modeling, and Research Prospects. *Biotechnol. Adv.* **2017**, *35* (4), 466–489.

Liang, Y.; Eudes, A.; Yogiswara, S.; Jing, B.; Benites, V.; Yamanaka, R.; Cheng, C.; Baidoo, E. E.; Mortimer, J.; Scheller, H. A Screening Method to Identify Efficient sgRNAs in Arabidopsis, Used in Conjunction with Cell-Specific Lignin Reduction. *Biotechnol. Biofuels* **2019**, *12* (1), 130.

Liu, J.; Song, Y.; Qiu, W. Oleaginous Microalgae Nannochloropsis as a New Model for Biofuel Production: Review & Analysis. *Renew. Sustain. Energy Rev.* **2017**, *72*, 154–162.

Liu, T.; Khosla, C. Genetic Engineering of *Escherichia coli* for Biofuel Production. *Annu. Review Genet.* **2010**, *44*, 53–69.

Martien, J.; Amador, D. Recent Applications of Metabolomics to Advance Microbial Biofuel Production. *Curr. Opin. Biotechnol.* **2017**, *43*, 118–126.

Melati, R.; Shimizu, F.; Oliveira, G.; Pagnocca, F.; de Souza, W.; Sant'Anna, C.; Brienzo, M. Key Factors Affecting the Recalcitrance and Conversion Process of Biomass. *BioEnergy Res.* **2019**, *12* (1), 1–20.

Meyer, K.; Damude, H.; Ripp, K.; Stecca, K. DGAT Genes from Oleaginous Organisms for Increased Seed Storage Lipid Production and Altered Fatty Acid Profiles in Oilseed Plants. Google Patents, 2017.

Miao, R.; Wegelius, A.; Durall, C.; Liang, F.; Khanna, N.; Lindblad, P. Engineering Cyanobacteria for Biofuel Production. In *Modern Topics in the Phototrophic Prokaryotes*; Springer, 2017; pp 351–393.

Montella, S.; Ventorino, V.; Lombard, V.; Henrissat, B.; Pepe, O.; Faraco, V. Discovery of Genes Coding for Carbohydrate-Active Enzyme by Metagenomic Analysis of Lignocellulosic Biomasses. *Sci. Rep.* **2017**, *7*, 42623.

Muthamilarasan, M.; Singh, N.; Prasad, M. Multi-Omics Approaches for Strategic Improvement of Stress Tolerance in Underutilized Crop Species: A Climate Change Perspective. *Adv. Genet.* **2019**, *10*.

Nething, D. Detection of Cellulose Synthase Antisense Transcripts Involved in Regulating Cell Wall Biosynthesis in Barley, Brachypodium and Arabidopsis; Ohio University, 2017.

Ogunwole, J.; Alabi, O.; Ugbabe, O.; Birhanu, B. Promoting Jatropha Agriculture for Sustainable Soil Capital Improvement: A Win-Win Technology for Rehabilitating Degraded Lands in Africa. In *New Frontiers in Natural Resources Management in Africa*; Springer, 2019; pp 27–39.

Ong, R.; Higbee, A.; Bottoms, S.; Dickinson, Q.; Xie, D.; Smith, S.; Serate, J.; Pohlmann, E.; Jones, A.; Coon, J. Inhibition of Microbial Biofuel Production in Drought-Stressed Switchgrass Hydrolysate. *Biotechnol. Biofuels* **2016**, *9* (1), 237.

Papanikolaou, S.; Aggelis, G. Sources of Microbial Oils with Emphasis to Mortierella (Umbelopsis) Isabellina Fungus. *World J. Microbiol. Biotechnol.* **2019**, *35* (4), 63.

Papp, T.; Nyilasi, I.; Csernetics, Á.; Nagy, G.; Takó, M.; Vágvölgyi, C. Improvement of Industrially Relevant Biological Activities in *Mucoromycotina fungi*. In *Gene Expression Systems in Fungi: Advancements and Applications*; Springer 2016; pp 97–118.

Passoth, V. Lipids of Yeasts and Filamentous Fungi and Their Importance for Biotechnology. In *Biotechnology of Yeasts and Filamentous Fungi*; Springer, 2017; pp 149–204.

Purdy, R. Incorporating Resilience and Sustainability in Renewable Energy Development: A Cameroon Case Study, 2019.

Qin, L.; Li, W.; Zhu, J.; Li, B.; Yuan, Y. Hydrolysis of Lignocellulosic Biomass to Sugars. In *Production of Platform Chemicals from Sustainable Resources*; Springer, 2017; pp 3–41.

Rodionova, M.; Poudyal, R.; Tiwari, I.; Voloshin, R.; Zharmukhamedov, S.; Nam, H.; Zayadan, B.; Bruce, B.; Hou, H.; Allakhverdiev, S. Biofuel Production: Challenges and Opportunities. *Int. J. Hydr. Energy* **2017**, *42* (12), 8450–8461.

Sánchez-Rodríguez, C.; Ketelaar, K.; Schneider, R.; Villalobos, J.; Somerville, C.; Persson, S.; Wallace, I. Brassinosteroid Insensitive Negatively Regulates Cellulose Synthesis in Arabidopsis by Phosphorylating Cellulose Synthase 1. *Proc. Natl. Acad. Sci.* **2017**, *114* (13), 3533–3538.

Saumya, S.; Aberami, J.; Sankar, P. Plastid Transformation–A Greener and Cleaner Technique for Overexpression of Proteins. *Res. J. Pharm. Technol.* **2019**, *12* (10), 5083–5090.

Shih, P.; Liang, Y.; Loqué, D. Biotechnology and Synthetic Biology Approaches for Metabolic Engineering of Bioenergy Crops. *Plant J.* **2016**, *87* (1), 103–117.

Silva, R. Análises de transcritoma e de metaboloma revelam que Qualea grandiflora Mart. possui um metabolismo Alumínio-dependente, **2017**.

Singh, V.; Yadav, V.; Mishra, V. Nanotechnology: An Application in Biofuel Production. In *Nanomaterials in Biofuels Research*; Springer, 2020; pp 143–160.

Sivasamy, A.; Cheah, K.; Fornasiero, P.; Kemausuor, F.; Zinoviev, S.; Miertus, S. Catalytic Applications in the Production of Biodiesel from Vegetable Oils. *ChemSusChem Chem. Sustain. Energy Mater.* **2009**, *2* (4), 278–300.

Soares, L.; Rabelo, C.; Delforno, T.; Silva, E.; Varesche, M. Experimental Design and Syntrophic Microbial Pathways for Biofuel Production from Sugarcane Bagasse under Thermophilic Condition. *Renew. Energy* **2019**, *140*, 852–861.

Soltanian, S.; Aghbashlo, M.; Almasi, F.; Hosseinzadeh, H.; Nizami, A.; Ok, Y.; Lam, S.; Tabatabaei, M. A Critical Review of the Effects of Pretreatment Methods on the Exergetic Aspects of Lignocellulosic Biofuels. *Energy Conv. Manage.* 2020.

Srivastava, A.; Villalobos, M.; Singh, R. Engineering Photosynthetic Microbes for Sustainable Bioenergy Production. In *Contemporary Environmental Issues and Challenges in Era of Climate Change*; Springer 2020; 183–198.

Stamenković, O.; Veličković, A.; Veljković, V. The Production of Biodiesel from Vegetable Oils by Ethanolysis: Current State and Perspectives. *Fuel* **2011**, *90* (11), 3141–3155.

Suganya, T.; Varman, M.; Masjuki, H.; Renganathan, S. Macroalgae and Microalgae as a Potential Source for Commercial Applications along with Biofuels Production: A Biorefinery Approach. *Renew. Sustain. Energy Rev.* **2016**, *55*, 909–941.

Tang, S.; Xu, T.; Peng, J.; Zhou, K.; Zhu, Y.; Zhou, W.; Cheng, H.; Zhou, H. Overexpression of an Endogenous Raw Starch Digesting Mesophilic α-amylase Gene in Bacillus Amyloliquefaciens Z3 by In Vitro Methylation Protocol. *J. Sci. Food Agric.* 2020.

Temme, K.; Tamsir, A.; Bloch, S.; Clark, R.; Emily, T. Methods and Compositions for Improving Plant Traits. Google Patents 2019.

Thanapimmetha, A.; Saisriyoot, M.; Khomlaem, C.; Chisti, Y.; Srinophakun, P. A Comparison of Methods of Ethanol Production from Sweet Sorghum Bagasse. *Biochem. Eng. J.* **2019**, *151*, 107352.

Tiwari, S.; Verma, T. Cellulose as a Potential Feedstock for Cellulose Enzyme Production. In *Approaches to Enhance Industrial Production of Fungal Cellulases*; Springer, 2019; pp 89–116.

Unkefer, C.; Sayre, R.; Magnuson, J.; Anderson, D.; Baxter, I.; Blaby, I.; Brown, J.; Carleton, M.; Cattolico, R.; Dale, T. Review of the Algal Biology Program within the National Alliance for Advanced Biofuels and Bioproducts. *Algal Res.* **2017**, *22*, 187–215.

Wargacki, A.; Leonard, E.; Win, M.; Regitsky, D.; Santos, C.; Kim, P.; Cooper, S.; Raisner, R.; Herman, A.; Sivitz, A. An Engineered Microbial Platform for Direct Biofuel Production from Brown Macroalgae. *Science* **2012**, *335* (6066), 308–313.

Wellmer, F.; Buchholz, P.; Gutzmer, J.; Hagelüken, C.; Herzig, P.; Littke, R.; Thauer, R. Current Status of Natural Resources—An Overview. In *Raw Materials for Future Energy Supply*; Springer, 2019; pp 107–144.

Weyens, N.; van der Lelie, D.; Taghavi, S.; Newman, L.; Vangronsveld, J. Exploiting Plant–Microbe Partnerships to Improve Biomass Production and Remediation. *Trends Biotechnol.* **2009,** *27* (10), 591–598.

Yan, F.; Dong, G.; Qiang, S.; Niu, Y.; Hu, C.; Meng, Y. Overexpression of Δ 12,Δ 15-Desaturases for Enhanced Lipids Synthesis in Yarrowia lipolytica. *Front. Microbiol.* **2020,** *11*, 289.

You, S.; Joo, Y.; Kang, D.; Shin, S.; Hyeon, J.; Woo, H.; Um, Y.; Park, C.; Han, S. Enhancing Fatty Acid Production of *Saccharomyces cerevisiae* as an Animal Feed Supplement. *J. Agric. Food Chem.* **2017,** *65* (50), 11029–11035.

Zabed, H.; Sahu, J.; Boyce, A.; Faruq, G. Fuel Ethanol Production from Lignocellulosic Biomass: An Overview on Feedstocks and Technological Approaches. *Renew. Sustain. Energy Rev.* **2016,** *66*, 751–774.

Zuccaro, G.; Pirozzi, D.; Yousuf, A. Lignocellulosic Biomass to Biodiesel. In *Lignocellulosic Biomass to Liquid Biofuels*; Elsevier, 2020; pp 127–167.

CHAPTER 12

Biosensors: A Biotechnological Tool for Monitoring Environmental Pollution

IRTEZA QAYOOM[1*], ZULAYKHA KHURSHID DIJOO[2], and
MEHVISH HAMEED[3]

[1]*Department of Environmental Science, Sri Pratap College,
Cluster University Srinagar, India*

[2]*Department of Environmental Sciences/Centre of Research for
Development, University of Kashmir, Srinagar, J&K, India*

[3]*College of Agricultural Engineering, Sher-e-Kashmir University of
Agricultural Sciences and Technology of Kashmir, Shalimar Campus,
Srinagar, J&K, India*

Corresponding author. E-mail: qirteza@gmail.com

ABSTRACT

Biosensors applicability lies in domains, such as defence, food analysis, agriculture, industrial processes, medical diagnosis, and environmental monitoring, etc. This chapter starts with a general introduction about the title followed by the classification of biosensors and applications in the area of environmental investigations and examining. The various approaches concerning diverse biocomponents, principle, transducers, and their application for different group of analytes and environmental contaminants have been deliberated upon.

12.1 INTRODUCTION

Environmental security has been conceded by a wide-ranging number of environmental contaminants as a consequence of industrial, land, agrarian, and other human undertakings. Environmental security is among the noteworthy apprehensions for the well-being of organisms which is threatened by a variety of pollutants negative to the surroundings. Therefore, monitoring of water, soil, and air pollutants is of absolute importance (Rogers et al., 2007). Different initiatives, regulatory measures, along with scientific and other public issues have been debated upon as well as implemented in order to limit and control the environmental pollution dangers, but it continues to remain an international challenge. Environmental monitoring is also among the various agendas of the International organizations because environmental pollution and the human health as well as socioeconomic expansion are intricately related (Justino et al., 2017). Countless anthropogenic practices have also contributed to the water pollution with biological micropollutants, such as viruses and bacteria, protozoa, and mycoplasmas that remain a major cause of death worldwide (Slifonova et al., 1993). For the general well-being and security of living systems, toxicants should be studied comprehensively for their identification and constant monitoring in environmental samples. Diverse methods have been well-thought out for the analysis of environmental pollutants. The conventional methods offer precise analysis of physiochemical properties of environmental samples with better accuracy and sensitivity, but require specialized laboratories with costly analytical instruments for pollutant analysis (Nigam and Shukla, 2015). Moreover, this system of analysis does not offer evidence on the bioavailability of pollutants and associated consequences on biological systems. To overcome restraints of the conventional methods, a parallel technique although complementary in nature is being assessed in which living systems are being worked for bioassays of environmental samples (Nigam and Shukla, 2015). Therefore, certain sensitive, rapid, and selective techniques have to be developed for analyzing pollutants for bioremediation process efficiently. Biosensors are an impressive alternative to traditional methods. Biosensors are the most popular environmental biotechnological tool used in the field of biomonitoring. These are used for monitoring biodegradability, toxicity, mutagenicity, concentration of hazardous substances, and concentration and pathogenicity of microorganism in wastes and in the environment (Singh, 2015). Biosensors are suitable for detecting and analyzing environmental pollutants because of its reliability, specificity, and sensitivity (Podola et al., 2004). Several definitions for biosensors were put forward and generally accepted. International Union of Pure and Applied

Chemistry (IUPAC) has well-defined biosensors as *"a self-contained integrated device that is capable of providing specific quantitative or semi-quantitative analytical information using a biological recognition element, which is retained in direct spatial contact with a transduction element"*(Thevenot et al., 1991). Turner (1987) has also defined biosensor as a "diagnostic tool which combines the specificity of a biological sensing element for the analytes of interest with a transducer to produce a signal proportional to target analytes concentration." The biosensor is not a single piece of instrument but a combination of a wide-ranging designed instrumentation (Mulchandani et al., 2002). Biosensors have noteworthy importance in environmental biotechnology due to its leads over conventional methods in the field of analysis. Biosensors have been used in the analysis of chemical compounds, such as pesticides, heavy metals (HMs), phenolic compounds, and persistent organic pollutants (e.g., PCB). These are also used for the evaluation of environmental quality factors (e.g., BOD). The use of biosensors in the areas of environmental toxicity, cytotoxicity, and Genotoxicity has been successfully proposed over recent years. This method combines molecular binding and knowledge of science and engineering on a common platform (Bhalla et al., 2016). The production of novel tools for effective environmental monitoring has increased significantly over the last decade. The earliest biosensor was developed by Clark and Lyons in 1956 for monitoring of oxygen. Later, Updike and Hicks developed an enzyme-based biosensor (Glucose biosensor) in 1967. It was made by using polyacrylamide immobilized to the surface of an oxygen electrode with enzyme glucose oxidase to enable quick and quantifiable analyses of glucose (Updike and Hicks, 1967; Clark and Lyons, 1962; Mascini, 2006).

Biosensors can also give information about the pollution site rapidly, important for environmental control plus monitoring. Furthermore, biosensor has an advantage over other analytical methods in that it is versatile and let investigators to analyze the in situ contaminant concentrations without need for any additional sample preparation. Biosensors can also be used to study the biological effects of compounds (e.g., toxicity) along with the identification of different compounds (Odobašić et al., 2019). Biosensors are commercially important and their importance is expected to grow with the advancement of technology (Wilson and Walker, 1994). Nanotechnology has also contributed to the development of biosensors important for the detection of environmental contaminants, thereby improving the analytical factors like sensitivity and thresholds of detection (Maduraiveeran and Jin, 2017).

12.2 BIOSENSOR CLASSIFICATION

A biosensor is an autonomous integrated tool which gives reliable, quantifiable, and analytical information, using a biochemical receptor intimately connected with a transducer element. It works on the principle of analyte binding with the desired biological material immobilized on the transducer to form a bound analyte which generates an electric signal detected by transducer. Biosensor is generally composed of three elements; a bioreceptor (biorecognition element), a transducer (electronic component), and a signal processing system (Sethi, 1994). Bioreceptor is a biological component that specifically recognizes the target analyte and interacts with it in such a manner so as to produce some physiochemical change detected by transducer (a measurable signal). The bioreceptors include cell organelle, DNA, whole cells, antibodies, microorganisms, enzymes, tissues (animal or plant), nucleic acids etc. (Xu and Ying, 2011). Transducer is a physical component which changes the response produced due to analyte bioreceptor interface to an electrochemical, optical or some physicochemical signal dependent on the type of transducer and transforms this signal into a measurable form. Choice of transducer varies with the type of interaction and the biological signal formed, for example, variation in pH, activation and/or inactivation of enzymes, and mass difference, etc. The specificity of a biosensor depends on the specificity of bioreceptor used.

On the basis of biorecognition process related to target analyte, biosensors are of two classes – biocatalytic and affinity sensors. Biocatalytic biosensors consist of a "biological constituent which catalyzes the chemical conversion of the analyte with which it interacts and detect the magnitude of the resulting changes, such as product formation, reactant disappearance, or inhibition of the reaction, which are correlated with the concentration of the analyte" (Marazuela and Moreno-Bondi, 2002). Affinity biosensors are centered on selective contact vis-à-vis the biological constituent and the analyte over their irreversible binding which results in a physicochemical change detectable via converter (Odobašić et al., 2019).

Biosensors are also categorized as per bioreceptors properties also. The Biological specificity-conferring mechanism of bioreceptors is involved in the detection of enzymes, microbial whole cells, florae, faunae, antibodies, proteins, or DNA pieces. However, the physicochemical type of transducers like electrochemical, optical, piezoelectric, calorimetric, or thermal, etc. are convenient for the detection of toxicants (Rodriguez-Mozaz et al., 2004; Wang et al., 2006; Salgado et al., 2011). The bioavailability extents are central to environmental examining plus risk assessment as they specify the biological consequences of the chemical like its toxicity, genotoxicity, carcinogenic nature, etc. instead of its sheer presence by means of detection by analytical tools (Ripp et al., 2010).

12.3 BIOSENSOR CLASSIFICATION IN LINE WITH BIORECEPTOR TYPE

12.3.1 ENZYME-BASED BIOSENSOR

Enzymes function as biocatalysts aiding the reactions that convert substrates into products. Enzymes are very selective toward a specific substrate which translates them into appropriate sensor substance. Due to its high sensitivity toward a particular substrate, enzyme-based biosensors check the action of enzyme in the company of the substrate (Ripp et al., 2010). The mechanism of detection of an analyte by enzyme-based biosensors takes place either directly by measuring the concentration of an analyte (or its products made in the course of enzymatic reaction) or by indirect monitoring wherein the enzyme inhibition by the analyte is determined (Rebollar-Pérez et al., 2015). A multitude of analytes for environmental valuation has been examined by this category of biosensor (Amine et al., 2006; Rodriguez-Mozaz et al., 2005, 2006). But indirect mode biosensors consist of scarce enzymes that are responsive to heavy metals (Turdean, 2011). The amount of an analyte is verified by calculating the fraction of inhibited enzymes after exposing to the inhibitor (Rebollar-Pérez et al., 2015). A novel example of enzyme-based biosensors with environmental application is the sulfur/sulfate-reducing bacterial cytochrome C_3 reductases which reduces heavy metals. This Cytochrome is immobilized on a glassy carbon electrode then its redox action is checked amperometrically. This is done in the presence of Cr^{+4} at the low detection limits of 0.2 cmg/L and instant response time (some minutes) (Michel et al., 2003). The linking of enzyme and transducing element vary according to immobilization technique

(Lojou and Bianco, 2006), and comprise of covalent bonding, adsorption, entrapment, or direct cross-linking. In enzyme-based biosensors, optical transducers are extensively utilized (Ripp et al., 2010).

12.3.2   ANTIBODY-BASED BIOSENSORS (IMMUNOSENSORS)

Immunosensors work on the principle of antigen and antibody interaction (North, 1985; Long et al., 2013). Surface immobilization of antibodies is an important step in the formation of biosensors, and may also cause loss of function in the case of nonoriented immobilization. Nonoriented immobilization results from arbitrary immobilization and steric hindrance due to high antibody density (Karyakin et al., 2000; Grieshaber et al., 2008). Biorecognition process consists of firm binding of antibody to antigens resulting in the formation of complexes. Among five classes of antibodies (IgA, IgD, IgE, IgG, and IgM), IgG is extensively employed for the detection of heavy metals, due to their superior affinity and specificity. Antibodies like monoclonal, polyclonal, or recombinant can be used as biorecognition component in biosensors. Monoclonal antibodies are specific as it binds to a specific antigen. While as polyclonal antibodies have affinity for numerous binding sites contrary to a lone antigen which causes it to bind strongly to the target. Recombinant antibodies are fashioned by genetic engineering technology (Ferrigno, 2016). High sensitivity plus specificity with marginal cross-reactivity are indispensable characteristics of the antibodies for recognition and measurement of targets precisely (Ferrigno, 2016). In order to immobilize antibodies on the sensor, surface covalent binding, noncovalent immobilization plus coupling interfaces are necessary as immobilization is the main phase that shapes the best working of an Immunosensor (Sharma et al., 2016). The activity of the antibodies can also be influenced by the reaction settings like temperature, pH, and ionic strength (Dzantiey and Zherdev, 2013). In electrochemical biosensors, a variation in the electric properties in the space between two electrodes occurs as a result of antigen–antibody responses (Katz and Willner, 2003).

12.3.3   PROTEIN-BASED BIOSENSORS

Protein-based biosensors are used as biorecognition element (bioreceptor). The proteins are immobilized on the transducer surface, for example, phytochelatins and metallothioneins (Cornelis et al., 2003). Interaction between analyte (metals) and protein in the biosensor takes place by means of complex formation. The subsequent deviations in the protein sheet are detected by determining the electrical capacity or impedance by the appropriate transducer. Protein biosensors are used to determine the bioavailable heavy metal concentrations (Odobašić et al., 2019). Bontidean et al. (1998) developed the protein-based biosensor using metal-binding proteins metallothioneins (SmtA) and regulator protein (MerR) for checking heavy metals. The former is a metal-binding protein that displays greatest response against amplified concentrations of cadmium or zinc, with more affinity for zinc, and minimum affinity for cadmium and copper. On the other hand, the latter is a metal regulatory protein in *P. aeruginosa* that offers better selectivity and specificity for mercury. A conformational modification takes place which is identified by a capacitive transducer using immobilized antibodies. These principles were utilized for detecting heavy metal ions (Bontidean et al., 1998).

12.3.4 WHOLE CELL-BASED BIOSENSORS (WCBBS)

Whole cells are also employed as a biocomponent in biosensor development. Whole cells have long been used for environmental applications, in particular for BOD monitoring (Nakamura, 2010). In whole cell-based biosensors microorganisms, protozoa, algae, fungi, plant cells, etc., are used which can be natural or genetically modified (Gu et al., 2004; Odobašić et al., 2019). WCBBs have advantages over enzyme-based biosensors as the former can be cultured, separated and purified without much trouble. These show added tolerance to changes in pH, temperature or ionic strength (Odobašić et al., 2019). To detect an analyte, a multiphase reaction takes place for the reason that one cell can consist of all the essential enzymes and cofactors. Biosensors like this can be renewed and well-looked-after effortlessly by letting cells to regrow in situ. In comparison to enzyme-based biosensors, WCBBs display intervention with nontargeted analytes plus a comparatively sluggish response (Odobašić et al., 2019). Conversely, recent advances in recombinant DNA technology have greatly benefited the designing of biosensors (Ron, 2007) and there is an improved stress on their usage in environmental contamination and toxicity monitoring. Frequently used Reporter gene for WCBBs design is lux (Billard and DuBow, 1998; Gu et al., 2004).

12.3.5   MICROBIAL BIOSENSORS

Microorganisms are ubiquitous and process a diverse number of chemical complexes. These microbes are incredibly appropriate for the detection of target substances and physiological pointers in several areas of study. Microbial biosensor can be an effective alternative to chemical analysis for on-site monitoring (Shin et al., 2005). Microorganisms vary in their aptness for a certain number of reasons, and can be made via cultivation in comparatively low-cost media. Furthermore, microbes are adept of identifying an extensive collection of chemicals; they are open to genetic alteration and modification according to the diverse reaction conditions (Lei et al., 2006; Yagi 2007). Lately, the use of rDNA technology has reformed the use of microbes by merging natural regulatory genes (i.e., those encoding a transcriptional regulator and a promoter/operator) with a reporter gene (Vollmer and Van Dyk 2004; Ron, 2007). In nonspecific biosensors, a reporter gene is merged to a constitutive promoter, and the toxicity of the target compound is determined as a reduction in the reporter protein activity. Even though microbial biosensors are established to be efficient means for in situ monitoring, however, these biosensors produce lesser reproducibility and sensitivity than chemical exploration (van der Meer et al., 2004).

12.4   CLASSIFICATION BY TRANSDUCER TYPE

Established around the principle used in transduction structures, biosensors may be distinguished as electrochemical, optical, piezoelectric, and calorimetric biosensors.

12.4.1   ELECTROCHEMICAL BIOSENSORS

These were the primary established and promoted biosensors. In electrochemical biosensors, the contact between the target analyte and the biomolecule (bioreceptor) gives rise to a chemical reaction in which ions and/or electrons are generated or consumed leading to variations in the electrical features of the analyte solution. In electrochemical biosensors, electric current or potential is generated which is sensed by transducer by giving an electrochemical signal which is correlated with the amount of analyte in the sample solution. Electrochemical biosensor necessitates marginal sample preparation and is highly sensitive even at low sample sizes however, the results suffer from weak reproducibility and constancy (Gautam et al., 2012).

Electrochemical biosensors are categorized depending on the form of quantified signal into potentiometric, amperometric, conductometric, and ion-selective field-effect transistors (ISFETs) biosensors. The fundamental blueprint of an electrochemical cell is usually delivered by diverse measurement principles (Wang et al., 2008). Potentiometric biosensors work by using ion-selective electrode (ISEs) in which an ion-selective membrane at the top is selective to target ions in the presence of interfering ions in the sample. Numerous potentiometric biosensors have been recognized in which pH electrode is joined with recombinant *E. coli* that consists of intracellular organophosphorus hydrolase at the cell surface, besides a wild-type organophosphorus metabolizing bacterium *Flavobacterium* sp. (Espinosa-Urgel et al., 2015; Lehmann et al., 2000). Similarly, a potentiometric oxygen electrode was established for the ethanol concentration measuring by making use of immobilized cells of *S. ellipsoideus* (Rogers and Gerlach, 1996). Amperometric biosensors are the commonly used while Amperometric biosensors show greater sensitivity and quicker results than potentiometric ones, however, they are less-selective and more vulnerable to electroactive species interferences (Martinkova et al., 2017; Monosik et al., 2012). Conductometric biosensors measure electrical conductivity in sample solutions in the presence of two electrodes and generate a biochemical reaction and work at reduced voltages also (Jaffrezic-Renault and Dzyadevych, 2008; Dzyadevych and Jaffrezic-Renault, 2014). ISFETs biosensors are appropriate for the detecting ions. The upsurge in sample activity causes rise in the potential of the gate electrode which is located in close vicinity to the analyte solution. The variation in electric potential is determined. 2, 4-Toluene diamine (2, 4-T) (Shanmugam et al., 2001) polychlorinated biphenyls (PCBs), triazines and various toxins, such as serin and soman (Keane et al., 2002) determined using these methods.

12.4.2 OPTICAL BIOSENSORS

In optical biosensors, transducer spots optical alterations in the input light due to interaction of the bioreceptor and the target analyte. The amplitude of these variations is correlated with the concentration of the analyte present. Optical biosensors do not respond to electromagnetic interference and require little instrumentation, and easy to operate, noninvasiveness, and are nonelectrical in nature (Odobašić et al., 2019). Surface plasmon resonance (SPR) biosensors operate by optical detection system in which the metal interface and dielectric excitation of electrons occur when light is incident on the metal surface producing electromagnetic waves (plasmons) (Wijaya et al., 2011; Asal et al., 2018). Fluorescence-based optical biosensors uncover the target atoms or molecules directly by determining the modification in the frequency of electromagnetic waves released by them. Detection can also be done by fluorescence labels or fluorescence energy transfer (FRET) (Odobašić et al., 2019. Luminescence-based biosensors are of two classes' chemiluminescent and bioluminescent optical biosensors. In these sensor systems, the triggered condition of the target atoms or molecules is acquired from exothermic chemical reaction, and on returning to the ground state, the excited species release light. When a chemical reaction like this takes place within a biological organism, then it is called bioluminescence (Odobašić et al., 2019). This method can detect COD in wastewaters by quantifying variations in bioluminescence signal.

12.4.3   PIEZOELECTRIC BIOSENSORS

In piezoelectric biosensors, a piezoelectric material is used as a transducer and is fixed to a biorecognition element. Within various natural and synthetic substances that exhibit piezoelectric effect, quartz crystals used (Monosik et al., 2012; Pohanka, 2017) due to their accessibility, chemical stability in aqueous solution and resistance to increased temperatures. The idea of this type of biosensors depends on the capability of a piezoelectric substance to generate an electrical potential when distorted under the applied mechanical stress, and vice versa, to elastically distort when subjected to an electric field.

12.4.4   THERMAL BIOSENSORS

Thermal biosensors are a class of biosensor in which the interactions between a recognition element and analyte result in a change of temperature.

It measures the heat change correlated with analyte concentration. Thermistors or thermopiles are used as thermal transducers in these devices (Wang et al., 2008; Perumal and Hashim, 2014). Few benefits of thermal biosensors are detection short of the need for labeling of reactants, not demanding regular recalibration, and no interventions by electrochemical and optical characteristics of the sample (Wang et al., 2008, Mohanty and Kougianos, 2006).

12.5 ENVIRONMENTAL APPLICATIONS OF BIOSENSORS

Various anthropogenic activities have resulted in the release of pollutants into the environment which causes environmental pollution. There has been an increase in the concentration of pollutants day by day which place a negative impact on the living system. Therefore, the identification of these pollutants is important. Biosensors are useful for the assessment of various types of pollutants in the environment (environmental research and risk control), although biosensors for air pollution is not much successful (Nigam and Shukla, 2015). The chief leads of biosensors over traditional techniques for environmental uses are the portability, smaller size, on-site follow-up, and the capability to determine contaminants in complex matrices with small sample preparation (Bilitewski and Turner, 2000; Karube et al., 1998). Therefore, biosensors have been proved to be an effective analytical tool in the analysis of environmental pollutants (Wang et al., 2014). However, bulk of the biosensors are limited for a particular toxicant or can be used for a restricted number of contaminants (Roda et al., 2001; Rodriguez-Mozaz et al., 2006; Rogers, 2006). Numerous biosensors have been efficaciously created for detecting pollutants like pesticides (Verma and Bhardwaj, 2015; Bucur et al., 2018; Cortina-Puig et al., 2010; Vargas-Bernal et al., 2012), phenols (Karim and Fakhruddin, 2012; Ozoner et al., 2011; Osma and Stoytcheva, 2014; Wang et al., 2018), HMs (Ma and Sang-Bing, 2017; Wong et al., 2007; Pal et al., 2009; Knecht and Sethi, 2009; Ilangovan et al., 2006; Turdean, 2011; Chen and Rosen, 2014), pathogens (Juronen et al., 2018; Oluwaseun et al., 2018; Justino et al., 2017; Peedel and Rinken, 2014; Viirlaid et al., 2015; Kuusk and Rinken, 2004), dioxins, PCBs etc. which are described below.

12.5.1   *BIOSENSORS FOR PESTICIDE DETECTION*

Pesticides are commonly present in the environment because of their excessive use in the agricultural fields to control pests and therefore increase crop productivity and crop yields (Verma and Bhardwaj 2015). The pesticide residues may contaminate the different components of the environment and may enter into the food chains, thus affecting various life forms. Pesticides are carcinogenic and may cause disruption to the working of nervous system and give birth to several immunological and respiratory diseases (Sassolas et al., 2012). Enzyme-based sensors, based on the enzyme inhibition and the constituents comprised in the enzymatic reactions are expended for the detection of pesticides (Sassolas et al., 2012). For detecting organophosphorus compounds and carbamate pesticides, enzyme-based biosensors, in which inhibition of acetylcholinesterase (AChE) and colin oxidase takes place, are frequently used (Rebollar-Pérez et al., 2015). The generally exercised transduction element for the detection of pesticide is electrochemical transducers because of its high sensitivity (Jaffrezic-Renault, 2001). The organophosphorus compounds moderately obstruct acetylcholine esterase action by phosphorylating its seriene group and becoming detectable by respective transducers. Inhibition of AChE in nerves can cause life-threatening condition due to the accumulation of the neurotransmitter acetylcholine (Verma and Bhardwaj, 2015). Numerous pesticides, such as trichlorfon, paraoxon, carbaryl, carbofuran, malaoxon, aldicarb, and diisopropyl fluorophosphates have been efficaciously identified by the biosensors based on AChE inhibition (Dzyadevych et al., 2005). Though the inhibition-based biosensors have high sensitivity but they lack specificity, as there are many other substances (carbamates, Heavy metals, etc.) that also inhibit these enzymes (Simonian et al., 1997; Ibrahim et al., 1998). Similarly, organophosphorus hydrolase (OPH) has been used for developing enzymatic biosensor for detecting pesticides. Lately, numerous OPH-based potentiometric, amperometric, and optical devices have been developed and introduced (Mulchandani et al., 1999; Viveros et al., 2006; Choi et al., 2010). Immunosensing technologies also have greater potential for rapid detection of pesticide remains in food and environment (Jiang et al., 2008). Photosynthetic biosensors can also be employed for detecting pesticides (Rodriguez-Mozaz et al., 2006). Sassolas et al. (2012) have reviewed some advances and novel developments in biosensors for the detection of pesticides like aptamers. Examples of some of the biosensors used in the detection of pesticides are listed in Table 12.1.

TABLE 12.1 Biosensors Developed for the Detection of Pesticides.

Pesticide	Enzyme/Recognition element	Transducer	Reference
Paraoxon	AChE	Amperometric	Andreescu (2002)
Paraoxon	AChE	Amperometric	Arduini et al. (2013)
Paraoxon	AChE and ChO$_x$	Optical (Colorimetric)	Guo et al. (2017)
Carbaryl	AChE	Amperometric	Bucur et al. (2006)
Carbaryl	AChE	Electrochemical (Impedimetric)	Gong et al. (2014)
Carbofuran	AChE	Amperometric	Bucur et al. (2006)
Carbofuran	AChE	Voltametric	Jeyapragasam and Saraswathi (2014)
Chlorpyrifos methyl oxon	AChE	Amperometric	Andreescu (2002)
Malaoxon	AChE	Amperometric	Jeanty et al. (2002)
Aldicarb	AChE	Amperometric	Arduini et al. (2006)
Trichlorfon	AChE	Amperometric	Li et al. (2007)
Diisopropyl fluorophosphates	AchE	Conductometric	Dzyadevych et al. (2005)
Triazine	Tyr	Amperometric	Campanella et al. (2002)
Paraoxon	Tyr	Amperometric	Campanella et al. (2007)
Methyl parathion	Tyr	Amperometric	De Albuquerque and Ferreira (2007)
Methyl parathion	AChE	Amperometric	Nunes et al. (2014)
Dichlorovos	AChE and ChO$_x$	Optical (fluorescence)	Meng et al. (2013)
Chloropyrifos	Tyrosinase	Electrochemical (Impedimetric)	Mayorga-Martinez et al. (2014)
Parathion	OPH	Amperometric	Tang et al. (2014)
Paraoxon	OPH	Optical	Cao et al. (2004)
Diazinon	OPH	Optical	White and Harmon (2005)
Atrazine	-	IS	Tran et al. (2012)
Endosulfan	-	IS	Liu et al. (2014a)
Atrazine	Antibodies (monoclonal)	Voltametric	Liu et al. (2014b)

AChE, Acetylcholinesterase; IS, Immunosensor; OPH, Organophosphorus hydrolase; Tyr, Tyrosine; Tyr, tyrosinase.

12.5.2   *BIOSENSORS FOR DETECTION OF HEAVY METALS*

Heavy metals (e.g., mercury, cadmium, zinc, cobalt, copper, lead etc.) pose a serious risk to the environment due to their toxicity and accretion potential in the environment. Enzyme-based electrochemical biosensors are the most frequent biosensors, developed for monitoring HMs and due to this different enzymes, such as urease, glucose oxidase, tyrosinase, peroxidase, etc. have been studied. Although some metals have biological significance (Rebollar-Pérez et al., 2015) but in some cases, they can inhibit the activity of enzymes (Rodriguez-Mozaz et al., 2006). Thus, heavy metal biosensors are designed on the notion of activation or inhibition of enzymes, for instance, zinc and copper can be identified by the activation of alkaline phosphatase and ascorbate oxidase-based biosensors, respectively (Satoh and Iijima, 1995). Additional heavy metals like cadmium, cobalt, copper, and nickel can be detected by the inhibition of glucose oxidase-based biosensor (Ghica et al., 2013). Conversely, these enzyme inhibition-based biosensors lack selectivity and typically analyze the total inhibition effect produced by heavy metals in the sample. To counter this, specific recombinant bacterial biosensors have been created, in which a reporter gene (for instance which codes for a luminescent protein) is merged with metal-specific inducible promoter (Petanen and Romantschuk, 2002), for instance, the bioavailable fraction of Cd, Zn, Hg, and Cr in soil (Rasmussen et al., 2000; Ivask et al., 2002) have been identified by this type of biosensor. Moreover, electrochemical aptasensors have also been depicted as favorable tools in HM monitoring because of the firm and effective interaction with very low levels of metal ions (Hayat and Marty, 2014; Wu et al., 2012; Zhang et al., 2018). Lately, single-chamber batch-mode cube microbial fuel cell (CMFC) and double-chamber microbial fuel cell (DMFC), and sediment microbial fuel cell (SMFC) were developed and are the latest electrobiochemical sensors for real-time HMs monitoring in wastewater effluent (Zhao et al., 2018). Optical biosensors also detect HMs due to the activity of a reporter gene controlled by inducible promoter (Rodriguez-Mozaz et al., 2006). In this approach called as "turn on assay," the reporter signal level rises as per the pollutant concentration. β-galactosidase (lacZ), green fluorescent protein (GFP) luciferase and (luc) are largely accepted reporter genes utilized in biosensor systems (Gutiérrez et al., 2015). For the detection of heavy metal, immunoassay method is also used, due to its high selectivity and sensitivity. Different biosensors developed for the identification of various types of heavy metals in environmental samples is reviewed in Table 12.2.

12.5.3 BIOSENSORS FOR DETECTION OF PHENOLIC COMPOUNDS

Phenols and their compounds are known toxic substances found in various processing and synthetic industries associated with pesticides, disinfectants, detergents, pharmaceuticals, polymers, and other products. As a consequence of their usage in many processes including petroleum and paper industry, phenols are among the most common organic impurities entering the marine system (Naghibi et al., 2003). Chlorophenols and nitrophenols which are the major degrading product of organophosphorus pesticides and chlorinated phenoxyacids, these compounds have been reported to cause severe toxicity and mutagenicity in plants and animals, and also affect physiologic processes, such as respiration, photosynthesis, and enzyme catalyzed reactions at low concentrations (Nigam and Shukla, 2015). For the determination of phenolic compounds, some effective enzymes, such as tyrosinase, peroxidases, and laccase have been used for the development of biosensors, tyrosinase being the most common. Several microorganisms can be utilized for the detection of phenolic compounds, such as *Pseudomonas, Arthrobacter, Rhodococcus, Trichosporon,* and *Moxraxella* (Karim and Fakhruddin, 2012). Table 12.3 encapsulates biosensors for detection of phenolic compounds.

12.5.4 BIOSENSORS FOR DETECTING SURFACTANTS

Surfactants are the basic active component present in most of the detergent products and constitute a wide group of organic pollutants (Taranova et al., 2002). The anionic surfactants are most widely used relative to cationic surfactants. Therefore, a number of biosensors have been developed only for the anionic surfactants, such as methylester, sodium dodecyl sulfate (SDS), alkane sulfonates, etc. For the determination of anionic surfactant-SDS, an amperometric biosensor was developed using *Pseudomonas rathonis* T (biological element, contains a plasmid for the degradation of surfactants) with a detection limit in the range of 0.25–0.75 mg/L (Reshetilov et al., 1997). Likewise, one more microbe-based biosensor has been considered for finding the anionic surfactants by utilizing bacterial strains of genera *Pseudomonas* and *Achromobacter* with detection limit

TABLE 12.2 Biosensors Developed for the Detection of Heavy Metals.

Heavy metals	Biosensing elements	Transducers	Limit of detection	References
Hg(II)	Urease	Amperometric	7.4×10^{-6} M	Kuralay et al. (2007)
Hg(II)	Urease	Potentiometric	0.05 µM	Yang et al. (2006)
Hg(II)	*Escherichia coli* S30 extract	Luminescence	5×10^{-9} M	Pellinen et al. (2004)
Hg(II)	Glucose oxidase	Amperometric	0.49 µg/L	Liu et al. (2009)
Pb(II)	Urease	Conductometric	0.1–10 mM	Ilangovan et al. (2014)
Cu(II)				
Cd(II)				
Cd(II)	Urease	Optical	10 µM	Tsai et al. (2003)
Cu(II)				
Cr(VI)	Cytochrome C_3 (*D. norvegicum*)	Amperometric	0.2 mg/L	Michel et al. (2006)
Hg(II)	Metallothionein SmA (*E. coli*)	Potentiometric	10^{-15} M	Bontidean et al. (2000)
Cu(II)				
Zn(II)				
Cd(II)				
Cu(II)	Glucose oxidase	Amperometric	0.2 µM	Ghica et al. (2013)
Cd(II)			2.4 µM	
Co(II)			2.1 µM	
Ni(II)			3.3 µM	
Hg(II)	Glucose oxidase	Amperometric	2.5 µM	Guascito et al. (2008)
Ag(I)			0.05 µM	
Cu(II)			5 µM	
Cd(II)			5 µM	
Hg(II)	Nucleic acids	Optical	1.2 nM	Long et al. (2013)

TABLE 12.3 Biosensors Developed for the Detection of Phenolic Compounds.

Phenolic compound	Biosensing element	Transducers	References
Phenol	*Pseudomonas putida* DSM50026	Amperometric	Timur et al. (2003)
Catechol	*Pseudomonas putida* DSM50026	Amperometric	Timur et al. (2003)
p-nitrophenol	*Arthrobacter*	Amperometric	Lei et al. (2003)
Chlorophenol	*Trichosporon beigelii*	Amperometric	Riedel et al. (1995)
Phenol	Tyrosinase	Electrochemical	Yildiz et al. (2007)
Phenol	Tyrosinase	Calorimetric	Russell and Burton (1999)
Chlorophenol	Tyrosinase	Optical	Freire et al. (2002)
Catechol	Tyrosinase	Electrochemical	Tembe et al. (2007)
m-cresol	Tyrosinase	Optical	Abdullah et al. (2005)
Catechol	Tyrosinase	Optical	Abdullah et al. (2005)

of 1 μM. The detection was due to reduced concentration of dissolved oxygen caused by cell respiration stimulated by sodium dodecyl sulfate (Taranova et al., 2002). Kucherenko et al. (2012) also developed a new biosensor method for the determination of surfactants in aqueous solutions based on the inhibition of an enzyme acetylcholinesterase.

12.5.5   BIOSENSORS FOR DETECTING BOD

Biological oxygen demand (BOD) is of utmost significance and a generally utilized parameter for describing the organic pollution of water and is estimated by determining the amount of oxygen required by aerobic microorganisms for degrading the organic matter in the wastewater. The conventional technique for calculating BOD involves taking a sample and incubating it for five days under standardized conditions. Although, the conventional method for calculating BOD has some advantages, but it is an extensive process and is not appropriate for on-line monitoring process, so to circumvent the disadvantage of the conventional method (BOD_5), it is important to develop an alternative method (Liu and Mattiasson, 2002). So, biosensors are an alternative for the estimation of BOD. The first microbial biosensor for the detection of BOD was introduced by Karube et al. (1977), it comprises of an oxygen electrode and a biofilm (appropriate microorganism immobilized on a porous cellulose membrane) and its principle is centered on the measurement of consumption of oxygen by microorganisms. Optical fiber biosensor based on biomaterial yeast has also been developed for monitoring low BOD in river waters (Dhewa, 2015; Rodriguez-Mozaz et al., 2006). Another optical biosensor was also developed for monitoring BOD level of sea water, with a response time of 3.2 min for 5 mg/L BOD (Jiang et al., 2006). Thermally killed cells (at 300°C) of complex microbial culture can also be used in designing of biosensor for monitoring BOD of wastewater (Tan and Lim, 2005).

12.5.6   BIOSENSORS FOR DETECTION OF PATHOGENS

Pathogens, for instance, viruses, bacteria, and protozoa are commonly found in the polluted and untreated waters which pose a serious threat to the environment and especially to human health (Rodriguez-Mozaz et al., 2004). The influx of wastewater from a nonpoint source and agricultural runoff into the water body led to the entry of fecal coliform bacteria (Alocilja and Radke, 2003) that can come into contact with humans via different routes and may have negative impact on human beings. Therefore, there is a need to detect these pathogens. As the conventional methods are time-consuming, biosensors have played an important role in the detection of these pathogens. These microorganisms would transform their metabolic redox reactions via oxidoreductase reactions into quantifiable electrical signals (Takayama et al., 1993) detected by transducer. To detect microbial content sample, transducer can either detect oxygen consumption or the appearance/disappearance of an electrochemically active metabolite (Ivnitski et al., 1999; Tenover, 1988; Highfield and Dougan, 1992; Kapperud et al., 1993; Sharpe, 1994; Tietjen and Fung, 1995; Feng, 1996; Zhai et al., 1997). A rapid and specific optical biosensor has been proposed to detect metabolically active *Legionella pneumophila* in complex environmental water samples based on SPR (Enrico et al., 2013; Foudeh et al., 2015). Aptamer-based biosensors have also been used for pathogen detection (Wang et al., 2012). The detection by DNA is more specific than immunosensor with enhanced sensitivity. Gene probes are already finding application in the detection of disease-causing microorganisms in water supplies, food, or in plants, animal or human tissues (Ivnitski et al., 1999). Table 12.4 summarizes the list of biosensors for the detection of pathogens.

12.6 OTHER ENVIRONMENTAL POLLUTANTS INCLUDE

12.6.1 NITROGEN COMPOUNDS

Nitrogen is commonly used for soil fertilization. However, nonstop consumption of these ions may have serious consequences on human health, particularly because they can react irreversibly with hemoglobin (Moorcroft et al., 2001). Amperometric biosensor has been developed for nitrite determination by immobilizing cytochrome C nitrate reductase (ccNiR) of *Desulfovibrio desulfuricans* and double-layered hydroxide holding anthraquinone-2-sulfonate that can sense nitrite concentration in the range of 0.015–2.35 μmol with a detection limit of 4 nM (Chen et al., 2007). Moreover, Khadro et al. (2008) also developed an enzymatic conductometric biosensor for the determination of nitrate in water, validated and used for natural water samples. This instrument was based on the

TABLE 12.4 Biosensors for the Detection of Pathogens.

Pathogens (Analyte)	Recognition element	Transducers	References
Escherichia coli	Polymerizable form of histidine	Optical (SPR)	Yilmaz et al. (2015)
Escherichia coli	Polymerizable form of histidine	Piezoelectric (QCM)	Yilmaz et al. (2015)
Escherichia coli	Antibodies (polyclonal)	Optical (electrochemiluminscence)	Chen et al. (2017)
Bacillus subtilis	Antibodies (polyclonal)	Electrochemical (Amperometric)	Yoo et al. (2017)
Legionella pneumophila	Antibody (polyclonal)	Electrochemical (Amperometric)	Martin et al. (2015)

QCM, Quartz crystal microbalance; SPR, Surface Plasmon Resonance.

methyl viologen mediator mixed with nitrate reductase from *Aspergillus niger* and Nafion® cation exchange polymer dissolved in a plasticized PVC membrane deposited on the sensitive surface of interdigitated electrodes. When stored in phosphate buffer pH 7.5 at 4°C, the sensor showed good stability over 2 months.

12.6.2 POLYCHLORINATED BIPHENYLS (PCBS)

Polychlorinated biphenyls (PCBs) are persistent organic pollutants present in the environment even though their production was banned several years ago. PCBs have several deleterious health effects including being carcinogenic, mutagenic, teratogenic, and immunotoxic (Chiu et al., 1995; Encarnação et al., 2019). Different biosensor configurations have been designed to determine the presence of PCBs in the environment quantitatively (Figure 12.1). Few of them are DNA biosensor with chronopotentiometric detection (Marrazza et al., 1999), immunosensors with electrochemical (Del Carlo et al., 1997), SPR (Shimomura et al., 2001) and fluorescence (Zhao et al., 1995; Endo et al., 2005) detection principles.

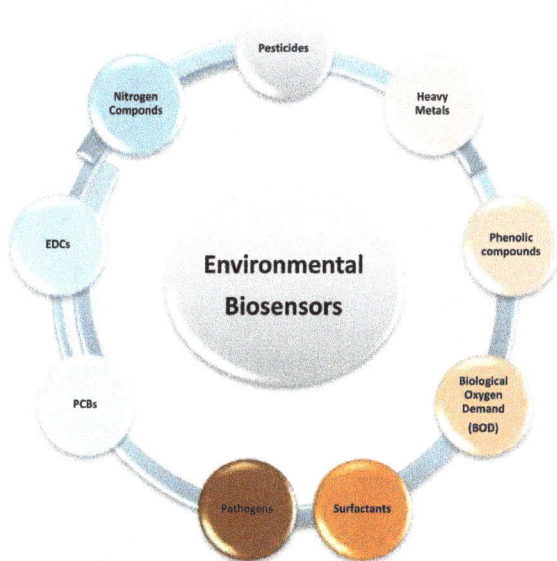

FIGURE 12.1 Biosensors and their potential roles for monitoring and detection of environmental contamination.

KEYWORDS

- biosensors
- biotechnology
- polychlorinated biphenyls
- pesticides
- heavy metals

REFERENCES

Abdullah, J.; Ahmad, M.; Karuppiah, N.; Heng, L. Y.; Sidek, H. Immobilization of Tyrosinase in Chitosan Film for an Optical Detection of Phenol. *Sens. Actuat.* B. **2005**, *114*, 604–609.

Akki, S. U.; Werth, C. J.; Silverman, S. K. Selective Aptamers for Detection of Estradiol and Ethinyl estradiol in Natural Waters. *Environ. Sci. Technol.* **2015**, *49*, 9905–9913.

Alocilja, E. C.; Radke, S. M. *Biosens. Bioelectron.* **2003**, *18*, 841–846.

Amine, A.; Mohammadi, H.; Bourais, I.; Palleschi, G. Enzyme Inhibition-Based Biosensors for Food Safety and Environmental Monitoring. *Biosens. Bioelectron.* **2006**, *21*, 1405–1423.

Andreescu, S. Screen-Printed Electrode Based on AChE for the Detection of Pesticides in Presence of Organic Solvents. *Talanta* **2002**, *57*, 169–176.

Arduini, F.; Guidone, S.; Amine, A.; Palleschi, G.; Moscone, D. Acetylcholinesterase Biosensor Based on Self-Assembled Monolayer-Modified Gold-Screen Printed Electrodes for Organophosphorus Insecticide Detection. *Sens. Actuat. B Chem.* **2013**, *179*, 201–208.

Arduini, F.; Ricci, F.; Tuta, C. S. et al., Detection of Carbamic and Organophosphorus Pesticides in Water Samples Using a Cholinesterase Biosensor Based on Prussian Blue-Modified Screen-Printed Electrode. *Anal. Chim. Acta.* **2006**, *580*, 155–162.

Asal, M.; Ozen, O.; Sahinler, M.; Polatoglu, I. Recent Developments in Enzyme, DNA and Immuno-Based Biosensors. *Sensors.* **2018**, *18*, 1924.

Bala, R.; Kumar, M.; Bansal, K.; Sharma, R. K.; Wangoo, N. Ultrasensitive Aptamer Biosensor for Malathion Detection Based on Cationic Polymer and Gold Nanoparticles. *Biosens. Bioelectron.* **2016**, *85*, 445–449.

Bhalla, N.; Jolly, P.; Formisano, N.; Estrela, P. Introduction to biosensors. *Essays Biochem.* **2016**, *60* (1), 1–8.

Bilitewski, U.; Turner, A. *Biosensors in Environmental Monitoring*; CRC Press: Boca Raton, FL, USA, 2000.

Bontidean, I.; Berggren, C.; Johansson, G.; Csöregi, E.; Mattiasson, B.; Lloyd, J. R.; Jakeman, K. J.; Brown, N. L. Detection of Heavy Metal Ions at Femtomolar Levels Using Protein-Based Biosensors. *Anal. Chem.* **1998**, *70*, 4162–4169.

Bontidean, I.; Lloyd, J. R.; Hobman, J. L. et al., Bacterial Metal-Resistance Proteins and Their Use in Biosensors for the Detection of Bioavailable Heavy Metals. *J. Inorg. Biochem.* **2000**, *79*, 225–229.

Bucur, B.; Fournier, D.; Danet, A.; Marty, J. L. Biosensors Based on Highly Sensitive Acetylcholinesterases for Enhanced Carbamate Insecticides Detection. *Anal. Chim. Acta.* **2006**, *562*, 115–121.

Bucur, B.; Munteanu, F. D.; Marty, J. L.; Vasilescu A. Advances in Enzyme-Based Biosensors for Pesticide Detection. *Biosensors.* **2018**, *8* (2), 1–28.

Campanella, L.; Dragone, R.; Lelo, D. et al. Tyrosinase Inhibition Organic Phase Biosensor for Triazinic and Benzotriazinic Pesticide Analysis (Part Two). *Anal. Bioanal. Chem.* **2002**, *384*, 915–921.

Campanella, L.; Lelo, D.; Martini, E.; Tomassetti, M. Organophosphorus and Carbamate Pesticide Analysis Using an Inhibition Tyrosinase Organic Phase Enzyme Sensor; Comparison by Butyrylcholinesterase + Choline Oxidase Opee and Application to Natural Waters. *Anal. Chim. Acta.* **2007**, *587*, 22–32.

Cao, X.; Mello, S. V.; Leblanc, R. M. et al., Detection of Paraoxon by Immobilized Organophosphorus Hydrolase in a Langmuir–Blodgett Film. *Coll. Surf. A Physicochem. Eng. Aspects.* **2004**, *250*, 349–356.

Chen, H.; Mousty, C.; Cosnier, S.; Silveira, C.; Moura, J. J. G.; Almeida, M. G. Highly Sensitive Nitrite Biosensor Based on the Electrical Wiring of Nitrite Reductase by [ZnCr-AQS] LDH. *Electrochem. Commun.* **2007**, *9*, 2240–2245.

Chen, J.; Rosen, B. P. Biosensors for Inorganic and Organic Arsenicals. *Biosensors* **2014**, *4*, 494–512.

Chen, S.; Chen, X.; Zhang, L.; Gao, J.; Ma, Q. Electrochemiluminescence Detection of *Escherichia coli* O157:H7 Based on a Novel Polydopamine Surface Imprinted Polymer Biosensor. *ACS Appl. Mater. Interf.* **2017**, *9*, 5430–5436.

Chiu, Y. W.; Carlson, R. E.; Marcus, K. L.; Karu, A. E. A Monoclonal Immunoassay for the Coplanar Polychlorinated Biphenyls. *Anal. Chem.* **1995**, 67 (21), 3829–3839.

Choi, B. G.; Park, H.; Park, T. J.; Yang, M. H.; Kim, J. S.; Jang, S. Y.; Heo, N. S.; Lee, S. Y.; Kong, J.; Hong, W. H. *ACS Nano* **2010**, *4*, 2910–2918.

Clark, L. C.; Lyons, C. Electrode Systems for Continuous Monitoring in Cardiovascular Surgery. *Ann. NY Acad. Sci.* **1962**, *102* (1), 29–45.

Cornelis, R.; Crews, H.; Caruso, J.; Heumann, K. *Handbook of Elemental Speciation: Techniques and Methodology*; John Wiley & Sons, Ltd: Chichester, **2003**.

Cortina-Puig, M.; Istamboulie, G.; Noguer, T.; Marty, J. L. Analysis of Pesticide Mixtures Using Intelligent Biosensors. In *Intelligent and Biosensors*; Somerset, V. S., Ed.; IntechOpen, UK, 2010. doi: 10.5772/7154

Ivnitski, D.; Abdel-Hamid, I.; Atanasov, P.; Wilkins, E. Biosensors for Detection of Pathogenic Bacteria. *Biosens. Bioelectron.* **1999**, *14*, 599–624.

Date, Y.; Aota, A.; Sasaki, K.; Namiki, Y.; Matsumoto, N.; Watanabe, Y.; Ohmura, N.; Matsue, T. Label-Free Impedimetric Immunoassay for Trace Levels of Polychlorinated Biphenyls in Insulating Oil. *Anal. Chem.* **2014**, *86* (6), 2989–2996.

De Albuquerque, Y. D. T.; Ferreira, L. F.; Amperometric Biosensing of Carbamate and Organophosphate Pesticides Utilizing Screen-Printed Tyrosinase-Modified Electrodes. *Anal. Chim. Acta.* **2007**, *596*, 210–221.

Del Carlo, M.; Lionti, I.; Taccini, M.; Cagnini, A.; Mascini, M. *Anal. Chim. Acta.* **1997**, *342*, 189–197.

Dempsey, E.; Diamond, D.; Collier, A. *Biosens. Bioelectron.* **2004**, *20*, 367–377, 35.

Dhewa,T. Biosensor for Environmental Monitoring: An Update. *Int. J. Environ. Res.* **2015**, *3* (2), 212–218. http://www.sci-encebeingjournal.com

DuBow, M. S. Bioluminescence-Based Assays for Detection and Characterization of Bacteria and Chemicals in Clinical Laboratories. *Clin. Biochem.* **1998**, *31*, 1–14.

Dzyadevych, S. V.; Jaffrezic-Renault, N. Conductometric Biosensors. In *Biological Identification*; Schaudies, R. P., Ed.; Woodhead Publishing: Cambridge, UK, **2014**, *6*, 153–193.

Dzyadevych, S. V.; Soldatkin, A. P.; Arkhypova, V. N. et al. Early-Warning Electrochemical Biosensor System for Environmental Monitoring Based on Enzyme Inhibition. *Sens. Actuat. B Chem.* **2005**, *105*, 81–87.

Encarnação, T.; Pais, A. A.; Campos, M. G.; Burrows, H. D. Endocrine Disrupting Chemicals: Impact on Human Health, Wildlife and the Environment. *Sci. Prog.* **2019**, *102* (1), 3–42.

Endo, T.; Okuyama, A.; Matsubara, Y.; Nishi, K.; Kobayashi, M.; Yamamura, S.; Morita, Y.; Takamura, Y.; Mizukami, H.; Tamiya, E. *Anal. Chim. Acta.* **2005**, *531*, 7–13.

Enrico, D. L.; Manera, M. G.; Montagna, G.; Cimaglia, F.; Chesa, M.; Poltronieri, P.; Santino, A.; Rella, R. SPR Based Immunosensor for Detection of Legionella Pneumophila in Water Samples. *Opt. Commun.* **2013**, *294*, 420–426.

Espinosa-Urgel, M. E.; Seranno, L.; Ramos, J. L.; Fernández Escamilla, A. M. Engineering Biological Approaches for Detection of Toxic Compounds: A New Microbial Biosensor Based on the Pseudomonas putida TtgR Repressor. *Mol. Biotechnol.* **2015**, *57*, 558–564.

Fan, L.; Zhao, G.; Shi, H.; Liu, M. A Simple and Label-Free Aptasensor Based on Nickel Hexacyanoferrate Nanoparticles as Signal Probe for Highly Sensitive Detection of 17β-estradiol. *Biosens. Bioelectron.* **2015**, *68*, 303–309.

Fan, L.; Zhao, G.; Shi, H.; Liu, M.; Wang, Y.; Ke, H. A Femtomolar Level and Highly Selective 17β-Estradiol Photoelectrochemical Aptasensor Applied in Environmental Water Samples Analysis. *Environ. Sci. Technol.* **2014**, *48*, 5754–5761.

Feng, P. Emergence of Rapid Methods for Identifying Microbial Pathogens in Food. *J. AOAC Int.* **1996**, *79* (3), 809–812.

Ferrigno, P. K. Non-Antibody Protein-Based Biosensors. *Essays Biochem.* **2016**, *60*, 19–25.

Foudeh, A. M.; Trigui, H.; Mendis, N.; Faucher, S. P.; Veres, T.; Tabrizian, M. Rapid and Specific SPRi Detection of *L. pneumophila* in Complex Environmental Water Samples. *Anal. Bioanal. Chem.* **2015**, *407*, 5541–5545.

Freire, R. S.; Durana, N.; Kubota, L. T. Electrochemical Biosensor-Based Devices for Continuous Phenols Monitoring in Environmental Matrices. *J. Braz. Chem. Soc.* **2002**, *13*, 119–123.

Gautam, P.; Suniti, S.; Prachi, K.; Amrita, D.; Madathil, B.; Nair, A. N. A Review on Recent Advances in Biosensors for Detection of Water Contamination. *Int. J. Environ. Sci.* **2012**, *2* (3), 1565–1574.

Ghica, M. E.; Carvalho, R. C.; Amine, A.; Brett, C. M. A. Glucose Oxidase Enzyme Inhibition Sensors for Heavy Metals at Carbon Film Electrodes Modified with Cobalt or Copper Hexacyanoferrate. *Sens. Actuat. B Chem.* **2013**, *178*, 270–278.

Gong, Z.; Guo, Y.; Sun, X.; Cao, Y.; Wang, X. Acetylcholinesterase Biosensor for Carbaryl Detection Based on Interdigitated Array Microelectrodes. *Bioprocess. Biosyst. Eng.* **2014**, *37*, 1929–1934.

Grieshaber, D.; MacKenzie, R.; Voeroes, J.; Reimhult, E. Electrochemical Biosensors-Sensor Principles and Architectures. *Sensors* **2008**, *8* (3), 1400–1458.

Gu, M. B.; Mitchell, R. J.; Kim, B. C. Whole-Cell-Based Biosensors for Environmental Biomonitoring and Application. *Adv. Biochem. Eng. /Biotechnol.* **2004**, *87*, 269–305.

Guascito, M. R.; Malitesta, C.; Mazzotta, E.; Turco, A. Inhibitive Determination of Metal Ions by an Amperometric Glucose Oxidase Biosensor. *Sens. Actuat. B Chem.* **2008**, *131*, 394–402.

Guo, L.; Li, Z.; Chen, H.; Wu, Y.; Chen, L.; Song, Z.; Lin, T. Colorimetric Biosensor for the Assay of Paraoxon in Environmental Water Samples Based on the Iodine-Starch Color Reaction. *Anal. Chim. Acta.* **2017**, *967*, 59–63.

Gutiérrez, J. C., Amaro, F., & Martín-González, A. Heavy Metal Whole-Cell Biosensors Using Eukaryotic Microorganisms: An Updated Critical Review. *Front. Microbiol.* **2015**, *6*, 48.

Hahn, T.; Tag, K.; Riedel, K.; Uhlig, S.; Baronian, K.; Kunze, G.; Gotthard, G. *Biosens. Bioelectron.* **2006**, *22*, 2078–2085.

Hayat, A.; Marty, J. L. Aptamer Based Electrochemical Sensors for Emerging Environmental Pollutants. *Front. Chem.* **2014**, *2*, 41.

Hayat, A.; Paniel, N.; Rhouati, A.; Marty, J. L.; Barthelmebs, L. Recent Advances in Ochratoxin A-Producing Fungi Detection Based on PCR Methods and Ochratoxin A Analysis in Food Matrices. *Food Control.* **2012**, *26*, 401–415.

He, M. Q.; Wang, K.; Wang, J.; Yu, Y. L.; He, R. H. A Sensitive Aptasensor Based on Molybdenum Carbide Nanotubes and Label-Free Aptamer for Detection of Bisphenol A. *Anal. Bioanal. Chem.* **2017**, *409*, 1797–1803.

Highfield, P. E.; Dougan, G. DNA Probes for Microbial Diagnosis. *Br. J. Biomed. Sci.* **1992**, *42*, 352–355.

Ibrahim, H. K. R.; Helmi, S.; Lewis, J.; Crane, M. *Bull. Environ. Contam. Toxicol.* **1998**, *60*, 448–455.

Ilangovan, R.; Daniel, D.; Krastanov, A. et al. Enzyme Based Biosensor for Heavy Metal Ions Determination. *Biotechnol. Biotechnol. Equip.* **2014**, *20*, 184–189.

Ilangovan, R.; Daniel, D.; Krastanov, A.; Zachariah, C.; Elizabeth, R. Enzyme Based Biosensor for Heavy Metal Ions Determination. *Biotechnol. Biotechnol. Equip.* **2006**, *20* (1), 184–189.

Ivask, A.; Virta, M.; Kahru, A. *Soil Biol. Biochem.* **2002**, *34*, 1439–1447.

Jaffrezic-Renault, N. New Trends in Biosensors for Organophosphorus Pesticides. *Sensors* **2001**, *1* (2), 60–74.

Jaffrezic-Renault, N.; Dzyadevych, S. V. Conductometric Microbiosensors for Environmental Monitoring. *Sensors.* **2008**, *8*, 2569–2588.

Jalalian, S. H.; Karimabadi, N.; Ramezani, M.; Abnous, K.; Taghdisi, S. M. Electrochemical and Optical Aptamer-Based Sensors for Detection of Tetracyclines. *Trends Food Sci. Technol.* **2018**, *73*, 45–57.

Jeanty, G.; Wojciechowska, A.; Marty J-L.; Trojanowicz, M. Flow-Injection Amperometric Determination of Pesticides on the Basis of Their Inhibition of Immobilized Acetylcholinesterases of Different Origin. *Anal. Bioanal. Chem.* **2002**, *373*, 691–695.

Jenison, R. D.; Gill, S. C.; Pardi, A.; Polisky, B. High-Resolution Molecular Discrimination by RNA. *Science* **1994**, *263*, 1425–1429.

Jeyapragasam, T.; Saraswathi, R. Electrochemical Biosensing of Carbofuran Based on Acetylcholinesterase Immobilizedontoironoxide-Chitosan Nanocomposite. *Sens. Actuat. B Chem.* **2014**, *191*, 681–687.

Jiang, X.; Li, D.; Xu, X.; Ying, Y.; Li, Y.; Ye, Z.; Wang, J. *Biosens. Bioelectron.* **2008**, *23*, 1577–1587.

Jiang, Y.; Xiao, L. L.; Zhao, L.; Chen, X.; Wang, X.; Wong, K. Y. Optical Biosensor for the Determination of BOD in Seawater. *Talanta.* **2006**, *70* (1), 97–103.

Juronen, D.; Kuusk, A.; Kivirand, K.; Rinken, A.; Rinken, T. Immunosensing System for Rapid Multiplex Detection of Mastitis-Causing Pathogens in Milk. *Talanta.* **2018**, *178*, 949–954.

Justino, C. I. L.; Duarte, A. C.; Rocha-Santos, T. A. P. Recent Progress in Biosensors for Environmental Monitoring: A Review. *Sensors* **2017**, *17* (12), 1–25. doi: 10.3390/s17122918

Kapperud, G.; Vardund, T.; Skjerve, E.; Horns, E.; Michaelsen, T. E. Detection of Pathogenic Yersinia Enterocolitica in Foods and Water by Immunomagnetic Separtaion, Nested Polymerase Chain Reaction, and Colorimetric Detection of Amplified DNA. *Appl. Environ. Microbiol.* **1993**, *59*, 2938–2944.

Karim, F.; Fakhruddin, A. N. M. Recent Advances in the Development of Biosensor for Phenol: A Review. *Rev. Environ. Sci. Biotechnol.* **2012**, *11*, 261–274.

Karube, I.; Mitsuda, S.; Matsunaga, T.; Suzuki, S. A Rapid Method for Estimation of BOD by Using Immobilized Microbial Cells. *J. Ferment. Technol.* **1977**, *55*, 243–248.

Karube, I.; Yano, K.; Sasaki, S.; Nomura Y.; Ikebukuro, K. *Ann. NY Acad. Sci.* **1998**, *864*, 23.

Karyakin, A. A.; Presnova, G. V.; Rubtsova, M. Y.; Egorov, A. M. Oriented Immobilization of Antibodies Onto the Gold Surfaces via Their Native Thiol Groups. *Analyt. Chem.* **2000**, *72* (16), 3805–3811.

Katz, E.; Willner, I. Probing Biomolecular Interactions at Conductive and Semiconductive Surfaces by Impedance Spectroscopy: Routes to Impedimetric Immunosensors, DNA–Sensors, and Enzyme Biosensors. *Electroanalysis.* **2003**, *15* (11), 913–947.

Keane, A.; Phoenix, P.; Ghoshal, S.; Lau, P. C. *J. Microbiol. Methods* **2002**, *49*, 103.

Khadro, B.; Namour, P.; Bessueille, F.; Leonard, D.; Jaffrezic-Renault, N. Enzymatic Conductometric Biosensor Based on PVC Membrane Containing Methyl Viologen/Nafion®/Nitrate Reductase for Determination of Nitrate in Natural Water Samples. *Sens. Mater.* **2008**, *20*, 267–279.

Khoshbin, Z.; Housaindokht, M. R.; Verdian, A.; Bozorgmehr, M. R. Simultaneous Detection and Determination of Mercury (II) and Lead (II) Ions through the Achievement of Novel Functional Nucleic Acid-Based Biosensors. *Biosens. Bioelectron.* **2018**, *116*, 130–147.

Knecht, M. R.; Sethi, M. Bioinspired Colorimetric Detection of Hg^{2+} and Pb^{2+} Heavy Metal Ions Using Au Nanoparticles. *Analyt. Bioanalyt. Chem.* **2009**, *394*, 33–46.

Kucherenko, I. S.; Soldatkin, O. O.; Arkhypova, V. M.; Dzyadevych, S. V; Soldatkin, A. P. A Novel Biosensor Method for Surfactant Determination Based on Acetylcholinesterase Inhibition. *Measure. Sci. Technol.* **2012**, *23* (6), 6.

Kuralay, F.; Ozyoruk, H.; Yıldız, A. Inhibitive Determination of Hg^{2+} Ion by an Amperometric Urea Biosensor Using Poly (Vinylferrocenium) Film. *Enzyme Microb. Technol.* **2007**, *40*, 1156–1159.

Kuusk, E.; Rinken, T. Transient Phase Calibration of Tyrosinase-Based Carbaryl Biosensor. *Enzyme Microb. Technol.* **2004**, *34* (7), 657–661.

Laschi, S.; Mascini, M.; Scortichini, G.; Fránek, M.; Mascini, M. Polychlorinated biphenyls (PCBs) Detection in Foods Amples Using an Electrochemical Immunosensor. *J. Agric. Food Chem.* **2003**, *51* (7), 1816–1822.

Lei, Y.; Chen, W.; Mulchandani, A. Microbial Biosensors. *Anal. Chim. Acta.* **2006**, *568*, 200–210.

Lehmann, M.; Riedel,K.; Alder, K.; Kunze, G. Amperometric Measurement of Copper Ion with a Deputy Substrate Using a Novel *Saccharomyces cerevisiae* sensor. *Biosens. Bioelectron.* **2000**, *15*, 211–219.

Lei, Y.; Mulchandani, P.; Chen, W.; Wang, J.; Mulchandani, A. A Microbial Biosensor for P-Nitrophenol Using Arthrobacter Sp. *Electroanalysis* **2003**, *15*, 1160–1164.

Li, X-H.; Xie, Z-H.; Min, H. et al. Amperometric Biosensor Based on immobilization acetylcholinesterase on Manganese Porphyrin Nanoparticles for Detection of Trichlorfon with Flow-Injection Analysis System. *Electroanalysis* **2007**, *19*, 2551–2557.

Liu, G.; Guo, W.; Song, D. A Multianalyte Electrochemical Immunosensor Based on Patterned Carbon Nanotubes Modified Substrates for Detection of Pesticides. *Biosens. Bioelectron.* **2014a**, *52*, 360–366.

Liu, X.; Li, W.-J.; Yang, Y.; Mao, L.-G.; Peng, Z. A Label-Free Electrochemical Immunosensor Based on Gold Nanoparticles for Direct Detection of Atrazine. *Sens. Actuat. B Chem.* **2014b**, *191*, 408–414.

Liu, J.; Xu, X.; Tang, L.; Zeng, G. Determination of Trace Mercury in Compost Extract by Inhibition Based Glucose Oxidase Biosensor. *Trans. Nonferrous Met. Soc. China* **2009**, *19*, 235–240.

Liu, J.; Mattiasson, B. Microbial BOD Sensors for Wastewater Analysis. *Water Res.* **2002**, *36* (15), 3786–3802.

Lojou, E.; Bianco, P. Application of the Electrochemical Concepts and Techniques to Amperometric Biosensor Devices. *J. Electroceram.* **2006**, *16* (1), 79–91.

Long, F.; Zhu, A.; Shi, H. Recent Advances in Optical Biosensors for Environmental Monitoring and Early Warning. *Sensors* **2013**, *13* (10), 13928–13948.

Long, F.; Zhu, A.; Shi, H.; Wang, H.; Liu, J. Rapid On-Site/In-Situ Detection of Heavy Metal Ions in Environmental Water Using a Structure-Switching DNA Optical Biosensor. *Sci. Rep.* **2013**, *3*, 2308.

Ma, H.; Sang-Bing, T. Design of Research on Performance of a New Iridium Coordination Compound for the Detection of Hg2+. *Int. J. Environ. Res. Public Health* **2017**, *14* (10), 1232.

Madianos, L.; Tsekenis, G.; Skotadis, E.; Patsiouras, L.; Tsoukalas, D. A Highly Sensitive Impedimetric Aptasensor for the Selective Detection of Acetamiprid and Atrazine Based on Microwires Formed by Platinum Nanoparticles. *Biosens. Bioelectron.* **2018**, *101*, 268–274.

Maduraiveeran, G.; Jin, W. Nanomaterials Based Electrochemical Sensor and Biosensor Platforms for Environmental Applications. *Trends Environ. Anal. Chem.* **2017**, *13*, 10–23.

Marazuela, M. D.; Moreno-Bondi, M. C. Fiber-Optic Biosensors—An Overview. *Analyt. Bioanalyt. Chem.* **2002**, *372*, 664–682.

Marrazza, G.; Chianella, I.; Mascini, M. *Anal Chim Acta.* **1999**, *387*, 297–307.

Martín, M.; Salazar, P.; Jiménez, C.; Lecuona, M.; Ramos, M. J.; Ocle, J.; Riche, R.; Villalonga, R.; Campuzano, S.; Pingarrón, J. M. et al. Rapid Legionella Pneumophila

Determination Based on Disposable-Shell Fe_3O_4@poly (Dopamine) Magnetic Nanoparticles Immunoplatform. *Anal. Chim. Acta.* **2015**, *887*, 51–58.

Martinkova, P.; Kostelnik, A.; Valek, T.; Pohanka, M. Main Streams in the Construction of Biosensors and Their Applications. *Int. J. Electrochem. Sci.* **2017**, *12*, 7386–7403.

Mascini, M. A Brief Story of Biosensor Technology. *Biotechnol. Appl. Photosynth. Proteins Biochips. Biosens. Biodevices.* **2006**, 4–10.

Mayorga-Martinez, C.; Pino, F.; Kurbanoglua, S.; Rivas, L.; Ozkan, S. A.; Merkoci, A. Iridium Oxide Nanoparticles Induced Dual Catalytic/Inhibition Based Detection of Phenol and Pesticide Compounds. *J. Mater. Chem. B.* **2014**, *2*, 2233–2239.

Meng, X.; Wei, J.; Ren, X.; Ren, J.; Tang, F. A Simple and Sensitive fluorescence Biosensor for Detection of Organophosphorus Pesticides Using H_2O_2-Sensitive Quantum Dots/Bi-Enzyme. *Biosens. Bioelectron.* **2013**, *47*, 402–407.

Michel, C.; Ouerd, A.; Battaglia-Brunet, F. et al. Cr (VI) Quantification Using an Amperometric Enzyme-Based Sensor: Interference and Physical and Chemical Factors Controlling the Biosensor Response in Ground Waters. *Biosens. Bioelectron.* **2006**, *22*, 285–290.

Michel, C.; Battaglia-Brunet, F.; Minh, C. T.; Bruschi, M.; Ignatiadis, I. Amperometric Cytochrome C3-Based Biosensor for Chromate Determination. *Biosens. Bioelectron.* **2003**, *19* (4), 345–352.

Mohammad, A. V.; Housaindokht, R.; Sheikhzadeh, E.; Pordeli, P.; Zaeri, Z. R.; Fard, F. J.; Masoumeh, N.; Mashreghi, M.; Haghparast, A.; Pour, A. Z.; Esmaeili, A. A.; Soleymani, S. A Sensitive Electrochemical Aptasensor Based on Single Wall Carbon Nanotube Modified Screen Printed Electrode for Detection of *Escherichia coli* O157:H7. *Adv. Mater. Lett.* **2018**, *9* (5), 369–374.

Mohanty, S. P.; Kougianos, E. Biosensors: A Tutorial Review. *IEEE Potent.* **2006**, *25* (2), 35–40.

Monosik, R.; Stredansky, M.; Sturdik, E. Biosensors—Classification, Characterization and New Trends. *Acta Chim. Slovaca* **2012**, *5* (1), 109–120.

Moorcroft, M. J.; Davis, J.; Compton, R. G. Detection and Determination of Nitrate and Nitrite: A Review. *Talanta* **2001**, *54*, 785–803.

Mulchandani, A.; Shengtian, P.; Chen, W. *Biotechnol. Prog.* **1999**, *15*, 130–134, 121.

Mulchandani, P.; Lei, Y.; Chen, W.; Wang, J. Microbial Biosensor for P-Nitrophenol Using Moraxella sp. *Anal. Chim. Acta.* **2002**, *470*, 79–86.

Murata, M.; Nakayama, M.; Irie, H.; Yakabe, K.; Fukuma, K.; Katayama, Y.; Maeda, M. *Anal. Sci.* **2001**, *17*, 34.

Naghibi, F.; Pourmorad, F.; Honary, S.; Shamsi, M. Decontamination of Water Polluted with Phenol Using Raphanus sativus Root. *Iran. J. Pharm. Res.* **2003**, *2*, 29–32. View at: Google Scholar.

Nakamura, H. *Anal. Methods* **2010**, *2*, 430.

Nigam, V. K.; Shukla, P. Enzyme Based Biosensors for Detection of Environmental Pollutants—A Review. *J. Microbiol. Biotechnol.* **2015**, *25* (11), 1773–1781.

North, J. R. Immunosensors: Antibody-Based Biosensors. *Trends Biotechnol.* **1985**, *3* (7), 180–186.

Nunes, G. S.; Lins, J. A. P.; Silva, F. G. S.; Araujo, L. C.; Silva, F. E. P. S.; Mendonça, C. D.; Badea, M.; Hayat, A.; Marty, J. L. Design of a Macroalgae Amperometric

Biosensor; Application to the Rapid Monitoring of Organophosphate Insecticides in an Agroecosystem. *Chemosphere* **2014**, *111*, 623–630.

Odobašić, A.; Šestan, I.; Begić, S. Biosensors for Determination of Heavy Metals in Waters, Biosensors for Environmental Monitoring, Toonika Rinken and Kairi Kivirand. *Intech Open* **2019**. doi: 10.5772/intechopen.84139

Oluwaseun, A. C.; Phazang, P.; Sarin, N. B. *Biosensors: A Fast-Growing Technology for Pathogen Detection in Agriculture and Food Sector, Biosensing Technologies for the Detection of Pathogens–A Prospective Way for Rapid Analysis*; Rinken, T., Kivirand, K., Eds.; *IntechOpen*: UK, 2018.

Osma, J. F.; Stoytcheva, M. *Biosensors: Recent Advances and Mathematical Challenges*; OmniaScience: Barcelona, Spain, 2014.

Ozoner, S. K.; Erhan, E.; Yilmaz, F. Enzyme Based Phenol Biosensors. In *Environmental Biosensors*; Somerset, V., Ed.; *IntechOpen*: UK, 2011.

Pal, P.; Bhattacharyay, D.; Mukhopadhyay, A.; Sarkar, P. The Detection of Mercury, Cadium, and Arsenic by the Deactivation of Urease on Rhodinized Carbon. *Environ. Eng. Sci.* **2009**, *26* (1), 25–32.

Paniel, N.; Baudart, J.; Hayat, A.; Barthelmebs, L. Aptasensor and Genosensor Methods for Detection of Microbes in Real World Samples. *Methods* **2013**, *64*, 229–240.

Peedel, D.; Rinken, T. Rapid Biosensing of *Staphylococcus aureus* Bacteria in Milk. *Analyt. Methods* **2014**, *6*, 2642–2647.

Pellinen, T.; Huovinen, T.; Karp, M. A Cell-Free Biosensor for the Detection of Transcriptional Inducers Using Firefly Luciferase as a Reporter. *Anal. Biochem.* **2004**, *330*, 52–57.

Perumal, V.; Hashim, U. Advances in Biosensors: Principle, Architecture and Applications. *J. Appl. Biomed.* **2014**, *12*, 1–5.

Petanen, T.; Romantschuk, M. Use of Bioluminescent Bacterial Biosensors as an Alternative Method for Measuring Heavy Metals in Soil Extracts. *Analyt. Chim Acta.* **2002**, *456*, 55.

Podola, B.; Nowack, E. C. M.; Melkonian, M. The Use of Multiple-Strain Algal Sensor Chips for the Detection and Identification of Volatile Organic Compound. *Biosens. Bioelectron.* **2004**, *19*, 1253–1260.

Pohanka, M. The Piezoelectric Biosensors: Principles and Applications, a Review. *Int. J. Electrochem. Sci.* **2017**, *12*, 496–506.

Rasmussen, L. D.; Sorensen, S. J.; Turner, R. R.; Barkay, T. *Soil Biol. Biochem.* **2000**, *32*, 639–646.

Rebollar-Pérez, G.; Campos-Terán, J.; Ornelas-Soto, N.; Méndez-Albores, A.; Torres, E. Biosensors Based on Oxidative Enzymes for Detection of Environmental Pollutants. *Biocatalysis.* **2015**, *1* (1).

Reshetilov, A. N.; Semenchuk, I. N.; Iliasov, P. V.; Taranova, L.A The Amperometric Biosensor for Detection of Sodium Dodecyl Sulfate. *Anal. Chim. Acta.* **1997**, *347*, 19–26.

Riedel, K.; Beyersdorf-Radeck, B.; Neumann, B.; Schaller, F. Microbial Sensors for Determination of Aromatics and Their Chloro Derivatives. Part III: Determination of Chlorinated phenols Using a Biosensor Containing *Trichosporon beigelii* (cutaneum). *Appl. Microbiol. Biotechnol.* **1995**, *43*, 7–9.

Ripp, S.; Diclaudio, M. L.; Sayler, G. S. Biosensors as Environmental Monitors. In *Environmental Microbiology*; Mitchell, R.; Gu, J., Eds.; 2nd ed.; Wiley-Blackwell: NJ, USA, 2010; pp 213–233,.

Roda, A.; Pasini, P.; Mirasoli, M.; Guardigli, M.; Russo, C.; Musiani, M. A Sensitive Determination of Urinary Mercury (II) by a Bioluminescent Transgenic Bacteria-Based Biosensor. *Anal. Lett.* **2001,** *34,* 29–41.

Rodriguez-Mozaz, S., de Alda, M. J. L.; Marco, M.-P.; Barceló, D. Biosensors for Environmental Monitoring: A Global Perspective. *Talanta* **2005,** *65,* 291–297.

Rodriguez-Mozaz, S.; De Alda, M. J. L.; Barcelo, D. Biosensors as Useful Tools for Environmental Analysis and Monitoring. *Anal. Bioanal. Chem.* **2006,** *386* (4), 1025–1041.

Rodriguez-Mozaz, S.; Marco, M. P.; Alda, M. J. L.; Barcel, D. Biosensors for Environmental Applications: Future Development Trends. *Pure Appl. Chem.* **2004,** *76,* 723–752.

Rogers, K. R. Recent Advances in Biosensor Techniques for Environmental Monitoring. *Anal. Chim. Acta.* **2006,** *568,* 222–231.

Rogers, K. R.; Gerlach, C. L. Environmental Biosensors—A Status Report. *Environ. Sci. Technol.* **1996,** *30,* 486–491.

Rogers, N. J.; Apte, S. C.; Batley, G. E.; Gadd, G. E.; Casey, P. S. Comparative Toxicity of Nanoparticulate ZnO, Bulk ZnO, and $ZnCl_2$ to a Freshwater Microalga *(Pseudokirchneriella subcapitata)*: The Importance of Particle Solubility. *Environ. Sci. Technol.* **2007,** *41* (24), 8484–8490.

Ron, E. Z. *Curr. Opin. Biotechnol.* **2007,** *18,* 252.

Russell, I. M.; Burton, S. G. The Development of an Immobilized Enzyme Bioprobe for the Detection of Phenolic Pollutants in Water. *Anal. Chim. Acta.* **1999,** *389,* 161–170.

Salgado, A. M.; Silva, L. M.; Melo, A. F. Biosensor for Environmental Applications. In *Environmental Biosensors*; Somerset, V., ed.; *InTech,* 2011.

Sassolas, A.; Prieto-Simón, B.; Marty, J.-L. Biosensors for Pesticide Detection: New Trends. *Am. J. Analyt. Chem.* **2012,** *3* (3), 210.

Satoh, I.; Iijima, Y. Multi-Ion Biosensor with Use of a Hybrid-Enzyme Membrane. *Sens. Actuat. B.* **1995,** *24* (1–3), 103–106.

Seeman, N. C. Biochemistry and Structural DNA Nanotechnology: An Evolving Symbiotic Relationship†. *Biochemistry* **2003,** *42,* 7259–7269.

Seeman, N. C. Nanomaterials Based on DNA. *Annu. Rev. Biochem.* **2010,** *79,* 65–87.

Sethi, R. S. Transducer Aspects of Biosensors. *Biosens. Bioelectron.* **1994,** *9,* 243–264.

Shanmugam, K.; Subramanayam, S.; Tarakad, S. V.; Kodandapani N.; D'Souza S. F.2,4-Toluene Diamines-Their Carcinogenity, Biodegradation, Analytical Techniques and an Approach towards Development of Biosensors. *Anal. Sci.* **2001,** *17,* 1369.

Sharma, S.; Byrne, H.; O'Kennedy, R. J. Antibodies and Antibody-Derived Analytical Biosensors. *Essays Biochem.* **2016,** *60,* 9–18.

Sharpe, A. N. Developments in Rapid Methods for Detection of Agents of Foodborne Disease. *Food Res. Int.* **1994,** *27,* 237–243.

Shimomura, M.; Nomura, Y.; Zhang, W.; Sakino, M.; Lee, K.-H; Ikebukuro, K.; Karube, I. Simple and Rapid Detection Method Using Surface Plasmon Resonance for Dioxins, Polychlorinated Biphenyl and Atrazine. *Anal. Chim. Acta.* **2001,** *434* (2), 223–230.

Shin, H. J.; Park, H. H.; Lim, W. K. Freeze-Dried Recombinant Bacteria for On-Site Detection of Phenolic Compounds by Color Change. *J. Biotechnol.* **2005,** *119,* 36–43.

Simonian, A. L.; Rainina, E. I.; Wild, J. R. *Analyt. Lett.* **1997,** *30,* 2453–2468.

Singh, R. L. *Principles and Applications of Environmental Biotechnology for a Sustainable Future*; Springer: Singapore, 2015; pp 1–487.

Slifonova, O.; Burlage, R.; Barkay, T. Bioluminescent Sensors for Detection of Bioavailable Hg(II) in the Environment. *Appl. Environ. Microbiol.* **1993**, *59*, 3083–3090.

Takayama, K.; Kurosaki, T.; Ikeda, T. Mediated Electrocatalysis at Biocatalyst Electrode Based on a Bacterium Gluconobacter Industrius. *J. Electroanalyt. Chem.* **1993**, *356*, 295–301.

Tan, F.; Cong, L.; Saucedo, N. M.; Gao, J.; Li, X.; Mulchandani, A. An Electrochemically Reduced Graphene Oxide Chemiresistive Sensor for Sensitive Detection of Hg(2+) Ion in Water Samples. *J. Hazard. Mater.* **2016**, *320*, 226–233.

Tan, T. C.; Lim, E. W. C. Thermally Killed Cells of Complex Microbial Culture for Biosensor Measurement of BOD of Wastewater. *Sens. Act. B.* **2005**, 0925-4005, *107* (2), 546–551.

Tan, T. C.; Lim, E. W. C. Thermally Killed Cells of Complex Microbial Culture for Biosensor Measurement of BOD of Wastewater. *Sens. Actuat. B Chem.* **2005**, *107*, 546–551.

Tang, J.; Xie, J.; Shao, N.; Yan, Y. The DNA Aptamers That Specifically Recognize Ricin Toxin Are Selected by Two In Vitro Selection Methods. *Electrophoresis.* **2006**, *27* (7), 1303–1311.

Tang, X.; Zhang, T.; Liang, B. et al. Sensitive Electrochemical Microbial Biosensor for p-Nitrophenylorganophosphates Based on Electrode Modified with Cell Surface-Displayed Organophosphorus Hydrolase and Ordered Mesopore Carbons. *Biosens. Bioelectron.* **2014a**, *60*, 137–142.

Tang, Y.; Ge, B.; Sen, D.; Yu, H. Z. Functional DNA Switches: Rational Design and Electrochemical Signaling. *Chem. Soc. Rev.* **2014b**, *43*, 518–529.

Taranova, L.; Semenchuk, I.; Manolov, T. et al. Bacteria-Degraders as the Base of an Amperometric Biosensor for Detection of Anionic Surfactants. *Biosens. Bioelectron.* **2002**, *17*, 635–640.

Tembe, S.; Inamder, S.; Haram, S.; Karvee, M.; D'Souza, S. F. Electrochemical Biosensor for Catechol Using Agarose-Guargum Entrapped Tyrosinase. *J. Biotechnol.* **2007**, *128*, 80–85.

Tenover, F. C. Diagnostic Deoxyribonucleic Acid Probes for Infectious Diseases. *Clin. Microbiol. Rev.* **1988**, *1*, 82.

Thévenot, D.; Toth, K.; Durst, R.; Wilson, G. Electrochemical Biosensors: Recommended Definitions and Classification. *Biosens. Bioelectron.* **2001**, *16*, 121–131.

Thevenot, D. R.; Toth, K.; Durst, R. A.; and Wilson, G. S., Electrochemical Biosensors: Recommended Definitions and Classifications. *Pure Appl. Chem.* **1991**, *71*, 2333–2348.

Tietjen, M.; Fung, D. Y. C. Salmonella and Food Safety. *Crit. Rev. Microb.* **1995**, *21*, 53–83.

Timur, S.; Pazarliog'lu, N.; Pilloton, R.; Telefoncu, A. Detection of Phenolic Compounds by Thick Film Sensors Based on *Pseudomonas putida*. *Talanta.* **2003**, *61*, 87–93.

Tran, H. V.; Yougnia, R.; Reisberg, S. et al. A Label-Free Electrochemical Immunosensor for Direct, Signal-on and Sensitive Pesticide Detection. *Biosens. Bioelectron.* **2012**, *31*, 62–68.

Tsai, H-C.; Doong, R-A.; Chiang, H-C.; Chen, K-T. Sol–Gel Derived Urease-Based Optical Biosensor for the Rapid Determination of Heavy Metals. *Anal. Chim. Acta.* **2003**, *481*, 75–84.

Turdean, G. L. Design and Development of Biosensors for the Detection of Heavy Metal Toxicity. *Int. J. Electrochem.* **2011**.

Turdean, G. L. Design and Development of Biosensors for the Detection of Heavy Metal Toxicity. *Int. J. Electrochem.* **2011**, 1–15. doi: 10.4061/2011/343125

Turner, A. P. F. *Biosensors-Fundamentals and Applications*; Turner, A. P. F.; Karube, I.; Wilson, G. S., Eds.; Oxford University Press, **1987**; p 5.

Updike, S. J.; Hicks, G. P. The Enzyme Electrod. *Nature* **1967**, *214* (5092), 986–988.

Van der Meer, J. R.; Tropel D.; Jaspers, M. Illuminating the Detection Chain of Bacterial Bioreporters. *Environ. Microbiol.* **2004**, *6*, 1005–1020.

Vargas-Bernal, R.; Rodríguez Miranda, E.; Herrera-Pérez, G. *Evolution and Expectations of Enzymatic Biosensors for Pesticides. Pesticides—Advances in Chemical and Botanical Pesticides*; Soundararajan, R. P., Ed.; *IntechOpen*: UK, 2012.

Verdian, A. Apta-Nanosensors for Detection and Quantitative Determination of Acetamiprid–A Pesticide Residue in Food and Environment. *Talanta* **2018**, *176*, 456–464.

Verma, N.; Bhardwaj, A. Biosensor Technology for Pesticides—A Review. *Appl. Biochem. Biotechnol.* **2015**, *175* (6), 3093–3119.

Verma, N.; Bhardwaj, A. Biosensor Technology for Pesticides—A Review. *Appl. Biochem. Biotechnol.* **2015**, *175*, 3093–3119.

Viirlaid, E.; Riiberg, R.; Mäeorg, U.; Rinken, T. Glyphosate Attachment on Amino Activated Carriers for Sample Stabilization and Concentration. *Agron. Res.* **2015**, *13* (4), 1152–1159.

Viveros, L.; Paliwal, S.; McCrae, D.; Wild, J.; Simonian, A. A. *Sens. Actuat. B.* **2006**, *115*, 150–157.

Vollmer, A. C.; Van Dyk, T. K. Stress Responsive Bacteria: Biosensors as Environmental Monitors. *Adv. Microb. Physiol.* **2004**, *49*, 131–174.

Wang, H.-Q.; Wu, Z.; Tang, L.-J.; Yu, R.-Q.; Jiang, J.-H. Fluorescence Protection Assay: A Novel Homogeneous Assay Platform toward Development of Aptamer Sensors for Protein Detection. *Nucl. Acids Res.* **2011**, *39*, e122.

Wang, X.; Dzyadevych, S. V.; Chovelon, J. M.; Jaffrezic-Renault, N.; Chen, L.; Xia, S.; Zhao, J. Conductometric Nitrate Biosensor Based on Methyl Viologen/Nafion®/Nitrate Reductase Interdigitated Electrodes. *Talanta.* **2006**, *69*, 450–455.

Wang, X.; Lu, X.; Chen, J. Development of Biosensor Technologies for Analysis of Environmental Contaminants. *Trends Environ. Anal. Chem.* **2014**, *2*, 25–32.

Wang, Y.; Xu, H.; Zhang, J.; Li, G. Electrochemical Sensors for Clinical Analysis. *Sensors* **2008**, *8*, 2043–2081.

Wang, Y.; Zhai, F.; Hasebe, Y.; Jia, H.; Zhang, Z. A highly Sensitive Electrochemical Biosensor for Phenol Derivatives Using a Graphene Oxide-Modified Tyrosinase Electrode. *Bioelectrochemistry* **2018**, *122*, 174–182.

Wang, Y-X.; Ye, Z-Z.; Si, C-Y.; Ying, Y-B. Application of Aptamer Based Biosensors for Detection of Pathogenic Microorganisms. *Chin. J. Anal. Chem.* **2012**, *40*, 634–642.

White, B. J.; Harmon, H. J. Optical Solid-State Detection of Organophosphates Using Organophosphorus Hydrolase. *Biosens. Bioelectron.* **2005**, *20*, 1977–1983.

Wijaya, E.; Lenaerts, C.; Maricot, S.; Hastanin, J.; Habraken, S.; Vilcot, J-P. Surface Plasmon Resonance-Based Biosensors: From the Development of Different SPR Structures to Novel Surface Functionalization Strategies. *Curr. Opin. Solid State Mater. Sci.* **2011**, *15*, 208–224.

Willner, I.; Zayats, M. Electronicaptamer-Based sensors. *Angew. Chem. Int. Ed. Engl.* **2007**, *46*, 6408–6418.

Wilson, K.; Walker, J. M. *Principles and Techniques of Practical Biochemistry*, 4th ed.; Cambridge University Press, 1994.

Wong, E. L. S.; Crow, E.; Gooding, J. J. The Electrochemical Detection of Cadmium Using Surface Immobilized DNA. *Electrochem. Commun.* **2007,** *9* (4), 845–849.

Wozei, E.; Hermanowicz, S. W.; Holman, H-Y. N. *Biosens. Bioelectron.* **2006,** *21,* 1654–1658.

Wu, Y.; Liu, L.; Zhan, S.; Wang, F.; Zhou, P. Ultrasensitive Aptamer Biosensor for Arsenic (III) Detection in Aqueous Solution Based on Surfactant-Induced Aggregation of Gold Nanoparticles. *Analyst* **2012,** *137* (18), 4171–4178.

Wu, Z.; Shen, H.; Hu, J.; Fu, Q.; Yao, C.; Yu, S.; Xiao, W.; Tang, Y. Aptamer-Based Fluorescence-Quenching Lateral Flow Strip for Rapid Detection of Mercury(II) Ion in Water Samples. *Anal. Bioanal. Chem.* **2017,** *409,* 5209–5216.

Xu, X.; Ying, Y. Microbial Biosensors for Environmental Monitoring and Food Analysis. *Food Rev. Int.* **2011,** *27* (3), 300–329.

Yagi, K. Applications of Whole-Cell Bacterial Sensors in Biotechnology and Environmental Science. *Appl. Microbiol. Biotechnol.* **2007,** *73,* 1251–1258.

Yang, Y.; Wang, Z.; Yang, M. et al. Inhibitive Determination of Mercury Ion Using a Renewable Urea Biosensor Based on Self-Assembled Gold Nanoparticles. *Sens. Actuat. B Chem.* **2006,** *114,* 1–8.

Yildirim, N.; Long, F.; Gao, C.; He, M.; Shi, H. C.; Gu, A. Z. Aptamer-Based Optical Biosensor for Rapid and Sensitive Detection of 17β-estradiol in Water Samples. *Environ. Sci. Technol.* **2012,** *46,* 3288–3294.

Yildiz, H. B.; Castillo, J.; Guschin, D. A.; Toppare, L.; Schuhmann, W. Phenol Biosensor Based on Electrochemically Controlled Tyrosinase in a Redox Polymer. *Microchimacta* **2007,** *159,* 27–34.

Yilmaz, E.; Majidi, D.; Ozgur, E.; Denizli, A. Whole Cell Imprinting Based *Escherichia coli* Sensors: A Study for SPR and QCM. *Sens. Actuat. B Chem.* **2015,** *209,* 714–721.

Yoo, M. S.; Shin, M.; Kim, Y.; Jang, M.; Choi, Y. E.; Park, S. J.; Choi, J.; Lee, J.; Park, C. Development of Electrochemical Biosensor for Detection of Pathogenic Microorganism in Asian Dust Events. *Chemosphere* **2017,** *175,* 269–274.

Zhai, J. H.; Cui, H.; Yang, R. F. DNA-Based Biosensors. *Biotechnol. Adv.* **1997,** *15* (1), 43–58.

Zhang, W.; Liu, Q. X.; Guo, Z. H.; Lin, J. S. Practical Application of Aptamer-Based Biosensors in Detection of Low Molecular Weight Pollutants in Water Sources. *Molecules* **2018,** *23* (2), 344.

Zhao, S.; Liu, P.; Niu, Y.; Chen, Z.; Khan, A.; Zhang, P.; Li, X. A Novel Early Warning System Based on a Sediment Microbial Fuel Cell for In Situ and Real Time Hexavalent Chromium Detection in Industrial Wastewater. *Sensors* **2018,** *18* (2), 642.

Zhao, C. Q.; Anis, N. A.; Rogers, K. R.; Kline, R. H.; Wright, J.; Eldefrawi, A. T.; Eldefrawi, M. E. *J. Agric. Food Chem.* **1995,** 2308–2315.

Zhihong, M.; Xiaohui, L.; Weiling, F. *Anal. Commun.* **1999,** *36,* 281–283.

CHAPTER 13

New Insights of Bacteriophages: Potential Tool for Wastewater Treatment

SAIMA HAMID* and MOHAMMAD YASEEN MIR

Centre of Research for Development, University of Kashmir, Srinagar, India

Corresponding author. E-mail: cord.babasaima4632@gmail.com

ABSTRACT

Over the last 40 years, less attention has been paid toward the improvement in wastewater treatment technologies. This issue can be solved by paying much attention toward the use of sophisticated biological organisms or tools. Among potential microorganism, bacteriophages have immense potential to use them as biotechnological tools for industrial and drinking water sources. A more detailed understanding of the nature and interactions of wastewater microbial communities is needed in order to implement phage biocontrol efficiently in wastewater treatment. Strategies to tackle both host specificities and host cell resistance, as well as safety issues concerning the emergence of pathogen via transduction, must also be set out. In fact, the future use of the phage to control foaming is an environmentally sustainable and cost-effective solution to avoid these dangerous bacteria like *V. Cholera, Shigella, Salmonella* from entering rivers, ponds, and lakes. In comparison to other methods currently employed in wastewater installations, it is inexpensive, quick and environmentally safe to use programmed phage's to regulate undesirable bacteria or prevent biofouling and foaming while maintaining additional useful ASP bacteria.

13.1   INTRODUCTION

Among the numerous species that do not damage organisms other than host bacterial cells in their respective established ecosystem are classified as bacteriophages or phages, which are viruses that infect bacteria and are responsible for adaptation of bacteria in their respective environs (Santos et al., 2014; Brussow and Kutter, 2005; Chibani-Chennoufi et al., 2004; Brussow and Hendrix, 2002; Bergh et al., 1989; McCallin et al., 2005). Bacteriophages are found in abundant quantities in soil etc. so they are considered as important organism in terms of gene alteration because not much has been explored from all the abundant phages, so there lays huge scope for understanding its genetic makeup. Researchers are using microorganisms as their tools to benefit society either in terms to make vaccines against diseases, food safety treatment methods by identifying those microbes which contain codes for proteins in their genomes, likewise bacteriophages are known to be considered as their potential biotechnological tool. As bacteriophages use other bacteria to replicate its genetic material inside the host to make number of copies which gives way to adaptive series of evolution for bacteria to stand with changing environ conditions and moreover genomic modifications, provide new platform for generation of new drugs and organism tools. In addition, the use of bacteriophageal particles in phage-based vaccination has emerged as one of the most effective prevention approaches, along with plant and animal gene transfer methods, phage-based collection of biological affinity molecules, bacterial biosensing tools, gene distribution, food biopreservation and protection, plant pathogens biocontrol, biofilm regulation, surface monitoring. Since certain phages are natural bactericidal microorganisms, they have been used as alternatives to antibiotics for over 90 years in the former Soviet Union and in Central Europe (Jassim and Limoges, 2014). The most recent findings have also shown that 70% of the world's bacteria now have an antibiotic resistance that has carelessly spread through fields and cattle and now includes infected rivers. As a result, certain serious infections cannot be treated and surgical procedures are more dangerous because of the prevalence of acquired medical bacteria that are antibiotic resistant. The subsequent ban on growth-funding antibiotics in several countries and the subsequent effects on human and animal welfare contribute to phages of potential biocontrol and multidrug resistant strain treatment with various bacteria (Jassim and Limoges, 2014).

13.2 PHAGE FUNCTIONS IN REGULATING BACTERIAL WATERBORNE PATHOGEN

In order to make a shift toward the use of microbes instead of chemicals against pathogenic organisms, phages were used to reduce the use of chemical agents on the basis of comparison which were made on merits and demerits of using biological organism as biotechnological agents like phages (Fujiwara et al., 2011; Sulakvelidze and Kutter, 2005; Loc-Carrillo and Abedon, 2011). It has been observed that bactericidal phages do not harm usual bacterial flora of soil, and have low rates of destroying bacterial biofilm which are important in terms of bioremediation (Jassim and Limoges, 2014). In testing water treatment system, in disinfection processes (Havelaar, 1987), typhimurium phages were also suggested for use as a diagnostical species of pathogenic human viruses. Phages have some desirable characteristics as therapeutic agents or biocontrol agents as they possess some specific properties of self-replication, mobility and are host-specific too, which enables them to adapt environmental conditions so easily (Jassim and Limoges, 2014; Sulakvelidze et al., 2001; Jassim and Limoges, 2014). Well regulated animal models have shown that the usage of engineered lytic phages can avoid infection or cure animals contaminated with such pathogenic bacteria and may be effective options for the treatment of drug-resistant infections (Abdulamir et al., 2014; Jassim and Limoges, 2014; Aldoori et al., 2015). A comprehensive methodology or protocol is necessary to implement the use of phages against specific host bacteria while considering its physicochemical parameters and importantly the dose application (Jassim and Limoges, 2013). Yet, the use of phages for the wide applications regarding humans are beyond knowledge to treat human infectious diseases (Potera, 2013; Jassim and Limoges, 2014). Although there is some wide applicability for use of Phages in food industry, even USFDA (2006) has approved the List of specific phage preparations. It is known that phages only infect and lyse bacterial cells and are harmless to mammalians (USFDA, 2006). This has led to the development of a phage related product which received regulatory approval from the FDA in 2011, as a natural antimicrobial for use in agrofood industry, treated as "generally regarded as safe" (GRAS) and by US-FSIS as safe for use in animals (Sillankorva et al., 2012; Klumpp and Loessner, 2013). As various bacteriophages are considered to cause bacterial virulence at their lysogenic or prophages stages or during transduction process for horizontal gene transfer via vectors, hence precautionary measures are needed to deal with such phages in the case of Human phage therapy (Verheust et al., 2010; Jassim and Limoges, 2014). The bacterial tolerance of Lysogenic phages is well known by supplying axial force in the bacterial pathogens leading to horizontal gene transfer to the production of the bacterial pathogen. The development of an efficient phage for the biocontrol system against medically important human and animal pathogens also first indicate the potential to produce many mutant strains which might be harmful. In order to avoid a bacterial resistance of phages, experts must focus on the modeling of the phage conditions in the natural habitats and must also discuss the use of phage biocontrol therapy in the wastewater treatment facilities.

13.3 FILAMENTOUS BACTERIOPHAGES *GORDONIA* IN UNTREATED ACTIVATED SLUDGE

In the sewage and industrial wastewater treatment facilities, after various biological processes, last step is to activate the left sludge which is converted into solid and less pathogenic, thus various microbes, such as bacteria, protozoa, virus, and metazoa are used which help to degrade organic pollutants into liquid mixture with low proportion of suspended solids. Although, some important filamentous bacteria in the presence of biopolymers which induce the formation of fluffy suspension solids contain the activated sludge as the critical component of this process (Jenkins et al., 2004, Martins et al., 2004; Wagner, 2002). Filamentous bacteria bind with biopolymers to form flocksbecause air bubbles trapped inside the sludge make it fluffy when present on aerators and make it difficult to handle such fluffy sludge (Martins et al., 2004; Petrovski et al., 2011). On the basis of using filamentous morphotypes, some high-throughput deep-sequencing-dependent population analysis, a versatile technique used in order to identify filamentous bacteria for sludge formation which were highly based on fluorescence *in situ* hybridization using rRNA-targeted probes (Eikelboom, 1975; Eikelboom, 1977; Eikelboom, 2002; Waarde et al., 2002; Guo et al., 2015; Ju et al., 2014). Accumulated studies have identified the different compositions of filamentous bacteria involved in foaming and have proposed some specific functions of long unbranched filaments of *Candidatus*, that is, *Microthrix parvicella* or short-branched filaments of actinomycetes containing mycolic acid, or Mycolata (Nielsen et al., 2009). Although, some of the traditional methods like surfactants, applications of chlorine, flow rate of return sludge, and regulation of aeration are commonly used to control foaming of sludge with least success rates (Seka et al., 2003). Hence, the bacteriophages have the potential to directly attack those filamentous bacteria which cause the fluffiness of sludge without affecting other nontarget beneficial microbes (Seka et al., 2003; Thomas et al., 2002). For many years, use of bacteriophages has created special research interest to use them as antimicrobial agents for the treatment of wastewater as they are also the part of same environs (Ewert and Paynter, 1980; Khan et al., 2002; Summers, 2001; Sulakvelidze et al., 2001). Most recent work into phage-based antimicrobials has focused on medicinal or agricultural pathogens of importance (Matsuzaki, 2005; Merril et al., 2003). The use of phages in environmental and/or industrial settings for the containment of issue bacteria also has tremendous potential to substitute ineffective and sometimes environmentally damaging biocides. Various studies have shown the isolation with therapeutic applications of filamentous bacteria against Phages, as most purposeful subject was the Mycobacteria species (Hatfull et al., 2010). *Haliscomeno-bacter hydrossis* is the filamentous bacteria which causes fluffy or bulky biomass and the Phages were reported to have neutralizing effects against filamentous bacteria (Kotay et al., 2011). *Gordonia, Tsukamurella* spp., *Nocardia, Rhodococcus* belong to the mycolata group which are related to Phages (Petrovski et al., 2010). GTE2 and GTE7 are the two known bacteriophages which have been shown to minimize their host level and to inhibit stable development of foam by their host bacteria in pure laboratory colonies studied in a laboratory foaming device (Petrovski et al., 2010). After those positive results, the next step in the phage-based foam control system is to determine whether Phage can control the foaming associated bacteria directly in the complex chemical and microbial environment of the activated sludge matrix. In addition, in terms of the size of each Phage host line against the plant-associated wastewater treatment associated hosts, the value of a phage for bacterial control must be determined. Filamentous bacteria were grown from several wastewater spray incidents and used to extract cognate phages that are capable of controlling these bacteria. The isolated phages were tested for use in reducing the levels of *Gordonia* in untreated activated sludge. Using the isolated *Gordonia* strains as hosts, wastewater-insulated phages have been able to control *Gordonia* rates in a given media. In fact, the application of these phages in activated sludge environments resulted in repeatable, major suppression of the *Gordonia* levels. This is unexpected considering the substantial variety and species richness found in the active sludge population of micro- and macroorganisms, with specific entity classes responsible for complex functions like floc formation, phosphorus elimination, nitrite oxidation, and denitrification (Gill and Young, 2011). Various studies have been conducted in which differences are to be anticipated with such a heterogeneous and diverse context composed of viruses, microbes, protozoa, and metazoans. The host *Gordonia* rates fluctuated at the beginning of the experiment (day zero to day 1) after immediate inoculation in the mixed sludge bioreactors likely representing the stage needed for the exogenously inserted *Gordonia* to be introduced into the microbial sludge environment. In contrast to regulation, a high degree of *Gordonia* G7 was observed on *Gordonia* and phage inoculation (day zero) in phage-treated environment. In the process used in the sludge scheme, this might have killed both indigenous and exogenous hosts. In the current laboratory-scale active sludge, this work offers information on phage-*Gordonia* population dynamics, conditions that are closer compared with successful studies performed in developed cultures to resembling real world treatment systems (Petrovski et al., 2011). In addition to the test of *Gordonia* levels in activated sludge, the findings at the end of the experimental phage procedure were more positive than those found at the end. Aerated, permanent crops compared with continuous flow aeration reactors that are similar to continuous organic enrichment cultures were the sludge air systems in this research. The control systems and the phage-treated sludge systems were only demonstrated in limited amounts at the end of the experiment in contrast to the recently collected sludge around less compact tubes. The poor health of flocs were obvious after long ventilation (nine days) in static cultivation. The sludge treated with a phage nevertheless performed a 97.3 mL/g sludge volumetric index, which is within the predicted scope of a safe sludge (70–150 mL/g) aeration basin. SVI is the best predictor for sludge settlement. A bulking SVI of greater than about 150 mL/g is often graded, while a bulking SVI of less than about 70 mL/g can leave a turbid, behind a turbid supernatant (Palm et al., 1980). The phage-treated sludge at the end of the experiment had decreased moisturization capacity in its supernatant after settlement relative to the control sludge with low SVI (54.9 mL/g, suggesting more full settlement of the sludge.

13.4 BACTERIOPHAGE ACTION MECHANISMS IN THE REGULATION OF BIOFILM

Biofouling is due to accumulation of microbial products (such as extracellular polymeric substances, EPS) on membranes occurs in wastewater treatment and membrane-based water processes as microbes can be found in every ecosystem (Wu and Fane, 2012). Membrane filtration processes have their own demerits as it needs high maintenance costs and needs to look upon other pre-set conditions of physicochemical parameters, such as chlorine, ozone, UV etc. (Al-Juboori and Yusaf, 2012; Matin et al., 2011). Moreover, researchers developed new antibiotic membranes through the modification of membrane surfaces and/or the incorporation of nanomaterials into the membrane matrix (Ng et al., 2013).

Various control strategies like dispersal by use of nitric oxide, inhibition of quorum sensing, enzymatic disruption of extracellular polymeric substances, have been developed as control strategies for biological-based membrane biofouling via disruption of biofilm by bacteriophages and inhibition of microbial attachment by energy uncoupling. As highlighted by Malaeb et al. (2013) and Siddiqui et al. (2015), biological-based strategies in the control of biofilm growth and membrane biofouling created r huge potential interest in particular as the continued growth of antibiotic resistant bacteria.

In various areas of research, such as agriculture, medicine and nutrition, the concept of bacteriophage therapy has been introduced (Chan et al., 2013; Nobrega et al., 2015). According to Campbell (2003) and Kingwell (2015), there are two stages in the lifecycle of bacteriophages that involve lytic and lysogenic processes that break down the bacterial cell wall and allow the replication of phage in the bacterial cells (host). The lytic bacteriophages are also classified as virulent bacteriophages, which in the infected cells, synthesize and construct new phage particles and then cause the lysis of the host cells. Recently, published information reveals that phages in effect, invade the neighboring fresh host cells. Bacteriophages are also classified as temperate bacteriophages, which are lysogenic bacteriophages, either experience a lytic process or incorporate their genome into the bacterial genome (i.e., prophage). The newly formed bacteriophages in turn attack different host cells adjacent to it. They are also known as the moderate bacteriophages, as they are the lysogenic bacteriophages which either undergo a lytic process or insert the genome into the bacterial (i.e., prophecy) genome. New bacteriophages can be released from the host cells and rented lytic processes under certain circumstances (such as UV light, mutagenic compounds, and unfavorable temperatures) (Campbell, 2003; Obeng et al., 2016). Recent research has shown that temperate phages can stimulate rapid reaction of bacterial hosts to fluctuating environmental conditions (Obeng et al., 2016). These natural characteristics of bacteriophages are helpful in controlling the production of biofilm, namely (1) bacteriophages replicate directly on an infection site and are closely associated with sustainable bacterial hosts which may achieve biofilm control in situ, (2) bacteriophages produce enzymes which can hydrolyze the biofilmpolymeric matrix, (3) bacteriophages are totally compatible with other biofouling control strategies, and (4) bacteriophages isolation and mass production is theoretically possible, enabling production on an industrial scale (Balcao et al., 2014; Campbell, 2003). Different models were extensively explored as bacterial population-sizing controllers that infect solely culture-formed biofilms. The natural host range recorded is, however, still small. Essentially, work in the biofouling alleviation principle of the bacteriophage-based membrane is still at an early stage of study. In one of the study of Prof Armon's group, it has been observed that on inoculating specific lytic bacteriophages that can infect *P. aeruginosa, A. Johnsoniiand B. Subtilis* into the feed waters at a concentration of 610 5 CFU/100 mL in a bench-scale UF system, The results of the of study revealed that the bacteriophages have been observed to minimize the formation of biofoulings on the membrane surfaces, which has allowed the membrane to permeate up to 40%–60% (Goldman et al., 2009) (Table 13.1).

13.5 PHAGE THERAPY

For the effectiveness of any phage therapy, it is necessary to isolate and to enrich the correct phage in order to generate adequate numbers for use. Phage enrichment usually includes the inoculation by a single-host strain of mixed environmental specimens and growth media. Plague assays are used to detect the production of phages in the host via lytic cycle in the original isolation of host after overnight incubation which helps to isolate only host-specific bacterium that can be culturable under in vitro conditions for about 1%–20% of what extracted from the environs (Andreotolla et al., 2002). However, advances in the culturing of bacteria have occurred in the last few years (Connon and Giovannoni, 2002; Rappe et al., 2002;

TABLE 13.1 Includes the List of Endogenous Bacteriophages in Membrane-Based Water and Wastewater Treatment Systems.

Type of bacteriophage	Feed water	Reactor scale	Removal efficiency	Membrane specification	Reference
Somatic coliphages and F-specific bacteriophage, and bacteriophages infecting B. fragilis	Sewage wastewater	Full-scale MBR	Hollow fiber PVDF (0.04 μm)	5.34 log removal for somatic coliphages; 3.5 log removal for F-specific bacteriophages; 3.8 for bacteriophages infecting B. fragilis	(Purnell et al., 2015)
Somatic coliphages and F-specific bacteriophages	River water	Pilot UF	Hollow fiber PVDF (0.04 μm)	3.8 log removal for somatic coliphages; 3 log removal for F-specific bacteriophages	(Ferrer et al., 2015)
F+ coliphage	Sewage wastewater	MBR Hollow fiber	Full-scale PVDF (0.04 μm)	5.4–7.1 log removal	(Chaudhry et al., 2015)
Somatic coliphages, Fspecific bacteriophages, and bacteriophages infecting B. fragilis	Sewage wastewater	Full-scale MBR	Flat sheet PE (0.4 μm)	4.4 log removal for somatic coliphages; 5.8 log removal for F-specific phages; 3.7–4.1 log removal for bacteriophages infecting	(De Luca et al., 2013)
Somatic and F-specific coliphages	Sewage wastewater	Full-scale MBR	Flat sheet PE (0.4 μm)	2.67–4.04 log removal for somatic coliphages; more than 4.58–6.0 log removal for F-specific phages	(Francy et al., 2012)
Somatic and F-specific coliphages	Sewage wastewater	Pilot-scale MBR	Flat sheet PE (0.4 μm)	2.6–5.6 log removal for both phages	(Marti et al., 2011)
Somatic and F-specific coliphages	Sewage wastewater	Pilot-scale MBR	Hollow fiber PVDF (0.04 μm)	3.1–5.8 log removal for somatic coliphages; 3.3–5.7 log removal for Fspecific phages	(Zhang and Farahbakhsh, 2007)
Somatic and F-specific coliphages	Sewage wastewater	Pilot-scale MBR	Flat sheet PE (0.4 μm)	3.08 log removal for somatic coliphages; 3.78 log removal for F-specific phages	(Ottoson et al., 2006)

TABLE 13.1 *(Continued)*

Type of bacteriophage	Feed water	Reactor scale	Removal efficiency	Membrane specification	Reference
Indigenous coliphages	Municipal wastewater	Pilot-scale MBR	Flat sheet PE (0.4 μm)	Flat sheet PE (0.4 μm)	(Oota et al., 2005)
Indigenous coliphages	Sewage wastewater	Bench-scale MBR	Flat sheet PE (0.4 μm)	2.3–5.9 log removal for Indigenous coliphages	(Ueda and Horan, 2000)
Male-specific (F+) coliphage	Sewage wastewater	Pilot-scale MBR	Hollow fibre PE (0.4 μm)	3.7 log removal	(Tam et al., 2007)

PAN, polyacrylonitrile; PE, polyethylene; PVDF, polyvinylidenefluoride.

O'Sullivan et al., 2004). However, in recent years, there have been advances in cultural growth (Connon and Giovannoni, 2002; O'Sullivan et al., 2004). During isolation of single-host bacterium like *Sphaerotilus natans*, the bacteriophages often give related efficiency of plating (EOP) of that alternative host, such as *Psuedomonas aeruginosa*. Multiple host isolation technique were employed for *E. coli* along with *P. aeruginosa* or *S. natans*, and it has been found that EOP was lower on the alternative host than the isolation host, and phage DNA was sensitive to both type I and II restriction endonucleases (Jensen et al., 1998). The phage lambda DNA underwent cleavage due to insensitivity toward restriction enzymes because the single host isolated DNA could not prevent it, thus multiple host isolation techniques have been found very useful over selection of single-host techniques. Hence, in order to achieve higher rates of efficiency, removal by adding appropriate quantity of bacteriophages in wastewater treatment facilities as since in 1950 onwards, large-scale phage development was performed (Marks and Sharp, 2000). After 1950, it has been found that at concentrations above 1013 PFU cm^3 in tanks of 150 L where *E. Coli* Phage A2 produced, was equal to 75 mg phage per culture and with increased aeration, lysis of phage after addition of bacterial host cultures were appropriate before CO_2 growth (Sergeant and Yeo, 1966). Previous studies indicate up to 200 mg L1 of the MS2 and T2 phases in smaller cultivation volumes, respectively (Seigel and Singer, 1953).

Enhanced biological phosphorus removal (EBPR) is a commonly used wastewater management method for the absorption of phosphorus and carbon. It works when the intracellular deposition of phosphate through polyphosphate-accumulating species (PAO) (Gu et al., 2008). PAO's metabolism is unusual because of its anaerobic carbon absorption combined with the high-energy release from accumulated intracellular polyphosphate in a phosphate stream. During the tricarboxylic acid process, the deposited carbon is then used aerobically to produce biomass while part is used for excess absorption of external phosphates and the regeneration of intracellular polyphosphate reserves (Mino et al., 1998).

The species *Accumulibacter, Betaproteobacteria, Candidatus has been identified* polyphosphate-accumulating species which are accountable for enhanced biological phosphorus removal (EBPR) (Gu et al., 2008; Barr, 2010). As per the reports of Martín (2006), the metagenomes were sequenced for the *Accumulibacter*-enriched EBPR sludges. These metagenomic sequences are important for understanding the EBPR system stage, for verifying the metabolic and biochemical pathways, and for promoting transcriptome, proteome, and metabolome analyses (He et al., 2010; Wilmes et al., 2008). A high degree of seamlessness between dominant strains of *Accumulibacter* was seen in further study of the metagenomes (García Martín, 2006). There was, however, a significant difference between extracellular polymeric substances in gene cassettes and frequently interspaced CRISPR elements that were shown to include the host protection against bacteriophageal predation (Barrangou et al., 2007; Kunin et al., 2008). EBPR systems function usually as flocular active sludge, whereby tiny biofilm aggregates (30–200 μm) are used to process wastewater in microorganisms (deKreuk et al., 2007). However, there was a recent concern in the usage, as aerobic granules, of activated sludge which are larger biofilm aggregates (200–2000 μm) (Liu and Tay, 2004; deKreuk et al., 2007). Faster settlement characteristics and higher concentration of biomass have been demonstrated for granular sludge relative to flocular sludge, rendering it operational and cost-effective (deKreuk et al., 2007). Aerobic granular sludge has still not been added in the WWTP, owing mainly to the inability to monitor pair and sustain granulating and EBPR processes satisfactorily. For the most part, large-sized triggered sludge systems effectively extract biological nutrients. However, some cases show declining and falling EBPR performance, both at laboratory and maximum stage, including favorable operating conditions (Crocetti et al., 2002; Thomas, 2003). The key issue of concern is sustaining steady output. The existence of glicogen-accumulating (GAO) species, which may compete with PAO for carbon substrate absorption, such as volatile fatty acids (VFA), is also a consequence of EBPR failure. GAO does not accumulate and process polyphosphates, so the primary molecule in the anaerobic carbon storages is turned into glycogen (Oehmen et al., 2005). One of the species, namely, *Competibacter phosphatis*, which belongs to group *Candidatus*, a deep-connected cluster in gammaproteobactéries (Crocetti et al., 2002; Kong et al., 2002), and *Defluvicoccus*-related species in Alphaproteobacteria are among the leader GAOs found so far (Wong et al., 2004).

13.6 CONCLUSION

Till date, various scientific reports has been published regarding many properties of bacteriophages, hence they are now considered as potential tool for wastewater treatment or have wide applicability in various environmental spheres like bioremediation. Although bacteriophages can serve as the basic platform in the field of genetic engineering, so that many benefits can be deprived out in order to deal with basic technologies with least benefits. As bacteriophages have specific host cells of bacteria, their harmful effects will be least if their genome will be manipulated in order to modify their genomic sequences to enhance their applicability in the areas like food, wastewater treatment, pollutant degradation, and so on. These viruses have one of the most advantages as they replicate their genetic material so rapidly inside the host bacteria and produce huge numbers to maintain their population in various environs. Such experiments on phages give insights into genome development, bacterial tolerance to changing environments, synthesis, and transcription of DNA, and theoretically deliver innovative drugs in biotechnology.

KEYWORDS

- **biotechnology**
- **bacteriophages**
- **bioremediation**
- **microbes**
- **biofilm**

REFERENCES

Abdulamir, A. S.; Jassim, S. A. A.; Abu Bakar, F. Novel Approach of Using a Cocktail of Designed Bacteriophages Against Gut Pathogenic *E. coli* for Bacterial Load Biocontrol. *Ann. Clin. Microbiol. Antimicrob.* **2014**, *13*, 39.

Abedon, S. T. Phage Therapy of Pulmonary Infections. *Bacteriophage* 2015, *5*, 1020260-1020260-13.

Abedon, S. T.; Thomas-Abedon, C. Phage Therapy Pharmacology. *Curr. Pharm. Biotechnol.* **2010**, *11*, 28–47.

Aldoori, A. A.; Mahdii, E. F.; Abbas, A. K.; Jassim, S. A. A. Bacteriophage Biocontrol Rescues Mice Bacteremic of Clinically Isolated Mastitis from Dairy Cows Associated with Methicillin-Resistant *Staphyloccocus aureus*. *Adv. Microbiol.* **2015**, *5*, 383–403.

Al-Juboori, R. A.; Yusaf, T. Biofouling in RO System: Me chanisms, Monitoring and Controlling. *Desalination* **2012**, *302*, 1–23.

Andreotolla, G.; Baldassarre, L.; Collivigarelli. C.; Pedrazzani, R.; Pricipi P.; Sorlini C. et al. A Comparison among Different Methods for Evaluating the Biomass Activity in Activated Sludge Systems: Preliminary Results. *Water Sci. Technol.* **2002**, *46*, 413–417.

Atterbury, R. J. Bacteriophage Biocontrol in Animals and Meat Products. *Microbial. Biotechnol.* **2009**, *2* (6), 601–612.

Balcao, V. M.; Glasser, C. A.; Chaud, M. V.; del Fiol, F. S.; Tubino, M.; Vila, M. M. D. C. Biomimetic Aqueous-Core Lipid Nanoballoons Integrating a Multiple Emulsion Formulation: A Suitable Housing System for Viable Lytic Bacteriophages. *Colloid Surf. B* **2014**, *123*, 478–485.

Balogh, B.; Jones, J. B.; Iriarte, F. B.; Momol, M. T. Phage Therapy for Plant Disease Control. *Curr. Pharm. Biotechnol.* **2010**, *11*, 48–57.

Barr, J. J.; Blackall, L. L.; Bond, P. L. Further Limitations of Phylogenetic Group-Specific Probes Used for Detection of Bacteria in Environmental Samples. *ISME J*, **2010**, *4*, 959–961.

Barrangou, R.; Fremaux, C.; Deveau, H. et al. CRISPR Provides Acquired Resistance Against Viruses in Prokaryotes. *Science* **2007**, *315*, 1709–1712.

Campbell, A. The Future of Bacteriophage Biology. *Nat. Rev. Genet.* **2003**, *4* (6), 471–477.

Catalão, M. J.; Gil, F.; Moniz-Pereira, J. Pimentel Diversity in Bacterial Lysis Systems: Bacteriophages Show the Way FEMS *Microbiol. Rev.* **2013**, *37*, 554–571.

Chan, B. K.; Abedon, S. T.; Loc-Carrillo, C. Phage Cocktails and the Future of Phage Therapy. *Future Microbiol* **2013**, *8* (6), 769–783.

Chan, B. K.; Abedon, S. T. Phage Therapy Pharmacology Phage Cocktails; Laskin, A. I., Sariaslani, S., Gadd, G. M., Eds., *Adv. Appl. Microbiol. 78*, Elsevier Academic Press Inc.: San Diego, 2012; pp 1–23.

Chaudhry, R. M.; Nelson, K. L.; Drewes, J. E. Mechanisms of Pathogenic Virus Removal in a Full-Scale Membrane Bioreactor. *Environ. Sci. Technol.* **2015**, *49* (5).

Connon, S. A.; Giovannoni, S. J. High-throughput Methods for Culturing Microorganisms in Very-Low-Nutrient Media Yield Diverse New Marine Isolates. *Appl. Environ. Microbiol.* **2002**, *68*, 3878–3885.

Crocetti, G. R.; Banfield, J. F.; Keller, J.; Bond, P. L.; Blackall, L. L. Glycogen-Accumulating Organisms in Laboratory-Scale and Full-Scale Wastewater Treatment Processes. *Microbiology* **2002**, *148*, 3353–3364.

Dąbrowska, K. Switała-Jeleń, K.; Opolski, A.; Górski, A. Possible Association between Phages, Hoc Protein, and the Immune System. *Arch. Virol.* **2006**, *151* (2), 209–215.

Dąbrowska, K.; Switala-Jelen, K.; Opolski, A.; Weber-Dabrowska, B. Bacteriophage Penetration in Vertebrates. *J. Appl. Microbiol.* **2005**, *98* (1), 7–13.

de los Reyes, M. F.; de los Reyes, F. L.; Hernandez, M.; Raskin, L. Quantification of Gordona Amarae Strains in Foaming Activated Sludge and Anaerobic Digester Systems with Oligonucleotide Hybridization Probes. *Appl. Environ. Microbiol.* **1998**, *64*, 2503–2512.

De Luca, G.; Sacchetti, R.; Leoni, E.; Zanetti, F. Removal of Indicator Bacteriophages from Municipal Wastewater by a Full-Scale Membrane Bioreactor and a Conventional Activated Sludge Process: Implications to Water Reuse. *Bioresour. Technol.* **2013**, *129*, 526–531.

deKreuk, M. K.; Kishida, N.; van Loosdrecht, M. C. M. Aerobic Granular Sludge—State of the Art. *Water Sci. Technol.* **2007**, *55*, 75–81.

Eikelboom, D. H. Filamentous Organisms Observed in Activated Sludge. *Water Res.* **1975**, *9*, 365–388.

Eikelboom, D. H. Identification of Filamentous Organisms in Bulking Activated Sludge. *Prog. Water Tech.* **1977**, *8*, 152–161.

Elkelboom, D. H.; Geurkink, B. Filamentous Micro-Organisms Observed in Industrial Activated Sludge Plants. *Water Sci. Technol.* **2002**, *46*, 535–542.

Ewert, D. L.; Paynter, M. J. Enumeration of Bacteriophages and Host Bacteria in Sewage and the Activated-Sludge Treatment Process. *Appl. Environ. Microbiol.* **1980**, *39*, 576–583.

Ferrer, O.; Casas, S.; Galvan, C.; Lucena, F.; Bosch, A.; Galofre, B.; Mesa, J.; Jofre, J.; Bernat, X. Direct Ultrafiltration Performance and Membrane Integrity Monitoring by Microbiological Analysis. *Water Res.* **2015**, *83*, 121–131.

Francy, D.; Stelzer, S.; E. Bushon, A. et al. Comparative Effectiveness of Membrane Bioreactors, Conventional Secondary Treatment, and Chlorine and UV Disinfection to Remove Microorganisms from Municipal Wastewaters. *Water Res.* **2012**, *46* (13), 4164–4178.

Fujiwara, A.; Fujisawa, M.; Hamasaki, R.; Kawasaki, T. Biocontrol of *Ralstonia solanacearum* by Treatment with Lytic Bacteriophages. *Appl. Environ. Microbiol.* **2011**, *77* (12), 4155–4162.

Garcıa, M. H.; Ivanova, N.; Kunin, V. et al. Metagenomic Analysis of Two Enhanced Biological Phosphorus Removal (EBPR) Sludge Communities. *Nat. Biotechnol.* **2006**, *24*, 1263–1269.

Gill, J. J. Practical and Theoretical Considerations for the Use of Bacteriophages in Food Systems. In *Bacteriophages in the Control of Food- and Waterborne Pathogens*; Sabour, P. M., Griffiths, M. W., Eds.; ASM Press: Washington, DC, 2010; pp 217–235.

Gill, J. J.; Young, R. *Therapeutic Applications of Phage Biology: History, Practice and Recommendations*; Caister Academic Press: Norfolk, UK, 2010.

Gill, J. J.; Young, R. *Therapeutic Applications of Phage Biology: History, Practice and Recommendations*; Caister Academic Press: Norfolk, UK, 2011.

Goldman, G.; Starosvetsky, J.; Armon, R. Inhibition of Biofilm Formation on UF Membrane by Use of Specific Bacteriophages. *J. Membr. Sci.* **2009**, *342* (1–2), 145–152.

Goodfellow, M.; Alderson, G.; Chun, J. Rhodococcal Systematics: Problems and Developments. *Antonie Van Leeuwenhoek*, **1998**, *74*, 3–20.

Gu, A. Z.; Saunders, A.; Neethling, J. B.; Stensel. H. D.; Blackall, L. L. Functionally Relevant Microorganisms to Enhanced Biological Phosphorus Removal Performance at Full-Scale Wastewater Treatment Plants in the United States. *Water Environ. Res.* **2008**, *80*, 688–698.

Guo, F.; Wang, Z. P.; Yu, K.; Zhang, T. Detailed Investigation of the Microbial Community in Foaming Activated Sludge Reveals Novel Foam Formers. *Sci. Rep.* **2015**, *5*, 7637.

Guo, F.; Zhang, T. Profiling Bulking and Foaming Bacteria in Activated Sludge by High Throughput Sequencing. *Water Res.* **2012**, *46*, 2772–2782.

Hagens, S.; Loessner, M. J. Bacteriophage for Biocontrol of Foodborne Pathogens: Calculations and Considerations. *Curr. Pharm. Biotechnol.* **2010**, *11*, 58–68.

Hatfull, G. F. et al. Comparative Genomic Analysis of 60 Mycobacteriophage Genomes: Genome Clustering, Gene Acquisition, and Gene Size. *J. Mol. Biol.* **2010**, *397*, 119–143.

Havelaar, A. H. Virus, Bacteriophages and Water Purification. *Vet Q* **1987**, *9* (4), 356–360.

He, S. M.; Kunin, V.; Haynes, M. et al. Metatranscriptomic Array Analysis of 'Candidatus Accumulibacter Phosphatis'-Enriched Enhanced Biological Phosphorus Removal Sludge. *Environ. Microbiol.* **2010,** *12*, 1205–1217.

Jassim, S. A. A.; Abdulamir, A. S.; Abu Bakar, F. Novel Phage-Based Bio-Processing of Pathogenic *Escherichia coli* and Its Biofilms. *World J. Microbiol. Biotechnol.* **2012,** *28*, 47–60.

Jassim, S. A. A.; Limoges, R. G. Impact of External Forces on Cyanophage–Host Interactions in Aquatic Ecosystems. *World J. Microbiol. Biotechnol.* **2013,** *29* (10), 1751–1762.

Jenkins, D.; Richard, M. G.; Daigger, G. T. Manual on the Causes and Control of Activated Sludge Bulking, Foaming, and Other Solids Separation Problems, ed. 3; CRC Press: Boca Raton, 2004. www.nature.com/scientificreports/ Scientific Reports *5*, 13754.

Jensen, E. C.; Schrader, H. S.; Rieland, B.; Thompson, T. L.; Lee, K. W.; Nickerson K. W. et al. Prevalence of Broad–Host Range Lytic Bacteriophages of *Sphaerotilus natans,* Escherichia Coliand Pseudomonas Aeruginosa. *Appl. Environ. Microbiol.* **1998,** *64*, 575–580.

Jones, J. B.; Vallad, G. E.; Iriarte, F. B.; Obradovic´, A. Considerations for Using Bacteriophages for Plant Disease Control. *Bacteriophage* **2012,** *2* (4), 208–214.

Ju, F.; Guo, F.; Ye, L.; Xia, Y.; Zhang, T. Metagenomic Analysis on Seasonal Microbial Variations of Activated Sludge from a Fullscale Wastewater Treatment Plant Over 4 Years. *Environ. Microbiol. Rep.* **2014,** *6*, 80–89.

Khairnar, K.; Pal, P.; Chandekar, R. H.; Paunikar, W. N. Isolation and Characterization of Bacteriophages Infecting Nocardioforms in Wastewater Treatment Plant. *Biotechnol. Res. Int.* **2014,** 151952.

Khan, M. A.; Satoh, H.; Katayama, H.; Kurisu, F.; Mino, T. Bacteriophages Isolated from Activated Sludge Processes and Their Polyvalency. *Water Res.* **2002,** *36*, 3364–3370.

Kingwell, K. Bacteriophage Therapies Re-Enter Clinical Trials. *Nat. Rev. Drug Discov.* **2015,** *14* (8), 515–516.

Klumpp, J.; Loessner, M. J. Listeria Phages Genomes, Evolution, and Application. *Bacteriophage* **2013,** *3* (3), 26861.

Kong, Y. H.; Ong, S. L.; Ng, W. J.; Liu, W. T. Diversity and Distribution of a Deeply Branching Novel Proteobacterial Group Found in Anaerobic–Aerobic Activated Sludge Processes. *Environ. Microbiol.* **2002,** *4*, 753–757.

Kotay, S. M.; Datta, T.; Choi, J.; Goel, R. Biocontrol of Biomass Bulking Caused by Haliscomenobacter Hydrossis Using a Newly Isolated Lytic Bacteriophage. *Water Res.* **2011,** *45*, 694–704.

Kragelund, C. Ecophysiology of Mycolic Acid-Containing Actinobacteria (Mycolata) in Activated Sludge Foams. *FEMS Microbiol Ecol* **2007,** *61*, 174–184.

Kunin, V.; He, S.; Warnecke, F. et al. A Bacterial Metapopulation Adapts Locally to Phage Predation Despite Global Dispersal. *Genome Res.* **2008,** *18*, 293–297.

Liu, Y.; Tay, J. H. State of the Art of Biogranulation Technology for Wastewater Treatment. *Biotechnol. Adv.* **2004,** *22*, 533–563.

Loc-Carrillo, C.; Abedon, S. T. Pros and Cons of Phage Therapy. *Bacteriophage* **2011,** *1*, 111–114.

Malaeb, L.; Le-Clech, P.; Vrouwenvelder, J. S.; Ayoub, G. M.; Saikaly, P. E. Do Biological-Based Strategies Hold Promise to Biofouling Control in MBRs? *Water Res.* **2013,** *47* (15), 5447–5463.

Marks, T.; Sharp, R. Bacteriophages and Biotechnology: A Review. *J. Chem. Technol. Biotechnol.* **2000,** *75,* 6–17.

Marti, E.; Monclus, H.; Jofre, J.; Rodriguez-Roda, I.; Comas, J.; Balcazar, J. L. Removal of Microbial Indicators from Municipal Wastewater by a Membrane Bioreactor (MBR). *Bioresour. Technol.* **2011,** *102* (8), 5004–5009.

Martins, A. M.; Pagilla, K.; Heijnen, J. J.; van, Loosdrecht, M. C. Filamentous Bulking Sludge–A Critical Review. *Water Res.* **2004,** *38,* 793–817.

Matin, A.; Khan, Z.; Zaidi, S. M. J.; Boyce, M. C. Biofouling in Reverse Osmosis Membranes for Seawater Desalination: Phenomena and Prevention. *Desalination* **2011,** 281.

Matsuzaki, S. Bacteriophage Therapy: A Revitalized Therapy Against Bacterial Infectious Diseases. *J. Infect. Chemother.* **2005,** *11,* 211–219.

Merril, C. R.; Scholl, D.; Adhya, S. L. The Prospect for Bacteriophage Therapy in Western Medicine. *Nat. Rev. Drug Discov.* **2003,** *2,* 489–497.

Mino, T.; van Loosdrecht, M. C.; Heijnen, J. J. Microbiology and Biochemistry of the Enhanced Biological Phosphorus Removal Process. *Water Res.* **1998,** *32,* 3193–3207.

Ng, L. Y.; Mohammad, A. W.; Leo, C. P.; Hilal, N. 2013. Polymeric Membranes Incorporated with Metal/Metal Oxide Nanoparticles: A Comprehensive Review. *Desalination* **2013,** *308,* 15–33.

Nobrega, F. L.; Costa, A. R.; Kluskens, L. D.; Azeredo, J. Revisiting Phage Therapy: New Applications for Old Resources. *Trends Microbiol.* **2015,** *23* (4), 185–191.

O'Sullivan, L. A.; Fuller, K. E.; Thomas, E. M.; Turley, C. M.; Fry, J. C. Distribution and Culturability of the Uncultivated dAGG58 Cluster T of the Bacteroidetes Phylum in Aquatic Environments. *FEMS Microbiol. Ecol.,* **2004,** *47,* 359–370.

Obeng, N.; Pratama, A. A.; van Elsas, J. D. The Significance of Mutualistic Phages for Bacterial Ecology and Evolution. *Trends Microbiol.* **2016,** *24* (6), 440–449.

Oehmen, A.; Saunders, A.M; Vives, M. T.; Yuan, Z. G.; Keller, J. Competition between Polyphosphate and Glycogen Accumulating Organisms in Enhanced Biological Phosphorus Removal Systems with Acetate and Propionate as Carbon Sources. *J. Biotechnol.* **2005,** *23,* 22–32.

Oota, S.; Murakami, T.; Takemura, K.; Noto, K. Evaluation of MBR Effluent Characteristics for Reuse Purposes. *Water Sci. Technol.* **2005,** *51* (6–7), 441–446.

Ottoson, J.; Hansen, A.; Björlenius, B.; Norder, H.; Stenström, T. A. Removal of Viruses, Parasitic Protozoa and Microbial Indicators in Conventional and Membrane Processes in a Wastewater Pilot Plant. *Water Res.* **2006,** *40* (7), 1449–1457.

Palm, J. C.; Jenkins, D.; Parker, D. S. Relationship between Organic Loading, Dissolved Oxygen Concentration and Sludge Settleability in the Completely-Mixed Activated Sludge Process. *J. Water Pollut. Control Federation,* **1980,** *52,* 2484.

Petrovski, S. AN examination of the Mechanisms for Stable Foam Formation in Activated Sludge Systems. *Water Res.* **2011,** *45,* 2146–2154.

Petrovski, S.; Seviour, R. J.; Tillett, D. Characterization of the Genome of the Polyvalent Lytic Bacteriophage GTE2, Which Has Potential for Biocontrol of Gordonia-, Rhodococcus-, and Nocardia-Stabilized Foams in Activated Sludge Plants. *Appl. Environ. Microbiol.* **2011a,** *77,* 3923–3929.

Petrovski, S.; Seviour, R. J.; Tillett, D. Genome Sequence and Characterization of the Tsukamurella Bacteriophage TPA2. *Appl. Environ. Microbiol.* **2011b,** *77,* 1389–1398.

Petrovski, S.; Seviour, R. J.; Tillett, D. Prevention of Gordonia and Nocardia Stabilized Foam Formation Using Bacteriophage GTE7. *Appl. Environ. Microbiol.* **2011c**, *77*, 7864–7867.

Purnell, S.; Ebdon, J.; Buck, A.; Tupper, M.; Taylor, H. Bacteriophage Removal in a Full-Scale Membrane Bioreactor (MBR)—Implications for Wastewater Reuse. *Water Res.* **2015**, *73*, 109–117.

Sabah, A.; Jassim, A. Natural Solution to Antibiotic Resistance: Bacteriophages 'The Living Drugs'. *World J. Microbiol. Biotechnol.* **2014**, *30*, 2153–2170.

Seigel, A.; Singer, S. J. The Preparation and Properties of Desoxypentosenucleic Acid of Bacteriophage T2. *Biochim. Biophys. Acta.* **1953**, *10*, 311–319.

Seka, M. A.; Hammes, F.; Verstraete, W. Predicting the Effects of Chlorine on the Micro-Organisms of Filamentous Bulking Activated Sludges. *Appl. Microbiol. Biotechnol.* **2003**, *61*, 562–568.

Sergeant, K.; Yeo, R. G. The Production of Bacteriophage A2. *Biotech. Bioeng.* **1966**, *8*, 195–215.

Siddiqui, M. F.; Rzechowicz, M.; Winters, H.; Zularisam, A. W.; Fane, A. G. Quorum Sensing Based Membrane Biofouling Control for Water Treatment: A Review. *J. Water Process Eng.* **2015**, *7*, 112–122.

Sulakvelidze, A.; Alavidze, Z.; Morris, J. G. Bacteriophage Therapy. *Antimicrob. Agents Chemother.* **2001**, *45* (3), 649–659.

Summer, E. J. et al. Genomic and Functional Analysis of Rhodococcus Equi Phages ReqiPepy6, ReqiPoco6, ReqiPine5 and ReqiDocB7. *Appl. Environ. Microbiol.* **2011**, *77*, 669–83.

Summers, W. C. Bacteriophage Therapy. *Annu. Rev. Microbiol.* **2001**, *55*, 437–445.

Tam, L.; Tang, S.; Lau, W. et al. A Pilot Study for the Wastewater Reclamation and Reuse with MBR/RO and MF/RO Systems. *Desalination* **2007**, *202*, 106–113.

Thomas, J. A.; Soddell, J. A.; Kurtboke, D. I. Fighting Foam with Phages? *Water Sci. Technol.* **2002**, *46*, 511–518.

USFDA. *Food Additives Permitted for Direct Addition to Food for Human Consumption; Bacteriophage Preparation*; FDA: Washington, DC, 2006. Publishing FDA Web. http://www.fda.gov/OHRMS/DOCKETS/98fr/cf0559.pdf (accessed 3 Aug 2006).

van der Waarde, J. et al. Molecular Monitoring of Bulking Sludge in Industrial Wastewater Treatment Plants. *Water Sci. Technol.* **2002**, *46*, 551–558.

Verheust, C.; Pauwels, K.; Mahillon, J.; Helinski, D. R.; Herman, P. Contained Use of Bacteriophages: Risk Assessment and Biosafety Recommendations. *Appl. Biosaf.* **2010**, *15* (1), 32–44.

Wagner, M. Microbial Community Composition and Function in Wastewater Treatment Plants. *Antonie Van Leeuwenhoek* **2002**, *81*, 665–680.

Wilmes, P.; Andersson, A. F.; Lefsrud, M. G. et al. Community Proteogenomics Highlights Microbial Strain-Variant Protein Expression within Activated Sludge Performing Enhanced Biological Phosphorus Removal. *ISME* **2008**, 853–864.

Withey, S.; Cartmell, E.; Avery, L. M.; Stephenson, T. Bacteriophages–Potential for Application in Wastewater Treatment Processes. *Sci. Total Environ.* **2005**, *339*, 1–18.

Withey, S.; Cartmell, E.; Avery, L. M.; Stephenson, T. Bacteriophages-Potential for Application in Wastewater Treatment Processes. *Sci. Total Environ.* **2005**, *339* (1–3), 1–18.

Wong, M. T.; Tan, F. M.; Ng, W. J.; Liu, W. T. Identification and Occurrence of Tetrad-Forming Alphaproteobacteria in Anaerobic–Aerobic Activated Sludge Processes. *Microbiology* **2004,** *150,* 3741–3748.

Wu, B.; Fane, A. G. Microbial Relevant Fouling in Membrane Bioreactors: Influencing Factors, Characterization, and Fouling Control. *Membranes* **2012,** *i,* 565–584.

Zhang, K.; Farahbakhsh, K. Removal of Native Coliphages and Coliform Bacteria from Municipal Wastewater by Various Wastewater Treatment Processes: Implications to Water Reuse. *Water Res.* **2007,** *41* (12), 2816–2824.

Zhang, T.; Shao, M. F.; Ye, L. Pyrosequencing Reveals Bacterial Diversity of Activated Sludge from 14 Sewage Treatment Plants. *ISME J.* **2012,** *6,* 1137–1147.

Index

For Product Safety Concerns and Information please contact our EU
representative GPSR@taylorandfrancis.com
Taylor & Francis Verlag GmbH, Kaufingerstraße 24, 80331 München, Germany